High Energy Cosmic Rays

Springer

Berlin
Heidelberg
New York
Hong Kong
London
Milan
Paris
Tokyo

Todor Stanev

High Energy Cosmic Rays

With 155 figures and 19 tables

Springer

Professor Todor Stanev
Bartol Research Institute
University of Delaware
Newark
Delaware, DE 19716
USA

The main image of the front cover design features a drawing of the muons from a Monte Carlo calculation of air shower development, both in the air (shown in blue) and below ground (shown in green). Below ground, the number of muons decreases with depth as the low-energy muons stop and decay. The picture also shows (in red) an air shower array on the surface, which detects the air shower electrons and muons, and an underground muon detector that can detect the highest energy muons which reach it. Image reproduced courtesy of the author.

SPRINGER–PRAXIS BOOKS IN ASTROPHYSICS AND ASTRONOMY
SUBJECT *ADVISORY EDITOR*: John Mason B.Sc., M.Sc., Ph.D.

ISBN 3-540-40653-0 Springer-Verlag Berlin Heidelberg New York

Springer-Verlag is a part of Springer Science + Business Media (springeronline.com)

Library of Congress Cataloging-in-Publication Data

Stanev, Todor.
    High energy cosmic rays / Todor Stanev.—2nd ed.
        p. cm.—(Springer-Praxis books in astrophysics and astronomy)
    Includes bibliographical references and index.
    ISBN 3-540-40653-0 (acid-free paper)
    1. Cosmic rays.   I. Title.   II. Series.
    QC485.S69 2003
    523.01′97223—dc22   2003064920

Bibliographic information published by Die Deutsche Bibliothek

Die Deutsche Bibliothek lists this publication in the Deutsche Nationalbibliografie;
detailed bibliographic data are available from the Internet at http://dnb.ddb.de

Typeset in LaTex by the author
Cover design: Jim Wilkie

Printed on acid-free paper

To Svetla, Petra and my father Stefan

Preface

This book discusses the processes and the astrophysical environment that lead to the acceleration of cosmic rays and govern their propagation through the Galaxy to the solar system, and through the solar system to the Earth. Most of these processes are also used in different methods of cosmic ray detection. The book also gives many samples of cosmic ray data and their physical interpretation.

I am grateful to many colleagues for their contribution to my understanding of cosmic rays starting from my early days in cosmic rays physics. Among them are B. Betev, M. Block, A. Franceschini, J.G. Learned, M.M. Shapiro, A.A. Watson and G. Yodh with whom I have worked on some specific topics.

Special acknowledgements are due to my frequent collaborators who enhance my knowledge of the subject. This long list includes most of all Tom Gaisser, and also Jaime Alvarez-Muñiz, Venya Berezinsky, Peter Biermann, Ralph Engel, Francis Halzen, Paolo Lipari, Raymond Protheroe, Jörg Rachen, David Seckel, Hristofor Vankov and Enrique Zas.

I thank my colleagues from the Laboratory for Particle Physics and Cosmology in Collège de France, and its director Prof. Marcel Froissart for their hospitality in 2001/2002 when the main body of the book was planned and the writing began.

It would be impossible to complete this book without the support and encouragement of my extended family.

Newark, DE, USA,   *Todor Stanev*
December 2003

Contents

Part II Contemporary Challenges

Part I

The Standard Model of Cosmic Rays

1 Overview

Cosmic rays are often defined as charged particles that reach the Earth from interstellar space. This definition describes correctly the majority of the cosmic ray particles which do consist of fully charged nuclei. At GeV energy the flux of hydrogen and helium nuclei dominate all other species. The chemical composition of cosmic rays extends to very high masses and we believe that cosmic rays include in various degrees all stable nuclei. In addition there is a steady flux of electrons which are also included in the above definition. Other, although not very common, components represent anti-matter – these are the antiprotons and positively charged electrons – positrons.

Neutral particles are obviously not included as a cosmic ray component. So are all kinds of particles that are not of interstellar origin. We will, however, deviate from the definition and deal with all types of neutral particles and occasionally with particles generated by the Sun, such as the solar neutrinos that provided the proof of the nuclear processes that power the Sun.

The way cosmic rays are approached in this book is similar to the list of topics discussed at the International Cosmic Ray Conference – a big scientific forum that meets bi-annually. At the last ten meetings the contributions to the meeting were divided into three sections: cosmic ray origin and Galactic phenomena, high energy phenomena and solar and heliospheric physics. We will discuss mostly the first two aspects of the field and touch the third one only when it also applies to the propagation of galactic cosmic rays in the heliosphere.

1.1 Where does the cosmic ray field belong?

This is a very difficult question that I have trouble answering when inquisitive co-travelers ask it of me in the airport. From the division of topics at the bi-annual meetings it is obvious that cosmic ray physics is a cross-disciplinary area where astrophysicists, high energy particle physicists and plasma physicists work together.

The field has been like that from the very beginning – the discovery of the cosmic ray radiation almost 100 years ago. At this time the research that led to the discovery was a cross between physics, material science and environmental studies. After the discovery of radioactivity it was noticed that the air is being ionized at a relatively high rate. The measurements showed that 10 to 20 ions were generated in a cubic centimeter of air every second. Was that inherent to the material or a product of the natural radioactivity of the Earth – this was the main question. Three types of radioactive rays were known at that time: α-rays (ionized He nuclei), β-rays (electrons) and γ-rays. Since the first two were very easy to shield from, γ-rays were suspected to be the ionizing agent.

The ionization was measured at different heights in towers, including the Eiffel tower, in attempts to figure out what the penetration power of these γ-rays is. The breakthrough occurred just before the First World War when Victor Hess in Austria and Kohlhörster in Germany decided to make measurements from balloons. In 1912 Hess flew in a balloon to altitudes of 5 km and discovered that, instead of decreasing, the ionization of the air strongly increases with altitude. The only explanation of his measurement, he believed, was that 'a radiation of very high penetrating power enters the atmosphere from above'. This marked the discovery of cosmic rays for which Hess received the Nobel prize in 1936.

Kohlhörster contributed a lot to these first measurements with his flights that reached altitude of 9 km. After the war the altitude dependence of the ionization became the topic of many measurements at different locations and altitudes. Maybe the biggest contribution was that of Millikan, which is ironic since he aimed at disproving the results of Hess and Kohlhörster. Millikan improved the detection technology and started measurements of the ionization with instruments that were lowered in mountain lakes at different depths. Since the total thickness of the atmosphere corresponds to only about 10 meters of water, Millikan believed that his measurements in water will determine better the absorption length of the cosmic radiation.

Figure 1.1 compares the results of Kohlhörster with two of the lake experiments of Millikan. Millikan first used 'cosmic rays' do describe the radiation and thus created the current name of the field. His idea was to reveal the origin of cosmic rays through the energy of the cosmic rays, which he believed are γ-rays from the nucleosynthesis of the common elements like helium and oxygen, ranging in energy from 30 to 250 MeV.

The experimental results did not help him, because (as is obvious from Fig. 1.1) cosmic rays have different absorption lengths in the atmosphere and in water. This is a result that we easily understand now, when we know that these are measurements of two different components of the atmospheric cosmic ray showers – the electromagnetic component in air and the penetrating muon component under water.

During the following twenty to thirty years cosmic ray research concentrated on the the high energy physics properties of cosmic rays. There were no other sources of high energy particles and most of the discoveries of new particles before 1950 were made in measurements of the cosmic ray interactions.

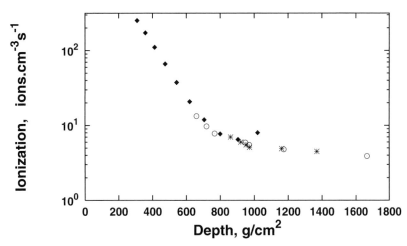

Fig. 1.1. The ionization as a function of the depth in the atmosphere. The diamonds are from the flight of Kohlhörster and the circles and asterisks are from the underwater measurements of Millikan.

Other related progress was the development of the quantum electrodynamics (QED) and the electromagnetic cascade theory that followed the discovery of cosmic ray induced showers in 1929 by Skobelzyn. These showers were first interpreted as a result of Compton scattering by the cosmic γ-rays. The theory of the electromagnetic interactions was soon developed and the theory of electromagnetic showers was fully developed by 1940.

The explosion of new discoveries became possible because of several new ways for direct detection of shower particles. Skobelzyn observed directly shower tracks in a cloud chamber. Particle ionized the material in the chamber and made the track visible. The amount of ionization could be measured by the thickness of the track. A similar way to see the tracks and to measure more precisely the amount of ionization was to use nuclear emulsion stacks. Layers of photo emulsion were stacked together and exposed to cosmic rays. The emulsion was later developed to see and match the tracks. It is a complicated and time-consuming method, but has unmatched accuracy and its principle is still used today.

In addition to viewing tracks, cosmic ray physicists learned how to count charged particles. The first successful device was the Geiger–Müller counter that gave a pulse after a charged particle passed through it. Counters make the particle detection much easier. They do not explicitly recognize the particle energy, but they could be shielded with amounts of matter that define the threshold energy of the particles to penetrate through the shielding.

The development of the experimental techniques made possible the important progress in the discovery of the nature of cosmic rays. The electrometers were replaced with counters of charged particles. When two counters were positioned one above the other both of them measured particles simultaneously, i.e. the measured particle had enough energy to penetrate through both counters. Even higher penetrating power was proved when several centimeters of lead or gold were placed between them. This proved that some of the cosmic ray particles are very different from γ-rays.

In a different type of experiments counters were put on the same level but at different distance from each other. The coincidences between the two counters demonstrated that some of the cosmic rays come to the surface of the Earth in groups – as atmospheric showers. The measurements of the rate of coincidences as a function of the distance between the counters and the application of the shower theory led to the conclusion that the primary particles that initiated the atmospheric showers have energy as high as 10^6 GeV.

In several years this was confirmed by nuclear emulsion experiments that were exposed to primary cosmic rays in high altitude balloon flights. Tracks of protons and heavier nuclei were discovered in the emulsion stacks which finally proved the true nature of cosmic rays. For a while cosmic ray experiments became the experimental side of nuclear physics.

Another track of the cosmic ray research in the same period was a study of the cosmic ray interactions with the geomagnetic field. Once the scientists knew that cosmic rays were positively charged nuclei, they figured out that the cosmic ray flux would depend on the strength of the magnetic field at the location of the experiment. This was confirmed, not without some controversy, in many scientific expeditions during which cosmic rays were measured at different geomagnetic latitudes.

The same type of experiments also discovered the 'east–west' effect – because of the direction of the geomagnetic field more primary cosmic rays come from the west than from the east. Positive particles coming from the west bend downwards – towards the atmosphere and the surface of the Earth, while the ones coming from the east bend away from the Earth. The effect is stronger and thus easier to measure at high geomagnetic latitude.

With the fast progress of the particle accelerators in the 1950s and 1960s cosmic rays lost their attraction for the majority of high energy physicists. Accelerators provided intense beams of known particle type and energy. A measurement that would last many years in cosmic rays could now be completed in hours. Experimental arrangements were much more sophisticated. They could be constructed to surround the known position of the particle interaction with the target. The results were precise because of the known primary energy and the better detection technique. There was only a relatively small group of physicists who were concerned with interactions well above the energy achievable by particle accelerators. The energy range kept increasing – as soon as a new accelerator was built the cosmic ray physicists had to jump to higher energy.

As squeezed as the field was, the results from the accelerator laboratories were God's gift for these remaining cosmic ray scientists. The characteristics of hadronic interactions became much better known and the analysis of the cosmic ray data improved significantly. The reputation of the field, however, decreased and this resulted in mass exodus of cosmic ray scientists to accelerator labs. At the same time there was a significant progress of the field in different directions.

The experimental study of the solar system led to the existence of what we now call 'space physics'. The magnetic fields in the heliosphere and the properties of the solar wind were studied in more and more detail. The behavior of the cosmic rays in the heliosphere became a major research topic. Results from measurements of the total cosmic ray intensity on the surface of the Earth at different geomagnetic latitudes (e.g. with different energy threshold) were compared to intensities measured by satellites and space missions and were related to the epoch of the solar cycle and the level of solar activity.

The possibility of studying directly primary cosmic rays with experiments mounted on balloons and satellites also led to a great precision of the knowledge of the chemical and isotopic composition of cosmic rays. The experimental progress inspired similar progress in the investigations of the formation of the measured chemical composition and its relation to the composition at the cosmic ray sources. The nuclear reaction cross-sections were measured in the laboratory by cosmic ray physicists and were applied to the cosmic ray interactions in their propagation in interstellar space. The detection of known unstable isotopes and the comparison of their fluxes allowed independent estimates of the cosmic ray containment time in the Galaxy.

At the same time there was rapid development in the theory of cosmic ray acceleration. Several models appeared almost simultaneously in the late 1970s that described the cosmic acceleration at astrophysical shocks. The combination of the results of these new developments led to the creation of what I call the *standard model of cosmic rays*.

The positive evolution in the development of the *standard model* still continues and the study of the cosmic ray propagation in the Galaxy and the heliosphere are now much more sophisticated and conclusive. A new development started about thirty years ago which eventually formulated the current status of the field. The early bird was the solar neutrino experiment set up by Ray Davis in the Homestake mine. Davis attempted to measure the neutrinos coming from the nuclear reactions at the Sun. By the early 1970s the results of this experiment attracted interest because of the missing solar neutrinos. At about the same time the extensive progress in particle theory developed scenarios in which the proton was not a stable particle and had a lifetime of the order of 10^{30} years, i.e. 20 orders of magnitude longer than the age of the Universe.

Huge experiments were built in the early 1980s to measure proton decay. They contained more than 10^{33} protons and were located deep underground to shield the reaction, which releases energy equal to one proton mass, from penetrating cosmic rays. Only limits on the proton lifetime exist now, but these experiments hinted that the muon neutrinos generated in the atmosphere by cosmic rays are also missing and extended the already existing hypothesis of neutrino oscillations.

As one will see in this book deep underground experiments can measure many different effects, only some of which are related to neutrino oscillations. The exciting set of topics started attracting many high energy accelerator physicists back to the cosmic ray field - although generally from the next age generation. Now deep underground physics is a fashionable field, with plans for new giant experiments that that only define the topics of the study and do not separate the physicists into accelerator or cosmic ray ones.

In the early 1980s, perhaps inspired by the advent of X-ray and γ-ray astronomy, cosmic ray physicists started looking at the exact direction from which high energy cosmic rays arrive at the Earth. The first suspicion that some very high energy particles come from the binary system Cygnus X-3 may, or may not, be correct, but the ambition for the development of cosmic ray astronomy that it created had a positive role for the development of the field. It coincided with the first measurements of TeV γ-rays, which on its own led to the current construction and operation of the third generation of telescopes.

During the last fifteen years many particle physicists became curious about the origin and nature of the highest energy cosmic rays. The next generation accelerator, the Large Hadronic Collider, will study particle interactions at equivalent laboratory energy of about 4×10^8 GeV, while particles of energy exceeding 10^{11} GeV have been detected in cosmic rays. How can Nature achieve higher energy than a perfectly engineered and fabricated machine? What are the objects that are capable of squeezing the energy of the fastest tennis ball into a volume less than 10^{-38} cm^3? A huge ground experiment is being built to observe these highest energy events and two more satellite experiments to detect them are now being considered.

Particles of such high energy have interactions in extragalactic space and may be deflected by magnetic fields, while high energy neutrinos could reach us from the edge of the Universe and point at their sources. Although such neutrinos are rare, their detection will create a new type of astronomy that does not observe electromagnetic waves. High energy neutrino astronomy underwent very rapid development during the last 10–15 years. Within a few years time two experiments, one in each hemisphere, and shielded by kilometers of water or ice will observe the whole sky for sources of high energy neutrinos. Their results will be complementary to those at all photon frequencies and to the detection of ultra high energy cosmic rays. There has never been a better chance to finally solve the problem of the origin of all cosmic rays.

I did not intentionally leave the reader wondering where the field of cosmic ray physics belonged in the classification of different fields. It certainly contains, contributes to and benefits from the research of many fields of physics and astrophysics. A better definition than an outline of its history and its ever-changing priorities is hardly possible.

1.2 Is progress in the cosmic ray field slow?

It certainly looks like that. Cosmic rays were discovered more than ninety years ago and we are still asking questions about their exact origin and even, at ultra high energy, about their nature. One can understand the degree of progress only after a careful examination of the achievements in the field and the current knowledge of cosmic rays.

We shall start with the energy range that cosmic rays cover. Figure 1.2 shows the energy spectrum of cosmic rays above 100 GeV per nucleus. This includes all charged nuclei. We only show this restricted energy range, because at lower energy the experiments are exact, but the instruments cannot cover the whole mass range. In that region the usual way of presenting the nuclei of different mass is in kinetic energy E_k per nucleon, and the conversion to energy per nucleus requires fitting and respectively introduces errors. The smooth cosmic ray spectrum becomes a wavy line.

At 100 GeV the difference between E_k and total energy per nucleon is the proton mass $m_p = 0.938$ GeV and the spectra measured in both units almost coincide. At lower energy they do not. In addition to the ten orders of magnitude shown in Fig. 1.2 cosmic rays include five more decades in kinetic energy.

The cosmic ray energy spectrum in Fig. 1.2 is an almost featureless power law spectrum with two transition regions where the slope of the spectrum changes. What is important for our discussion here is that the number of particles above 100 GeV is higher than that above 10^{11} GeV by sixteen orders of magnitude. The units of integral flux at three energies quote the approximate number of particles above that energy that hit the atmosphere. Six particles per square kilometer per minute is relevant for energy of about 10^7 GeV. A typical instrument of area 1 m^2 will have to wait for two years to detect a single particle above that energy. Even the measurements of the air showers initiated by these particles in the atmosphere last for many years. The low flux of cosmic rays at very high energy is one of the reasons for the difficult progress in their understanding.

Another reason is the necessity for particle and nuclear physics input in the analysis of indirect cosmic ray experiments, such as air showers. The first accelerator that studied hadronic interactions at energies applicable to Fig. 1.2, ISR, started working in 1971, more than thirty years after the existence of cosmic ray particles of that energy was shown by Pierre Auger. And ISR studied proton–proton not nucleus–nucleus collisions. The adequate

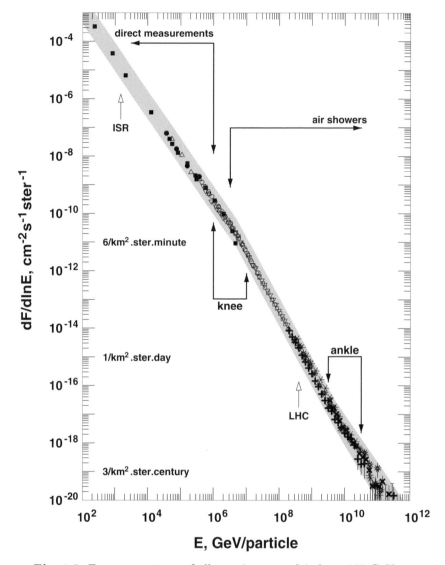

Fig. 1.2. Energy spectrum of all cosmic ray nuclei above 100 GeV.

analysis of the events and the reconstruction of the primary particle characteristics at high energy was, and still remains, difficult and model dependent.

The argument, however, works both ways. The next, and possibly last,
accelerator at record energy, the Large Hadronic Collider, will reach the
equivalent laboratory energy of approximately 4×10^8 GeV. The only way
to measure the evolution of the hadronic interactions at higher energy and
test their understanding and the theoretical models is through studies of more
energetic cosmic ray interactions.

If we come back to Fig. 1.2, we can roughly identify the two positions at which the spectral index changes and the three energy ranges they define. The region between 10^6 and 10^7 GeV where the cosmic ray spectrum becomes steeper is called the 'knee'. Below the 'knee' the number of particles decreases by a factor of 50 when the threshold energy is increased by ten. Above the 'knee' this factor is about 100. At higher energy the spectrum becomes again flatter at the 'ankle'. We believe that cosmic rays below the 'knee' are accelerated at supernova remnants, that particles of energy between the 'knee' and the 'ankle' come from some other galactic sources (possibly nonstandard supernova remnants) and that the highest energy particles are of extragalactic origin.

To develop a solid theory of the origin of cosmic rays in the whole energy range one needs detailed information about all possible sources and their environments. We have to know the structure and the strength of the magnetic fields and their extension. In spite of the huge progress in astrophysics during the past twenty or thirty years such information is only available for a small number of objects. The conditions in interstellar and extragalactic space are even less known.

This argument can also be turned around. The understanding of the details of cosmic ray diffusion in the Galaxy, and respectively the chemical composition of cosmic rays at their sources will lead to additional progress in astrophysics. The solution of these problems is done through trial and error – build a model and test it, understand why it does not work and then build a better model. This is a slow process.

The process of research contains periods of frustration and dissatisfaction, but I would not agree that the progress in cosmic ray physics is slower than in many other fields. The field involves relatively few scientists and its progress is balanced with the progress in the other disciplines that it requires input from and that use its results.

1.3 Main topics for future research

All predictions for the future of a scientific field are very uncertain. I will take the risk and attempt to outline the regions where the need for quick progress is urgent and the chances for such progress are good.

The first region is probably the measurement of the the cosmic ray chemical composition in the region of the 'knee'. At GeV energy hydrogen and helium nuclei dominate the cosmic ray spectrum but there are indications that heavier nuclei have energy spectra flatter than hydrogen. The tendency is that the average mass of the cosmic ray nuclei increases with energy. This is a widely accepted fact although the details of the energy dependence vary from analysis to analysis.

If, as we currently believe, the 'knee' of the cosmic ray spectrum marks the limiting energy for a class of cosmic ray accelerators, the composition should become heavier when that limit is approached. The reason is that the limiting high energy for a nucleus is proportional to its charge. The proton spectrum from an accelerator will cut off at energy 26 times lower than the cutoff of iron. If a new type of accelerator takes over at higher energy one should expect that, at least at the beginning, hydrogen and helium nuclei will again dominate and the average mass will decrease.

The exact solution of this problem cannot be achieved in theoretical investigations. It needs better experimental data. These consist of two different types. There is a chance now that a sophisticated modern detector could fly at the International Space Station and measure directly the cosmic ray chemical composition up to 10^6 GeV. Such data will provide an overlap with measurements of air showers on the ground. This overlap can be used to improve the analysis of the air shower experiments, which are the only experiments that can reach the highest cosmic ray energies.

Considerable progress has already been made in the use of standard hadronic interaction models and shower simulation codes. Because of this we can now judge objectively the differences between the various experimental approaches and results.

A second topic of high current interest is the end of the cosmic ray spectrum. This is the region marked with 3 particles per square kilometer per steradian per century in Fig. 1.2. One needs experiments with an effective area of thousands of square kilometers to collect reasonably good statistics. Such experiments are now becoming operational or are in the the process of construction. They will easily triple the world statistics during their first year of operation. It is very important that these new experiments not only measure the energy, but also investigate the type of the primary cosmic ray particles. This will give us an important clue for the processes that create the highest energy particles in Nature.

There are also ambitious projects for the construction of space-based air shower experiments that will have effective areas in millions of square kilometers. Some of these projects may be realized on a timescale of 10 years or even less.

The third topic is the creation of a new type of astronomy – the high energy neutrino astronomy. Optical astronomy is possible only when the astrophysical objects are not obscured by large amounts of matter. Higher energy photons penetrate matter more easily, but at an energy above 1000 GeV start interacting with the isotropic optical and infrared background and, at still higher energy, with the microwave background. Only neutrinos can penetrate the Universe without interactions at energies higher than 10^5 GeV.

There is also a different reason for the development of neutrino astronomy. Everything we touch consists of protons, neutrons and electrons. The number of protons in cosmic rays is about 100 times higher than the number of electrons. Astrophysics, however, considers almost exclusively electromagnetic processes, and does not account for the very likely presence of nucleons and hadronic interactions. Photons of all energies will be created in both electromagnetic and hadronic processes. Neutrinos belong exclusively to hadronic and nuclear interactions. Detection of high energy neutrinos of astrophysical origin will reveal the importance of hadronic interactions in astrophysical objects.

The construction of the first high energy neutrino telescope was started about ten years ago. The observational technique was proved and now a cubic kilometers detector is funded for construction at the South Pole. Another large neutrino telescope is in construction in the Mediterranean. We are close to the realization of a new observational technique that will help us understand the Universe better.

1.4 How is this book organized

The idea for this book originated seven years ago. I was asked if I was interested in writing a book on cosmic rays. My first reaction was that there were many good books on this subject and I would hardly be able to write a better one.

I should list here the books I admire and which have contributed much to what I know about cosmic rays. The first one I want to acknowledge is the book of Rossi [1]. Rossi presents the 1952 knowledge of cosmic rays and particle physics in a very clear and exact way. The book contains a still very useful description of electromagnetic cascade processes. The book of Ginzburg & Syrovatskii [2] set the stage for the current standard theory of cosmic rays. The book of Hayakawa [3] discusses both the physics of cosmic ray detection and the astrophysics aspects of cosmic ray research. Some parts of the book are now outdated. The book of Hillas [4] contains elegant original derivations of many cosmic ray properties. It also contains reprints (and translations) of 14 important classical cosmic ray articles.

The book of Berezinsky et al. [5] covers all astrophysical aspects of cosmic rays in a consistent manner. The book of Gaisser [6] has become very popular because it derives the basic equations that govern the development of cosmic ray cascades and gives the reader the instruments to calculate the fluxes of secondary particles. The book of Longair [7] uses cosmic ray examples to illustrate many processes in high energy astrophysics. Grieder [8] has collected a large amount of cosmic ray data and presents them with a clear brief explanation of the relevant processes. Schlickeiser recently published [9] a book that contains exact definitions and plasma processes descriptions. And for those who want to learn more about solar energetics and solar neutrinos I recommend the book of Bahcall [10].

This book is divided in two parts. Part I describes the *standard model of cosmic rays* and gives some of the data that helped in its construction. It starts with a short introduction to the interactions that are important for understanding the processes that are included in the model. The process description is limited to low and moderate energies and does not include their high energy extensions.

Chapter 3 starts with some knowledge of the astrophysical processes that precede the cosmic ray acceleration. It briefly describes solar energetics, stellar evolution, supernova explosions and supernova remnants. I also include data on solar and supernova neutrinos – the results of observations that were crucial for the confirmation of the theoretical predictions for stellar evolutions. Some of the subsections are typed in *italic*. They contain more detailed information than the general flow of the subject requires and can be skipped by a reader who has no interest in these topics. Sections in *italics* appear in all chapters of Part I. The second section of Chap. 3 deals with the acceleration of charged particles. The description is not very detailed – it only aims to familiarize the reader with the basic acceleration scenario.

Chapter 4 is dedicated to the passage of the cosmic rays through the Galaxy. It gives some basic information about the conditions in interstellar space, introduces a model of the galactic magnetic field and charged particle propagation in random magnetic fields. The next topic is the change of the chemical composition of the cosmic rays in their diffusion through the Galaxy. Cosmic ray pathlength and escape time are estimated on the basis of their isotopic composition. An additional source of information about the cosmic ray diffusion in the Galaxy, the isotropic flux of GeV γ-rays generated by cosmic rays, is discussed at the end of this chapter.

Chapter 5 is the first of three chapters that describe the experimental data taken at the top of the atmosphere, at the Earth's surface and underground. It starts with a brief description of the different types of detectors and of the changes in cosmic rays that are caused by heliospheric and geomagnetic effects. Then it gives the general picture of the cosmic ray spectrum and composition as measured by satellite and balloon experiments, followed by more detailed data on separate nuclear components. The chapter ends with a discussion of the fluxes of cosmic ray electrons and antiprotons.

Chapter 6 deals with cosmic ray processes in the atmosphere. It starts with the structure of the atmosphere and gives some of the analytic solutions for the atmospheric fluxes of secondary particles. The fluxes of atmospheric muons are then given and compared to Monte Carlo calculations.

Chapter 7 is devoted to the muons and neutrinos that are detected in underground experiments. It discusses high energy muon energy loss and propagation through rock and water and gives the exact formulae for the different energy loss processes. The next topic is the production and detection of atmospheric neutrinos. The section gives the neutrino interaction cross-sections as a function of the neutrino type and energy and discusses experimental results from different methods of atmospheric neutrino detection. The chapter ends with an introduction to neutrino oscillations that covers both atmospheric and solar neutrinos.

The second part of the book describes the contemporary challenges in cosmic ray research. It covers the three topics introduced in Sect. 1.3. Chapter 8 starts with a general discussion of cascades on the basis of the electromagnetic cascade theory. The accuracy of the theoretical approaches is discussed in comparison to Monte Carlo calculations. Attention then is drawn to hadronic showers at high energy. First a simple shower phenomenology is discussed and different detection methods and experimental arrangements are introduced together with the extension of the hadronic interaction models to high energy. The chapter ends with the methods for analysis of air showers and the data for the cosmic ray spectrum and composition at and above the 'knee' of the cosmic ray spectrum.

Chapter 9 discusses the end of the cosmic ray spectrum. It starts with an introduction to the energy loss of the cosmic ray nuclei in propagation in extragalactic space. Individual giant air shower detectors that have been built to explore the end of the cosmic ray spectrum and their results are followed by models for production of ultra high energy cosmic rays. The next topic is ultra high energy cosmic ray astronomy – the arrival directions of these particles are discussed together with the results of proton scattering in galactic and possible extragalactic magnetic fields. The chapter ends with a description of future giant air shower detectors.

The last chapter discusses the topic of high energy neutrino astronomy. High energy γ-ray astronomy results and analyses are used to identify potential sources and signal strengths in galactic and extragalactic sources. Individual sources are described and predictions are given for the isotropic high energy neutrino flux. Neutrinos from the propagation of ultra high energy protons are introduced to justify the relation between ultra high energy cosmic rays and high energy neutrinos. The book ends with a discussion of the possible methods for neutrino detection and of projects for different types of neutrino telescopes.

The reader should know that all the figures in this book have been generated by the author.

2 Cosmic ray interactions

This book is not a book on high energy physics and particle interactions. We have, however, to give the reader some information on the structure of matter and the interactions between its building blocks, because these are necessary for the understanding of the phenomena of cosmic ray acceleration, propagation in the Universe, and detection.

This chapter gives a simple introduction to our understanding of the structure of matter and of the different interactions that cosmic rays undergo in their propagation from their sources to us. The description of the interactions is brief and biased toward higher energy particles, with an energy of about 1 GeV and higher, which are the main subject of our interest. Three types of interactions are discussed:

- electromagnetic interactions of charged particles, which in are mostly important for the propagation of electrons and photons;
- inelastic hadronic interactions, that are important for the production of secondary particle fluxes;
- nuclear interactions, when heavier nuclei are split into lighter ones, that are mostly important for changes of the chemical and isotopic composition of accelerated cosmic ray nuclei.

We will not discuss the weak interactions in this section. Discussion and formulae for the interactions of neutrinos will be given in Sect. 7.2.1.

2.1 Components and structure of matter

The progress in understanding the structure of matter is intimately and naturally linked to the exploration of smaller and smaller dimension. Rutherford's experiments revealed the existence of the atomic nucleus which takes up a very small fraction of the volume of the atom. Nuclei consist of their components, protons (p) and neutrons (n). The electrons (e^-) are negatively charged particles that orbit the positively charged nucleus to complete the atom.

With the exceptionally rapid development of the experimental particle physics in the last fifty years the number of such 'elementary' particles became very large. The last issue of the *Review of Particle Properties* [11], that keeps

track of all results in the field, lists 143 particles the existence of which
is firmly established and whose properties are well known. Most of these
particles are not stable. They decay with a very short lifetime. The long-living
π^+ decays in 0.026 μs and most other lifetimes are shorter by many orders
of magnitude. There are also as many particle 'candidates', the properties of
which are not well known and which do not satisfy the conditions for fully
established 'elementary' particles.

A few of these particles, such as the electron and electron neutrino (ν_e)
are truly elementary, i.e. they are indeed among the building blocks of matter. Both the electron and the neutrino are *leptons*. Others, the *hadrons*, are
combinations of smaller blocks, *quarks*, which have never been observed individually, in isolation. Their properties have been derived from the properties
of the hadrons they build up because of the conservation of the quantum
numbers that these particles carry. A third type of particles is called gauge
bosons. These carry the forces between the hadrons and the leptons.

Table 2.1 gives information about the properties of some of the quarks
and leptons. Each quark and lepton has its antiparticle. A proton consists of

Table 2.1. Basic building blocks of matter.

Name	Baryon number	Lepton number	Charge
Quarks:			
up (u)	1/3	0	2/3
down (d)	1/3	0	−1/3
Leptons:			
electron (e^-)	0	1	−1
electron neutrino (ν_e)	0	1	0

two up quarks and one down quark. Its structure is *uud*. Consequently it has
a baryon number of 1 $(1/3 + 1/3 + 1/3)$ and charge +1 $(2/3 + 2/3 - 1/3)$.
All charges are measured in units of the electron charge. A neutron consists
of one up quark and two down quarks (*udd*). It is neutral $(2/3 - 1/3 - 1/3)$
and has a baryon number 1. The antiparticle of the proton, the antiproton
consists of $\bar{u}\bar{u}\bar{d}$. It has a charge of -1 and baryon number -1. All baryons are
strongly interacting particles. Other hadrons, which also interact strongly, are
the mesons. Mesons consist of a quark–antiquark combination. The positive
pion π^+ is a $(u\bar{d})$ combination and has a charge of 1 $(2/3 - (-1/3))$ and
baryon number 0 $(1/3 - 1/3)$.

Quarks have also an additional quantum number specific to them, *color*,
which allows the combinations of identical quarks, which otherwise would
have been forbidden by Fermi's statistics. The classical example for that is

the doubly charged baryon Λ^{++} which consists of three up quarks (uuu) of different colors.

2.1.1 Strong, electromagnetic and weak interactions

The particles shown in Table 2.1 represent only one of the three families of quarks and leptons. Table 2.2 gives the names of all quarks and leptons and the types of their interactions. All charged particles have electromagnetic

Table 2.2. Quarks, leptons, interactions they participate in, and the force carriers.

Name	The three families			Interactions	Gauge boson
Quarks	u (up)	c (charm)	t (top)	strong &	g
	d (down)	s (strange)	b (beauty)	EM	γ
Leptons	e (electron)	μ (muon)	τ (tau)	EM &	γ
	ν_e	ν_μ	ν_τ	weak	$W^\pm, Z$

interactions. Hadrons have also strong interactions and neutral leptons have only weak interactions. These three types of interactions reflect the strength and extension of the corresponding forces. The strong force has a short range of the order of the radius of a proton (1 fm = 10^{-13} cm) and strength α_s of 1. The force is carried by gluons (g). The electromagnetic force is carried by γ-rays and its coupling constant α has strength lower by two orders of magnitude. The weak force is carried by the intermediate vector bosons $W^\pm$ and the neutral Z and has a coupling α_W of the order of $10^{-6}\alpha_s$.

These features are also reflected in the corresponding particle decays. Weak decays, like $\pi^+ \rightarrow \mu^+\nu_\mu$ have lifetimes in excess of 10^{-12} seconds. Note the conservation of the quantum numbers in the decay. The sum of the lepton numbers of the decay products is 0, as is that of the parent π^+. Electromagnetic decays ($\pi^0 \rightarrow \gamma\gamma$) have lifetimes shorter than 10^{-16} s, while decays guided by the strong force have lifetimes of the order of 10^{-23} s.

2.1.2 Units of energy and interaction strength

The basic unit of energy in particle physics and cosmic ray physics is the electronvolt (eV). This is the kinetic energy gained by an electron by passing through a potential difference of 1 V. Different appropriate energy measures are obtained by scaling the eV in threefold order of magnitude units, i.e. a kiloelectronvolt (KeV) is 10^3 eV, megaelectronvolt (MeV) is 10^6 eV, (giga) GeV = 10^9 eV, (tera) TeV = 10^{12} eV, (peta) PeV = 10^{15} eV, (eta) EeV = 10^{18} eV and (zeta) ZeV = 10^{21} eV. The total particle energy and the kinetic energy $E_k = E - mc^2$ are measured in the same units. Particle momenta $p = (E^2 - m^2c^4)^{1/2}$ are measured in eV/c.

The interaction strength is measured by the interaction cross-section σ, which is expressed in units of area. The basic unit is the barn $= 10^{-24}$ cm^2. Common units are the millibarn, 1 mb $= 10^{-3}$ b and the microbarn, 1 μb $= 10^{-6}$ b. Cross-sections are usually given per one nucleon (or nucleus) of target. If a particle has interaction cross-section of $\sigma = 1$ mb, its mean free path in a medium of nucleon density $\rho = 10^3$ cm^{-3} is $\lambda = (\sigma\rho)^{-1} = 10^{24}$ cm. If the density ρ is in terms of g/cm^3 then the mean free path is calculated in terms of the column density g/cm^2, i.e. $\lambda = [(N_A/A)\rho\sigma]^{-1}$ g/cm^2, where N_A is Avogadro's number and A is the mass number of the target.

Other quantities that we will use soon are the particle Lorentz factor $\gamma = E_{tot}/mc^2$ that is the ratio of the total particle energy and its velocity $\beta = v/c$ in terms of the speed of light. These quantities are related in the following way:

$$\gamma = \frac{1}{\sqrt{1-\beta^2}}$$

2.2 Electromagnetic processes in matter

Most of the information given in this section is directly applicable only for electrons. Later, in Chap. 7 we discuss the electromagnetic interactions and the energy loss of muons.

2.2.1 Coulomb scattering

The basis of all elecromagnetic interactions is the Coulomb scattering between electric charges. The force between two point charges q_1 and q_2 at distance R from each other is

$$F = \frac{q_1 q_2}{R^2} n \;, \tag{2.1}$$

where n is the unit vector from one of the charges to the other.

Experimentally this process was studied by Rutherford in which he discovered the structure of the atom [12]. Rutherford bombarded heavy nuclei with α particles (He nuclei) and measured the angular deflection of the projectile nuclei.

In the Coulomb field the particle trajectory changes. The deflection angle ϑ depends on the impact parameter b between the two charges and the velocity and mass of the particles that carry them. The impact parameter is the closest distance between the two particles with charges q_1 and q_2. The deflection (scattering) angle is

$$\tan\frac{\vartheta}{2} = \frac{zZe^2}{Mv^2b} \;. \tag{2.2}$$

Equation (2.2) expresses the charges in terms of the charge of the electron
e. In the case of a projectile electron with charge $z = 1$ one can write the
differential cross-section for scattering as

$$\frac{d\sigma}{d\Omega} = \frac{b}{\sin\vartheta}\frac{db}{d\vartheta} = \frac{Z^2}{4}r_e^2 sin^{-2}\frac{\vartheta}{2}, \qquad (2.3)$$

where $r_e = e^2/(m_e c^2)$ is the *classical* radius of the electron and Z is the
charge of the medium.

The fact that the electrons change their direction suggests that there
is transfer of energy between the two particles. This is expressed through
the momentum transfer in terms of the momentum of the projectile particle,
which here is an electron. The momentum transfer q is related to the electron
momentum before the scattering p and the scattering angle ϑ as

$$q = 2p\sin\frac{\vartheta}{2}. \qquad (2.4)$$

There are two consequences from the scattering: the projectile particle
changes its direction and its energy is changed in the scattering process.

The formulae above are strictly valid for point-like charges moving with
nonrelativistic velocity. Several corrections have to be introduced for relativistic particles and for more realistic scattering conditions.

The correction for relativistic particles, that was introduced by Mott, adds
the term $(1 - \beta^2 \sin^2\frac{\vartheta}{2})$ to the differential cross-section in (2.3).

Another correction is needed to account for the size of the target nucleus.
This is the nuclear formfactor, that accounts for the distribution of charge
inside the nucleus.

A third important one is for the screening of the nuclear field by the atomic
electrons. The basic correction for screening is in the form $(1+\frac{1}{Za})^{-2}$, where
the screening radius a is defined as the exponent of the potential decrease with
distance. The screening radius is often approximated as $a = (\hbar^2/me^2)Z^{-1/3}$.

2.2.2 Ionization loss

Charged particles traveling through matter lose energy on excitation and
ionization of its atoms. The energy loss per unit of column depth (in units of
MeV per g/cm^2) is:

$$\frac{dE}{dx} = -\frac{N_A Z}{A}\frac{2\pi(ze^2)^2}{Mv^2}\left[\ln\frac{2Mv^2\gamma^2W}{I^2} - 2\beta^2\right], \qquad (2.5)$$

where Z is the atomic number of the medium, A is its mass number and I is
its average ionization potential. N_A is Avogadro's number. ze is the charge
of the the particle, v is its velocity, and M - its mass. γ and β characterize

particle energy and momentum. This expression is obtained by Hayakawa [3] by integrating the formulae of Bethe [13] and Bloch [14] to the maximum energy loss W. The ionization loss is thus proportional to a constant L that includes the charge and atomic number of the medium.

$$L \equiv \frac{2\pi N_A Z}{A} \left(\frac{e^2}{mc^2}\right)^2 mc^2 = 0.0765 \left(\frac{2Z}{A}\right) \text{MeV(g/cm}^2)^{-1} . \quad (2.6)$$

In rarefied media the ionization energy loss increases logarithmically with the particle energy. In (2.5) this is expressed through the γ^2 term. In denser media this increase is suppressed (the density effect), which is accounted for by introducing the term $-\delta$ in (2.5). The energy loss on ionization can then be written in a simplified form as

$$\frac{dE}{dx} = -L\frac{Z^2}{\beta^2} \left(B + 0.69 + 2\ln\gamma\beta + \ln W - 2\beta^2 - \delta\right) \text{MeV(g/cm}^2)^{-1} ,$$

$$(2.7)$$

where $B \equiv \ln(mc^2/I^2)$, $W \simeq E/2$ and $\delta = 2\ln\gamma\beta$ plus a correction C depending on the particle energy and on the properties of the medium. The $2\ln\gamma\beta$ term compensates for the logarithmic increase of the ionization loss. The parameters guiding the ionization energy loss including the density effect are given in Table 2.3 for particles with momenta $\gg M$ as calculated by Sternheimer [15]. C values in Table 2.3 are only correct at high energy when the ionization loss is almost constant. The low energy values can be found in Hayakawa's book [3] or in the original paper. We give values for a sample of materials, which are important for cosmic ray propagation in the interstellar medium (H, He), in the atmosphere (N, O), and in particle detectors (C, Fe).

Table 2.3. Parameters guiding the energy loss on ionization in different media [3].

Element	I, eV	L	B	C
Hydrogen	21.8	0.152	21.07	−9.50
Helium	44.0	0.077	19.39	−2.13
Carbon	77.8	0.077	18.25	−3.22
Nitrogen	90.9	0.077	17.94	−10.68
Oxygen	104	0.077	17.67	−10.80
Iron	286	0.072	15.32	−4.62

Equation (2.7) gives the average energy loss on ionization. In fact, the energy loss has significant fluctuations, especially when the thickness of the target is small.

2.2.3 Cherenkov light

A small fraction of the energy loss is emitted in the form of Cherenkov radiation. The name is after its discoverer, P. Cherenkov. Cherenkov light is emitted when a particle moves in a medium with velocity βc greater than the phase velocity of the light c/n, where n is the refraction index of the medium.

This requirement sets a threshold energy for the emission of Cherenkov light that depends on the value of the refraction index. The refraction index of air at sea level is $n = 1.0003$ and the threshold energy is $E_{thr} > m\sqrt{1 - 1/n^2}$ (m is the particle mass) ~ 21 MeV for electrons. The refraction index in water is 1.33 which gives much lower threshold energy, about 1 MeV for electrons.

The Cherenkov light is emitted on a cone around the particle trajectory. The cone opening angle is

$$\cos\theta = \frac{1}{\beta n} + q, \tag{2.8}$$

where q is a quantum correction factor with small practical importance. The maximum opening angle is achieved at high energy when $\beta = 1$ and $\cos\theta$ is inversely proportional to the refraction index of the medium.

The intensity of the radiation per unit pathlength is proportional to the square of the particle charge and is

$$dN/dL = z^2 \frac{\alpha}{\hbar c}\left[1 - \frac{1}{\beta^2 n^2}\right], \tag{2.9}$$

where $\alpha(\hbar c)^{-1} = 370$ eV^{-1} cm^{-1}. The second factor in (2.9) introduces an energy dependence in the threshold energy range. The wavelength distribution of the emitted Cherenkov light is proportional to λ^{-2} and is distributed in the visible and UV range.

2.2.4 Compton scattering

The process in which photons interact with the atomic electrons and transfer a fraction of their energy to the electrons is the Compton scattering. The differential cross-section for Compton scattering of a photon of energy k is given by

$$\sigma_C(k, k') = 2\pi r_e^2 \frac{1}{k'}\frac{1}{q}\left[1 + \left(\frac{k'}{k}\right)^2 - \frac{2(q+1)}{q^2} + \frac{1+2q}{q^2}\frac{k'}{k} + \frac{1}{q^2}\frac{k}{k'}\right], \tag{2.10}$$

where k' is the photon energy after the scattering and q is the primary photon energy in units of electron mass – $q \equiv k/mc^2$.

The total Compton scattering cross-section can be integrated from (2.10) to

$$\sigma_C(k) = \frac{\pi r_e^2}{q} \times \left[\left(1 - \frac{2(q+1)}{q^2}\right) \ln(2q+1) + \frac{1}{2} + \frac{4}{q} - \frac{1}{2(2q+1)^2} \right].$$

$$(2.11)$$

At low q values, $q \ll 1$, σ_C approaches the Thomson cross-section σ_T ($= 8\pi r_e^2/3 = 665$ mb) and decreases with increasing energy. For k much greater than the electron mass the total cross-section is well represented by the much simpler formula

$$\sigma_C \simeq \sigma_T \frac{3}{8q} \left(\ln 2q + \frac{1}{2} \right) \qquad (2.12)$$

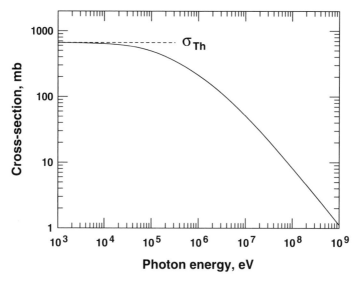

Fig. 2.1. Cross-section for Compton scattering as a function of the photon energy.

The angle between the primary and the secondary photon θ can be expressed as a function of the energies of the photon before and after the scattering.

$$\cos\theta = 1 - \frac{mc^2(k/k' - 1)}{k}, \qquad (2.13)$$

which makes Compton scattering a convenient physical base for construction of particle detectors. If the scattered photon and electron are both detected this gives not only the energy of the primary photon but also its direction.

2.2.5 Bremsstrahlung

Charged particles also interact with the electromagnetic field of the atomic nuclei and generate photons. The process is called bremsstrahlung. The energy loss on bremsstrahlung is

$$\frac{dE}{dx} = -\frac{N}{A} \int_0^{E-mc^2} \sigma_{br}(E,k) k \, dk \, , \tag{2.14}$$

where E is the energy of the charged particle and k is the energy of the emitted photon. The bremsstrahlung cross-section σ_{br} for electrons is given as a function of E and k as [16]

$$\sigma_{br} = \frac{4Z^2 \alpha r_e^2}{k} F(E,k) \, , \tag{2.15}$$

where $r_e = e^2/\hbar c$ is the classical radius of the electron and α is the fine splitting constant. This cross-section is calculated by Bethe and Heitler [16]. The function $F(E,k)$ depends on the screening parameter ξ, which expresses the screening of the nuclear field by the atomic electrons.

$$\xi \equiv 100 Mc^2 \frac{k}{E} \frac{1}{E-k} Z^{-1/3}. \tag{2.16}$$

The screening parameter ξ is inversely proportional to the energy of the charged particle and is proportional to the ratio of the electron energy before and after the process. As a function of the ratio of the photon to the electron energy $u = k/E$

$$F(E,k) = \left[4(1-u)/3 + u^2\right] \ln Z^{-1/3} + (1-u)/9 \tag{2.17}$$

in the case of vanishing $\xi \simeq 0$, which generally describes interactions of high energy electrons. For large ξ values (in the no screening regime)

$$F(E,k) = \left[4(1-u)/3 + u^2\right] \left[\ln\left(\frac{2E}{mc^2} \frac{1-u}{u}\right) - 1/2\right] \tag{2.18}$$

Figure 2.2 shows the form of $F(E,k)$ in these two regimes for electron energy of 100 MeV. The correct representation would be the use of the full screening formula for small k values with the no screening formula for k approaching the electron energy and intermediate formulae in between.

Because of the term $1/k$ the differential cross-section for bremsstrahlung becomes infinite when k approaches 0 and creates the 'infrared catastrophe'. This, however, integrates out when the total energy loss is calculated. For vanishing ξ the energy loss is

$$\frac{dE}{dx} = \frac{4NZ}{A} \alpha r_e^2 E \left[\ln 191 Z^{-1/3} + 1/18\right] \, . \tag{2.19}$$

The correction term $1/18$ in (2.19) comes from the interactions with the fields of the atomic electrons. The interaction cross-section has the the same form as (2.15) with the Z^2 term replaced by Z. The total bremsstrahlung cross-section is thus proportional to $Z(Z+1)$. This nuclear formfactor varies

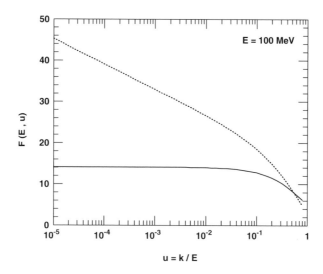

Fig. 2.2. $F(E,k)$ from (2.17) (solid line) and (2.18) (dotted line) for electron energy 100 MeV.

slightly from this general form and for practical reasons is better represented as $Z(Z + 1 \pm \epsilon)$.

At a certain energy ε_0 the energy loss for bremsstrahlung equals the ionization energy loss. ε_0 is called critical energy and decreases with the charge of the medium Z.

The general form of the energy loss in (2.19) allows the introduction of the radiation length X_0 which gives the average amount of matter for bremsstrahlung energy loss.

$$X_0 \equiv \left[\frac{4NZ(Z+1)}{A} \alpha r_e^2 \ln\left(191 Z^{-1/3}\right) \right]^{-1} \tag{2.20}$$

Approximate values for X_0 can be easily calculated with the formula recommended by Hayakawa [3]

$$X_0 \simeq 10^3 \times \frac{A}{6Z(Z+1)} \text{ g.cm}^{-2} . \tag{2.21}$$

More exact values of X_0 and ε_0 in different targets are given in Table 2.4, which is partially extracted from the *Review of Particle Properties* [11].

For mixtures of different elements the radiation length is calculated as a weighted sum of the radiation lengths for the different components

$$\frac{1}{X_0} = \sum_i \frac{w_i}{X_0^i} , \tag{2.22}$$

Table 2.4. Radiation lengths and critical energies for the most common elements. The radiation length values are from [11] and the critical energies from [3].

Element	Z	A	X_0, g/cm^2	ε_0, MeV
Hydrogen	1	1.01	61.28	350.
Helium	2	4.00	94.52	250.
Carbon	6	12.01	42.70	79.
Nitrogen	7	14.01	37.99	85.
Oxygen	8	16.00	34.24	75.
Silicon	14	28.09	28.08	37.5
Iron	26	55.85	13.84	20.7

where w_i is the fraction of weight of the component element in the compound mixture. If we assume, for example, that the atmosphere consists only of 25% O nuclei and 75% N nuclei, and has an average atomic weight of 14.5, (2.22) will give us $1/X_0 = 14 \times 0.25/34.24 + 16 \times 0.75/37.99 = 0.02712$ and we obtain $X_0(\text{air}) = 36.9$ g/cm^2. There is still some ambiguity about the exact value of the radiation length in air. The particle data book [11] gives $X_0 = 36.66$ g/cm^2 for sea level and temperature of 20°C. In cosmic ray physics an accepted number is 37.1 g/cm^2.

2.2.6 Creation of electron–positron pairs

It is convenient to discuss the creation of pairs after the bremsstrahlung because physically it is the inverse process. The pair production cross-section σ_{pair} can be calculated by substituting the electron and the positron of the pair for the electron before and after bremsstrahlung. The cross-section then becomes

$$\sigma_{pair}(k, E) = \sigma_{br}(E, k)\frac{E^2}{k^2} = \frac{4Z^2\alpha r_e^2}{k} G(k, E) ,$$

where k is the energy of the primary photon. The function $G(k, E)$ is of the order 1 and can be expressed as a function of the ratio $v = E/k$. E is the energy of one of the members of the pair. The shape of $G(k, E)$ also depends on the screening parameter ξ. For the case of full screening

$$G(k, v) = [1 + 4v(v - 1)/3]\ln\left(191Z^{-1/3}\right) - v(1 - v)/9 . \qquad (2.23)$$

For no screening

$$G(k, v) = [1 + 4v(v - 1)/3] \times \left[\ln\left(\frac{2k}{mc^2}\right)v(1 - v) - 1/2\right] . \qquad (2.24)$$

Figure 2.3 shows the function $G(k, v)$ for primary proton energy $k = 100$ MeV in the two extreme cases of full screening and no screening, which at this energy are very different. It is thus extremely important to use the correct expressions as the incorrect application of the simple equation (2.23) can lead to errors of one order of magnitude.

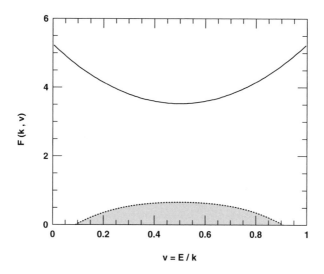

Fig. 2.3. $G(E, k)$ from (2.23) (solid line) and (2.24) (dotted line, shaded areas) for primary photon energy 100 MeV.

The total pair production cross-section can be directly integrated

$$\sigma_{pair}(k) = \int_{mc^2}^{k-mc^2} \sigma_{pp}(k, E) \, dE \qquad (2.25)$$

$$= 4Z^2 \alpha r_e^2 \frac{\ln\left(191 Z^{-1/3}\right)}{9} - \frac{1}{54}$$

in the case of vanishing ξ. The correction term $1/54$ comes from pair production in the field of the atomic electrons.

2.3 Electromagnetic collisions on magnetic and photon fields

2.3.1 Synchrotron radiation

Synrotron radiation is a very important energy loss process for charged particles in the presence of magnetic fields. In astrophysics it is often called magnetic bremsstrahlung. An electron moving in magnetic field B with an angle θ to the field direction loses energy to synchrotron radiation at a rate

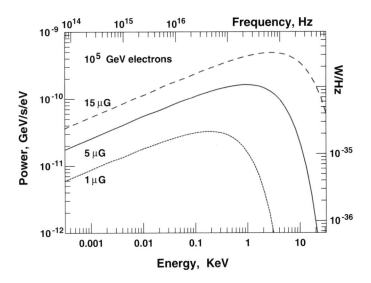

Fig. 2.4. Power spectrum of synchrotron radiation emitted by 10^5 GeV electrons in 1, 5, and 15 μG fields.

$$-\frac{dE}{dt} = 2\sigma_T c\gamma^2 U_B \beta^2 \sin^2\theta \, , \qquad (2.26)$$

where β and γ are the velocity and Lorentz factor of the electron, U_B is the energy density of the magnetic field ($= B^2/8\pi$) and σ_T is the Thomson cross-section. Note that the synchrotron energy loss is proportional to the square of the particle Lorentz factor and is thus inversely proportional to the square of the particle mass for the same total energy E_{tot}. A proton loses only $(m_e/m_p)^2 \simeq 3 \times 10^{-7}$ times as much energy as an electron of the same E_{tot}.

For an ensemble of electrons that are scattered randomly in all directions one could calculate the energy loss averaged over all pitch angles, which is

$$\left\langle -\frac{dE}{dt} \right\rangle = \frac{4}{3}\sigma_T c\gamma^2 U_B \qquad (2.27)$$

for relativistic electrons with $\beta \simeq 1$. Expressed in particle physics units the average energy loss becomes

$$-\frac{dE}{dt} = 3.79 \times 10^{-6} \left(\frac{B}{\text{gauss}}\right)^2 \left(\frac{E_e}{\text{GeV}}\right)^2 \text{ GeV/s} \, . \qquad (2.28)$$

The characteristic frequency of the radiated photons is the critical frequency

$$\nu_c = \frac{3}{4\pi}\gamma^2 \frac{eB}{m_e c}\sin\theta = 1.61 \times 10^{13} \left(\frac{B}{\text{gauss}}\right)\left(\frac{E}{\text{GeV}}\right)^2 \text{ Hz} \, . \qquad (2.29)$$

Expressed as a fraction of the electron energy the critical frequency is proportional to the product of the energy and the magnetic field value $\nu_c/E_e \propto E_e \times B$. The higher the energy and the magnetic field, the harder is the spectrum of the radiated photons.

The emissivity of a relativistic electron of energy E_e averaged over all pitch angles cannot be expressed in a final analytic form and is given by the integral

$$ j(E,\nu)\,d\nu = \frac{\sqrt{3}e^3 B}{m_e c^2} \int_0^\pi d\theta \, \sin^2 \theta/2(\nu/\nu_c)\,d\nu \int_{\nu/\nu_c}^\infty K_{5/3}(\eta)\,d\eta \,, \qquad (2.30) $$

where $K_{5/3}$ is the Bessel function of order 5/3. The number spectrum of the synchrotron radiation peaks at $0.29\nu_c$. Figure 2.4 shows the power spectrum of synchrotron radiation of a 10^5 GeV electron in 1, 5 and 15 μG fields. Note the shift of the critical frequency ν_c with the increase of the strength of the magnetic field.

Synchrotron radiation plays a very important role in astrophysics as described in the book [68] of Ginzburg. It is a major contributor to the non-thermal emission spectra of all astrophysical sources in a very wide frequency range stretching from radio waves to X-rays. The physical picture of the synchrotron emission is described in the book of Longair [7] and the full derivation of all relevant formulae is given by Blumenthal & Gould [69].

In spite of the complicated expression for the spectrum of photons radiated by a single electron the spectrum emitted by electrons with a power law distribution is well defined. If the differential power law index of the electron spectrum is α the integral spectral index of their synchrotron emission is $(\alpha - 1)/2$.

2.3.2 Inverse Compton effect

The formulae for Compton effect, equations (2.10) and (2.11), are also applicable to the inverse Compton effect, the interaction when an electron interacts with a photon from an ambient photon field, loses energy and boosts the photon. The formulae can be used when the energy of the primary electron E (which is assumed to be $m_e c^2$ in (2.10)) and the ambient photon ϵ are used to represent the photon energy in the electron rest frame, i.e.

$$ k = \frac{\epsilon E}{m_e c^2}(1 - \beta \cos\theta) \,, \qquad (2.31) $$

where $\cos\theta$ is the angle between the photon and the electron in the photon frame and β is the electron velocity in units of c.

The inverse Compton scattering is a very important process for the production of very high energy γ-rays when accelerated electrons collide with photons of the microwave background radiation of other ambient fields. In the Thomson regime ($\epsilon E \ll (m_e c^2)^2$) the cross-section is approximatelly

σ_T and the average energy of the boosted photon $E_\gamma = \epsilon(E/m_ec^2)^2$. For $\epsilon E \gg (m_ec^2)^2$ the average E_γ approaches the electron energy E. The cross-section is then much smaller.

Using these estimates we can also write the energy loss formulae for the electrons. In the Thomson regime

$$-\frac{dE}{dx} = \sigma_T U \left(\frac{E}{m_ec^2}\right)^2 , \qquad (2.32)$$

where U is the energy density of the photon field. In the high energy (Klein–Nishina) regime

$$-\frac{dE}{dx} = \frac{3}{8}\sigma_T U \left(\frac{m_ec^2}{\epsilon}\right)^2 \ln\left(\frac{2\epsilon E}{m_e^2 c^4}\right) . \qquad (2.33)$$

An important production mechanism for very high energy γ-rays combines the synchrotron radiation with the inverse Compton effect. High energy electrons first lose energy on synchrotron radiation and then boost the synchrotron photons to TeV by inverse Compton effect.

2.4 Inelastic hadronic interactions below 1000 GeV

This book will deal a lot with the interactions of hadrons in different energy ranges. At high energy these interactions are well understood and fairly well described by the Quantum Chromo Dynamics theory, which is based on the principles outlined in Sect. 2.1. At relatively low energies, though, the theory does not work and we have to rely on the phenomenological description of the particle interactions. This is what we will discuss in this section. Later, in part II, we shall discuss the way interactions are understood in terms of QCD. Let us start with several general definitions.

A particle of mass M that moves with a velocity βc cm/s is fully characterized by a four–vector $\mathbf{p}$. The components of $\mathbf{p}$ are the particle energy E and the particle momentum $\mathbf{p}\,(p_x, p_y, p_z) - \mathbf{p}^2 = E^2 - |\mathbf{p}|^2 = m^2$. The relative particle velocity $\beta = p/E$ and its Lorentz factor $\gamma = (1 - \beta^2)^{1/2} = E/m$.

It is very convenient to discuss hadronic interactions in the center of mass (CM) system, in which the momenta of the two interaction particles are collinear, have the same magnitude and point in opposite directions. The total CM energy of the interaction, $\sqrt{s}$ is

$$\sqrt{s} = (\mathbf{p}_1 + \mathbf{p}_2)^2 = \left[(E_1 + E_2)^2 - (\mathbf{p}_1 - \mathbf{p}_2)^2\right]^{1/2} . \qquad (2.34)$$

When one of the interacting particles is at rest

$$\sqrt{s} = (m_1^2 + m_2^2 + 2m_2 E_1^{Lab})^{1/2}, \qquad (2.35)$$

where E_1^{Lab} is the energy of the incident particle 1 in the rest frame of the particle 2. This is the laboratory (Lab) frame.

An inelastic interaction is by definition such an interaction where at least one new (secondary) particle is created. From (2.34) it follows directly that $\sqrt{s}$ has to be large enough to accommodate the mass of the secondary particle. If two protons collide and produce a neutral pion, $p + p \longrightarrow p + p + \pi^0$, the CM energy has to be larger than $\sqrt{s} > 2m_p + m_\pi = 2.01$ GeV. The minimum energy of the incident proton in the Lab frame is then $E_1 = s/2m_p - m_p = 1.22$ GeV. This is the absolute minimum energy required for the production of a neutral pion in the interactions of two protons.

The cross-section for inelastic interactions depends on the incident particle energy. Figure 2.5 shows the direct data on the pp inelastic cross-section measured in accelerator experiments from the compilation of Ref. [17]. Although the errors in the threshold region are significant, the data show a quick rise to ~ 30 mb in the threshold region between 1 and 2 GeV, and then a smooth logarithmic increase at higher energy. The measurement of the in-

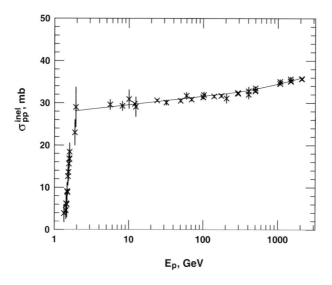

Fig. 2.5. Results from direct accelerator measurements of the pp inelastic cross-section. The data points are from the compilation of Ref. [17] and the line is the fit from Ref. [18].

elastic cross-section are very difficult, because they require detectors with a full 4π coverage of the interaction region. Much more certain estimates are obtained by separate measurements of the total σ_{pp}^{tot} and the elastic σ_{pp}^{el} cross-sections. σ_{pp}^{inel} is the difference between the total and the elastic cross-sections and its energy dependence is established very well by fits of all available measurements.

Δ resonance. Two-body decays. Lorentz transformation

The threshold energy range in Fig. 2.5 is dominated by the production of resonances. A resonance is a hadronic state with defined quantum numbers and quark content and mass depending on the amount of energy available for its production. The most common (with highest production cross-section) resonance is $\Delta(1232)$ with an average mass of 1.232 GeV and width $\Gamma = 115$ MeV.

The width of the Δ resonance is defined by the elastic cross section of its production in $p\pi$ (or $p\gamma$ collisions. It is described by the Breit–Wigner formula

$$\sigma_\Delta \propto \frac{\Gamma^2/4}{(E - m_\Delta)^2 + \Gamma^2/4} \, . \tag{2.36}$$

In pp interactions the Breit–Wigner formula reflects the amount of energy available for Δ production in the CM system. The width Γ is defined so that the production cross–section decreases by a factor of 2 at $E = m_\Delta + \Gamma/2$ from its maximum value at $E = m_\Delta$, as shown in Fig. 2.6.

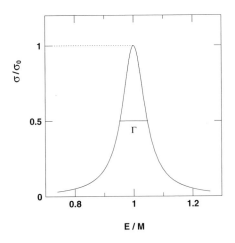

Fig. 2.6. The elastic cross-section for resonance production as a function of the ratio between the incident energy E and the resonance mass M as described by the Breit–Wigner formula.

In a two-body decay the two final state particles share the available energy (the mass of the decaying particle) according to the value of their own mass because their momenta are equal in absolute value. In the decay $M \longrightarrow m_1 + m_2 \ |\boldsymbol{p_1}| = |\boldsymbol{p_2}| = p$, where

$$p = \frac{\left[(M^2 - (m_1 + m_2)^2)(M^2 - (m_1 - m_2)^2)\right]^{1/2}}{2M} \; . \qquad (2.37)$$

In the frame of the decaying particle the energies $E_{1,2}$ are respectively $E_{1,2} = (M^2 + m_{1,2}^2 - m_{2,1}^2)/2M$. In the case of $\Delta \longrightarrow p + \pi^0$ decay the momenta of both the proton and the π^0 are $p^\Delta = 0.227$ GeV/c and the energies in the Δ frame are $E_p^\Delta = 0.965$ GeV and $E_\pi^\Delta = 0.267$ GeV.

The two decay products will have the same energies and momenta in the center of mass system only if the Δ were stationary in CM. Otherwise the energies and the longitudinal components of the momenta (in the direction of the Δ velocity have to be Lorentz transformed using the Lorentz factor of the Δ in the CM system $\gamma_\Delta = E_\Delta^{CM}/m_\Delta$. The transformation is:

$$E_p^{CM} = \gamma_\Delta(E_p^\Delta + \beta p^\Delta \cos\theta) \quad E_\pi^{CM} = \gamma_\Delta(E_\pi^\Delta - \beta p^\Delta \cos\theta) \; , \qquad (2.38)$$

where θ is the angle between the proton direction in the Δ frame and the Δ direction in the CM frame. The minus sign in the transformation of the pion energy appears because the proton and the pion move in opposite directions in the Δ frame. The longitudinal momenta $p_\parallel$ are transformed as

$$p_{p,\parallel}^{CM} = = \gamma_\Delta(p^\Delta \cos\theta + \beta E_p^\Delta); \qquad (2.39)$$
$$p_{\pi,\parallel}^{CM} = = \gamma_\Delta(-p^\Delta \cos\theta + \beta E_\pi^\Delta)$$

and the transverse momenta $p_\perp$ (normal to $p_\parallel$) are not changed.

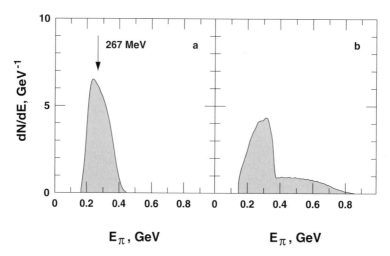

Fig. 2.7. Energy distribution of pions from Δ decay in the CM (a) and the Lab (b) systems. The Δ's are generated in pp collisions of Lab energy of 2 GeV.

Another Lorentz transformation is necessary to obtain the energy of the two decay products in the Lab system. The Lorentz factor of the CM system in the Lab is $\gamma_{CM} = (E_1^{Lab} + m_2)/\sqrt{s}$. Figure 2.7 shows the energy distributions of pions generated in $\Delta \longrightarrow p + \pi^0$ decays in pp interactions at $E_1^{Lab} = 2$ GeV. The arrow in the left-hand panel shows the pion energy in the Δ frame. The pion energy distribution in the CM frame is generated by the Δ velocity in the CM frame. The distribution in the Lab system is still much wider and shows two different components that are generated by Δ resonances moving forwards and backwards in the CM frame.

$\Delta(1232)$ is the lightest resonance that dominates the inelastic cross-section at its energy threshold. It is followed by a number of heavier resonances the production of which require higher CM energy. Each one of these resonances has lower cross-section than $\Delta(1232)$. The sum of all resonances in the *second resonance region* (heavier than $\Delta(1232)$), however, is at least equally important close to the threshold for inelastic interactions. The masses, widths and decay channels for all identified resonances are listed in the Particle Data Book [11].

2.4.1 Secondary particles spectra, average multiplicity and inelasticity

In the resonance region the multiplicity of the secondary particles in the final state (when all short-lived particles have already decayed) is fixed since every resonance has a well defined set of decays branches. At higher energy, $\sqrt{s} \approx 2.5$ GeV the resonance production is no longer dominant and the interactions are dominated by multiparticle production. The inelastic interaction at this and higher energy are described by a combination of parameters which parametrize the energy spectra and the multiplicity of the secondary particles.

The typical accelerator experiments in this energy range do not measure all secondaries, rather the particles of given type that are emitted with given momentum at certain angle to the direction of the incident particle beam. The general assumption is that the probability for the production of a particle with longitudinal momentum $p_\parallel$ and transverse momentum $p_\perp$ can be factorized in terms of the two components of the momentum as

$$\frac{d^2\sigma}{dp_\parallel dp_\perp} = \sigma^{inel} f(p_\parallel) \times g(p_\perp). \tag{2.40}$$

Experimental data that study the secondary particle production in different (overlapping or complementary) regions of the parameter space could be analyzed together in terms of σ^{inel} and the functions f and g.

A very important hypothesis of the analysis of accelerator data in this energy range is that with increasing energy, when $\sqrt{s}$ becomes significantly higher than the masses on the particles involved in the inelastic interactions, the cross-section σ^{inel} will become constant and the functions f and g will not depend on $\sqrt{s}$. There are different versions of this *scaling* hypothesis.

Feynman scaling, introduced by R. Feynman [19] postulates scaling in terms of $x_F = p^{CM}/(\sqrt{s}/2)$ where $\sqrt{s}/2$ is the maximum momentum that a particle can have in the center of mass system. In terms of x_F the momentum distribution of the secondary mesons in pp collisions have $\frac{1}{x}(1 - x_F)^n$ spectrum, with the power n for pions $n_\pi \approx 4$. The $p_\perp$ distribution approaches an energy independent shape with $\langle p_\perp^\pi \rangle = 0.34$ GeV/c.

Another version is that of *radial scaling* [20], which uses the scaling variable $x_R = E^{CM}/(2\sqrt{s})$. Hillas [18] uses the radial scaling hypothesis to parametrize the particle production data in pp collisions at Lab energy from 10 to 2000 GeV in the laboratory system. This parametrization is very useful for the description of cosmic ray collisions. The general form, after a transformation in the Lab system and integration over $p_\perp$ is

$$x\frac{dn}{dx} = f(x) \times H(E) , \tag{2.41}$$

where E is measured in the Lab system and $x = E/E_0$, where E_0 is the incident proton energy in the Lab.

For π^+, π^-, and π^0

$$f^\pi(x) = 1.22(1 - x)^{3.5} + 0.98 \exp{(-18x)} \tag{2.42}$$

and

$$H^\pi = \left(1 + \frac{0.4}{E - 0.14}\right)^{-1} , \tag{2.43}$$

where the energy is in GeV. Similar expressions describe the production of kaons and nucleon–antinucleon ($N\bar{N}$) pairs. The protons in this model have flat x distribution. The average transverse momenta for kaons and for the leading (fastest, most energetic) nucleon in this fit are respectively 0.40 and 0.50 GeV/c.

Equations (2.41) and (2.42) describe very well the experimental data. The agreement is also good for the forward hemisphere after a transformation in the CM system. Since the parametrization is aimed for use in cosmic ray calculations in the Lab, where the particles from the backward CM hemisphere are not essential, the discrepancies in that part of the phase space are not very important. Figure 2.8 shows the x distributions for pions and kaons as fitted by Hillas. Reference [18] also gives the fit to σ_{pp}^{inel} which is shown in Fig. 2.5 with a solid line. The parametrization as a function of the incident proton energy in the Lab is

$$\sigma_{pp}^{inel} = 32.2 \left(1 + 0.237 \ln \frac{E}{200\,\text{GeV}} + 0.01 \ln^2 \frac{E}{200\text{GeV}} \, \vartheta \, \frac{E}{200\text{GeV}}\right) \text{mb} , \tag{2.44}$$

where ϑ is the Heaviside function. As can be seen in Fig. 2.5, equation (2.44) provides a very good fit to the measured cross-section above $\sqrt{s} = 2$ GeV.

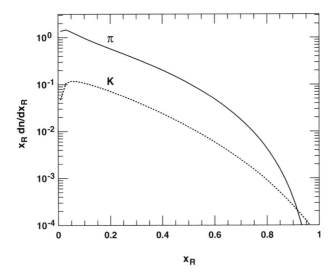

Fig. 2.8. Energy spectra of the secondary pions (solid line) and kaons (dots) calculated for incident proton energy of 100 GeV from the parametrizations of Ref. [18] in terms of the laboratory radial scaling parameter x_R (see (2.42) and (2.43)).

Integrating from 0 to 1 the expression $f(x) \times H(E)\,dx/x$ gives the average multiplicity of pions produced in pp interactions. Equations (2.42) and (2.43) do not work exactly (because they are intended for use in cosmic ray experiments and do not represent exactly the backward hemisphere) but similar parametrizations in the CM system give very good representations of the observed particle multiplicity. The experimental data on the multiplicity of secondary particles is summarized in several papers [21] and [22] that give parametrizations of the multiplicity of charged secondary particles as a function of the interaction energy. Albini et al. [21] express the charged multiplicity as

$$\langle n^{ch} \rangle = 1.17 + 0.30 \log s + 0.13 \log^2 s \,, \qquad (2.45)$$

while Thome et al. [22] write

$$\langle n^{ch} \rangle = 0.88 + 0.44 \log s + 0.118 \log^2 s. \qquad (2.46)$$

The two expressions agree very well in the region of the fits, as could be seen in Fig. 2.9. The figure also shows the average multiplicity of charged pions and kaons as given by Antinucci et al. [23], who parametrized the energy dependence of the multiplicity of different types of secondaries. Since the measurements of different secondaries have not been made in the same energy ranges, the sum of all secondaries from Ref. [23] only roughly agrees with the expressions 2.45 and 2.46.

A very important general parameter of the inelastic collisions is the coefficient of inelasticity K_{inel}. By definition this is the fraction of the primary

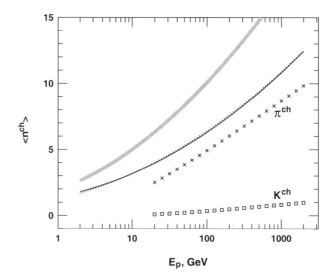

Fig. 2.9. Average charged multiplicity as parametrized by Ref. [21] (solid line) and by Ref. [22] (dotted line). The symbols show the average multiplicity of charged pions and kaons from the summary of Antinucci et al [23]. The thick lighter curve shows the average charge multiplicity in proton interactions on air nuclei.

energy E that a particle has conserved after an inelastic interaction, i.e. $K_{inel} = 1. - \sum_i E_s^i / E$, where the sum is over all secondary particles i generated in the interaction. In the Hillas model the leading secondary nucleons (which are fragments of the primary nucleons) have a flat distribution and $K_{inel} = 0.5$. This is an accepted number for pp interactions at moderate energies. There is a decrease of K_{inel} at $\sqrt{s}$ values above ~ 50 GeV. The distribution of leading nucleons is, however, not flat. About 18% of the inelastic cross-section is *diffractive* where the leading nucleon takes most of the energy and very few secondaries are produced. One half of the diffractive interactions are diffractions of the target nucleon. In the Lab system such interactions could be considered as non-diffractive. For cosmic ray purposes the diffractive cross-section is 9% of the total inelastic cross-section. Summing over the distribution of leading nucleons one again obtains $K_{inel} \simeq 0.5$. Sometimes the elasticity coefficient $K_{el} = 1 - K_{inel}$ is used instead of K_{inel}.

2.4.2 Kinematic variables and invariant cross-section

Experimental data taken at a certain angle with respect to the incident particle beam are represented as functions of a pair of variables that express the longitudinal and transverse characteristics of the detected particles. Such sets could be $(p_\parallel^{Lab}, \vartheta^{Lab})$, which is used in counter experiments, the already in-

troduced Feynman scaling variable $x_F = 2_{\parallel}/\sqrt{s}$ and $p_{\perp}^2$ (or $p_{\parallel}^{Lab}$, $p_{\perp}^2$) or a combination of those, the particle rapidity y. The rapidity is defined as:

$$y = \frac{1}{2}\ln\frac{E+p_{\parallel}}{E-p_{\parallel}} = \ln\frac{E+p_{\parallel}}{m_{\perp}}, \qquad (2.47)$$

where $m_{\perp} = \sqrt{m^2+p_{\perp}^2}$. Rapidity is very easy to transform from one system to another because the transformation only adds a constant to the rapidity value, i.e. $y^{Lab} = y^{CM} + \zeta$, where $\zeta = \ln\sqrt{s}/m_p$ for pp collisions. The energy and longitudinal momentum of a particle in any frame are expressed as a function of the rapidity as

$$E = m_{\perp}\cosh y; \; p_{\parallel} = m_{\perp}\sinh y. \qquad (2.48)$$

The maximum rapidity of a particle in the CM system is $y_{max}^{CM} = \ln\sqrt{s}/m$. When the momenta of the secondary particle are unknown and only the angle ϑ can be observed the kinematic variable is the pseudo rapidity $\eta = -\ln\tan\vartheta/2$. The pseudorapidity $\eta \simeq y$ for momenta much greater than the particle mass and angles $\vartheta \gg 1/\gamma$, where γ is the particle Lorentz factor. It is closely related to the old cosmic rays variable $\log_{10}\tan\vartheta^{Lab}$.

The production of particles with certain longitudinal and transverse characteristics is usually expressed in terms of the Lorentz invariant cross-section $2E\frac{d\sigma}{d^3p}$. In terms of the other kinematic variables the invariant cross-section is written as

$$\begin{aligned}
2E\frac{d\sigma}{d^3p} &= \frac{2E}{\pi}\frac{d\sigma}{p_{\parallel}dp_{\perp}^2} \qquad (2.49)\\
&= \frac{2x_0}{\pi}\frac{d\sigma}{dxdp_{\perp}^2} \text{ with } x_0 = \sqrt{x^2+2m_{\perp}/\sqrt{s}}\\
&= \frac{\pi}{2}\frac{d\sigma}{dy\,dp_{\perp}^2}\\
&= \frac{2E}{p^2}\frac{d\sigma}{dp\,d\Omega}.
\end{aligned}$$

2.5 Nuclear fragmentation

Nuclei are complicated systems of protons and neutrons that are held together by a multitude of forces. The simplest model of a nucleus, the liquid drop model, treats it as a fluid consisting of these nucleons. It is easy to derive the nuclear binding energy E_b of a nucleus of mass A consisting of N neutrons and Z protons in this model. It is given by the difference between the masses of the constituent nucleons and the nucleus itself.

$$\frac{E_b}{c^2} = \Delta M_A = Zm_p + Nm_n - M_A \tag{2.50}$$

The order of magnitude for the average binding energy is 20 MeV per nucleon. Rachen [24] has represented the semi-empirical Weizsäcker mass formula in terms of the nuclear mass A in the liquid drop model for stable nuclei as

$$E_b(A) = A\left[15.8 - 18.3A^{-1/3} - 0.18A^{2/3} + 1.3 \times 10^{-3}A^{4/3} - 6.4 \times 10^{-6}A^2\right] \tag{2.51}$$

From energy conservation one can also calculate what is the energy required to separate a fragment F containing N_F neutrons and Z_F protons from the nucleus A. This is the separation energy equal to the difference in the binding energy of the original nucleus and the two nuclei in the final state.

$$E_s = E_b(N, Z) - E_b(N_F, Z_F) - E_b(N - N_F, Z - Z_F) \tag{2.52}$$

This energy would be the threshold for the reaction $A \to F$ if the protons did not carry electric charge. The charge of the nucleus decreases E_s by an amount E_s^C, which represent the Coulomb barrier.

$$E_s^C \simeq \frac{Z_F(Z - Z_F)}{(A - F)^{1/3}} \tag{2.53}$$

The term in the denominator is the estimate of the radius of the remnant nucleus, after separating F.

The total energy needed to separate the fragment from the nucleus will be $E_s^{tot} = E_s - E_s^C$. This gives some ground rules for estimating the probability for separating single nucleons from a nucleus. That probability depends strongly on E_s^C because E_s is the same if we use the parametrization 2.51, which is based on the average charge of stable nuclei of mass A. It is generally easier to separate a proton from a nucleus than it is to separate a neutron. This is indeed true for relatively light nuclei with equal number of protons and neutrons, but not for heavy nuclei with many more neutrons than protons because the nucleus as a whole is positively charged.

The process is much more complicated than our simple treatment here. The account for the quantum effects show that the threshold energy is actually lower because of tunneling. A detailed treatment should also include the Fermi motion of the nucleons inside the nucleus.

Generally one could distinguish two types of nucleons in a nuclear collision. The 'participants' have direct collisions with nucleons from the other colliding nucleus. The 'spectators' are the rest of the nucleons, which do not participate directly in the collision. They are, however, also excited by the collision gaining energy from collisions with some of the participant nucleons and from the surface deformation of the 'pre–fragment' nucleus which contains all spectators. The 'pre–fragment' nucleus relaxes by evaporation in smaller fragments. For heavy nuclei ($A > 60$) an empirical distribution exists for the masses of the fragments in the final state. The distribution is [25]

$$P(x) \propto 0.1/x^{2.5} + \exp(3.7x) ,\tag{2.54}$$

where x is the ratio of the fragment mass to the mass of the original nucleus. $P(x)$ has a minimum at about $x = 1/3$. The branch of the distribution left of that minimum is called multifragmentation, because the parent nucleus fragments in many light fragments including numerous constituent nucleons. The part of the distribution above $x = 1/3$ describes the spallation, i.e. fragmentation processes which result in a large fragment and the emission of several constituent nucleons. Figure 2.10 shows the distribution of fragments for iron nuclei interacting on carbon target.

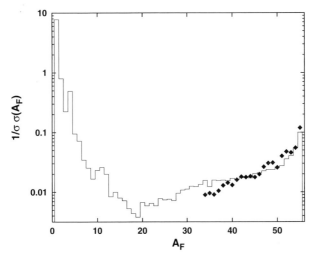

Fig. 2.10. Mass distribution of the fragments in Fe collisions on C target. The data are from Ref. [26] and the calculation is from Ref. [27].

The partial cross-section for the fragmentation of nucleus A into nucleus of mass A' could be written as $\sigma_{A\rightarrow A'} = a\sigma_0 P(A'/A)$ and is called the mass changing cross-section. The normalization is chosen so that the sum over all fragmentation probabilities P_j equals 1. The total mass changing cross-section for a nucleus of mass A at a given energy is σ_0. The charge changing cross-sections are similarly defined. The total cross-sections are proportional to $A^{2/3}$. The total cross-sections as well as the particle cross-section have very strong energy dependence at kinetic energies comparable to the nucleon mass. At higher energy, ~ 5 GeV per nucleon the cross-sections saturate and become nearly constant. Silberberg and collaborators [28] give the following parametrizations for the total mass changing cross-section as a function of the nuclear mass

$$\sigma = 45A^{0.7}\left[1 + 0.016 \sin\left(5.3 - 2.63 \ln A\right)\right] \text{mb}. \qquad (2.55)$$

The energy dependence of the cross-sections are defined with respect to the constant cross-section at energies above several GeV/nucleon. The parametrizations as a function of the kinetic energy per nucleon is

$$\sigma(E_k) = \sigma_{HE}\left[1 - 0.62 \exp(-E_k/200\text{MeV}) \times \sin\left(10.9(E_k/\text{MeV})^{-0.28}\right)\right], \qquad (2.56)$$

where σ_{HE} is the constant high energy value. The cross-sections peak in the region of the giant resonance, about 20 MeV/nucleon, reaching 60% above the high energy value, then reach a minimum at energy about 200 MeV and then slowly grow to reach the constant high energy value at about $E_k = 2$ GeV.

Nuclear fragmentation is a very complicated process which cannot be fully described in analytic terms. We have to rely on measurements of the partial and the total cross-sections. The data set of such measurements is large but still inadequate for the description of all possible fragmentation channels of all existing nuclei. It is therefore important to interpolate between different measured cross-sections using basic nuclear physics knowledge and the trends observed in the data samples. A very valuable compilation and interpolation procedure was developed by Silberberg & Tsao [29], which is widely used in studies of the cosmic ray propagation. More recently there have been extensive new cross-section measurements [30, 31] that will help significantly in the understanding of the formation of the chemical composition of the cosmic rays.

3 The birth of cosmic rays

Cosmic rays are an essential part of the Universe. Their origin is interrelated with the main processes and the dynamics of star formation, stellar evolution, supernova explosions and to the state and conditions of the interstellar matter in the Galaxy. This chapter is intended to guide the reader through some of these processes that lead to the acceleration of galactic cosmic rays.

This is also justified by the fact that the detection of the neutrinos emitted by the Sun and by the explosion of the supernova 1987a are the beginning of the neutrino astronomy, a new branch of astronomy that is subject of high current scientific interest. We give a brief account of these results.

The knowledge of stellar evolution is based in many respects on the observations of the Sun, which has been studied much more closely than any other star. We therefore use the nuclear processes and the stellar evolution models created for the Sun to give an idea of the principles of stellar evolution.

The explosion of SN1987a is the only supernova explosion that has been observed by modern science. We use it as an example for all other supernova explosions and the development of supernova remnants, where we believe that charged particles become cosmic rays.

3.1 Stellar energetics. The pp chain

Stars are very massive, and thus have a very strong gravitational field. If the gravitational attraction inside a star were not balanced it would immediately collapse. The condition for stability of a star is the existence of hydrostatic equilibrium, i.e. everywhere inside the star the gravitational force should be balanced by a force that acts outwards, that we will very generally call internal pressure. If we imagine that a star consists of static spherical shells of radius r, the equation of hydrostatic equilibrium is

$$\frac{dP(r)}{dr} = -\frac{GM(r)\rho(r)}{r^2} , \tag{3.1}$$

where $M(r)$ is the stellar mass enclosed inside the r shell, and $P(r)$ and $\rho(r)$ are the pressure and density on the shell. The pressure is dominated by the thermal motion of electrons and ions, which is proportional to the fourth power of the temperature T. There is also a contribution from any outward flow inside the star, usually photons (including visual light) created inside the star and moving towards its surface.

Stars also radiate energy at a high rate; they are very luminous. If the pressure inside the star were only thermal this would require a temperature gradient which gives rise to an energy flux from the central region of the star towards its exterior. In general, each spherical shell contributes certain amount of energy to the total stellar luminosity. If we characterize the energy as of mechanical or nuclear origin, the rate at which energy is generated on the shell at distance r from the center of the star is

$$\frac{dL_r}{dr} = 4\pi r^2 \times \rho \times \left(\varepsilon_{nucl} - T \frac{dS}{dt} \right), \tag{3.2}$$

where ε_{nucl} is the rate at which nuclear energy is generated. The term $T\, dS/dT$ expresses the rate of generation of mechanical energy through the stellar entropy S and the temperature T. For stars like our Sun, that are on the *main sequence* track of stellar evolution, nuclear energy dominates.

This became obvious in the late 1920s when the age of the Earth was first measured and scientists realized how old many rocks on Earth are. If the gravitational contraction and the corresponding heating were the main energy source of the Sun, then its lifetime would be determined by the gravitational timescale, which is

$$t_{gravity} = G_N M_\odot^2 / R_\odot L_\odot \simeq 10^7 \text{ years.} \tag{3.3}$$

$M_\odot$ here is the solar mass, $R_\odot$ is the solar radius and $L_\odot$ is the solar luminosity. Some rocks on Earth were found to be older than ten million years and the assumption that Earth is older than the Sun seemed very unlikely. The only energy source that can support the solar luminosity on a longer timescale is nuclear energy. If only a small fraction of the solar mass ϵ were converted to energy in nuclear burning of protons into α-particles, the nuclear timescale would be much longer.

$$t_{nucl} = \epsilon \times M_\odot c^2 / L_\odot \simeq 10^{10} \text{ years} \tag{3.4}$$

for $0.1 \times \epsilon = 7 \times 10^{-4}$. The coefficient 0.1 indicates the fraction of the solar mass that is exhausted while the Sun is being powered by this process or, in astronomical terms, stays on the main sequence branch of stellar evolution. The energy for that period is supplied by burning of 0.7% of the solar mass.

Nuclear energy is created because the mass of a He nucleus is lower than the mass of the four protons involved in its creation by 24.7 MeV. A fraction of this mass difference is liberated and converted to thermal energy by every fusion reaction. Detailed models for this fusion chain, called the *pp* chain, have been developed by J.N. Bahcall and collaborators. I will follow Bahcall's discussion [10] in my brief account of it.

The *pp* chain consists of the following reactions:

1a	p + p	$\longrightarrow \, ^2\text{H} + \text{e}^+ + \nu_e$
1b	p + e$^-$ + p	$\longrightarrow \, ^2\text{H} + \nu_e$
2	$^2\text{H} + \text{p}$	$\longrightarrow \, ^3\text{He} + \gamma$
3a	$^3\text{He} + \, ^3\text{He}$	$\longrightarrow \, ^4\text{He} + 2 \, \text{p}$
3b	$^3\text{He} + \, ^4\text{He}$	$\longrightarrow \, ^7\text{Be} + \gamma$
3c	$^3\text{He} + \text{p}$	$\longrightarrow \, ^4\text{He} + \text{e}+ + \nu_e$
4a	$^7\text{Be} + \text{p}$	$\longrightarrow \, ^7\text{Li} + \nu_e$
4b	$^7\text{Be} + \text{p}$	$\longrightarrow \, ^8\text{B} + \gamma$
5a	$^7\text{Li} + \text{p}$	$\longrightarrow 2 \times \, ^4\text{He}$
5b	^8B	$\longrightarrow \, ^8\text{B}^* + \text{e}^+ + \nu_e$
6b	$^8\text{B}^*$	$\longrightarrow 2 \times \, ^4\text{He}$

The letters attached to some of the reactions mean that these are competing reactions that lead to different branches. Reaction 3a, for example, is followed by 4a and 5a, while 3b is followed by 4b, 5b, and 6b. All three branches end with the production of ^{4}He. Each of the individual reactions requires a minimum temperature and pressure, and therefore may only happen within a maximum distance from the center of the Sun.

Five of the reactions listed above generate an electron neutrino ν_e in the final state. The jargon for neutrinos from these processes is *pp* (1a), *pep* (1a), *hep* (3c), *Be* (4a) and *B* (5b). The maximum energy for *pp* neutrinos, the most abundant ones, is 0.42 MeV; for *pep* 1.44 MeV; for *B* 15 MeV; and for *hep* is not exactly known but is about 18 MeV. *Be* neutrinos are emitted in two lines of 0.383 and 0.861 MeV. The detection of these neutrinos is the only way to prove experimentally that nuclear energy is powering the Sun and all main sequence stars.

3.1.1 Solar neutrinos

A detailed calculation of the solar neutrino luminosity requires the exact knowledge of the cross-section of all processes listed above and their energy dependence. One could, however, roughly estimate the solar neutrino flux in a much simpler way. The most common reaction is 1a, which in addition to the three particles in the final state liberates about 10 MeV in thermal energy. If this were the only energy source of the Sun, the reaction rate could be estimated from the total solar luminosity as $L_\odot/10$ MeV $\simeq 10^{38}$ per second. The neutrino flux at Earth is then approximately $F_\nu \simeq 2.5 \times 10^{38}/4\pi d^2 \simeq 10^{11}$ cm^{-2}s^{-1}, where d is the average distance to the Sun, one astronomical unit (AU). This number is surprisingly close (better than a factor of two) to the total number of neutrinos produced in the *pp* cycle.

The real calculation is a complicated iterative procedure which consists a chain of successive stellar evolution calculations. The whole procedure is called the standard solar model (SSM). The iterations continue until a satisfactory agreement with all measured solar parameters is achieved. These include the solar luminosity, mass and age, as well as much less obvious data, such as the strength of the p modes of solar oscillations. 98.4% of the solar neutrino luminosity is produced by the pp cycle and the rest comes from the CNO cycle that fuses ^{4}He nuclei into medium heavy carbon, nitrogen and oxygen nuclei. The SSM gives detailed information about the fluxes and the energy spectra of neutrinos from different channels and the regions of the Sun where they are produced. The neutrino flux from the pp reaction is 6×10^{10} cm^{-2} s^{-1}.

There are two basic methods for detection of solar neutrinos – the radio chemical method, that uses inverse β-decay, and one based on neutrino–electron interactions.

Inverse β-decay is a process, that is the opposite of nuclear decay, i.e. electron neutrinos interact with a nucleus A of charge Z (i.e. contains Z protons and $N = A - Z$ neutrons), which converts to the state $A'(Z+1, N-1)$ and creates an electron e$^-$. If the target nucleus is such that the A' state is unstable and not very common, one can count the number of A' decays and thus measure the number of solar neutrino interactions. Decays are identified by the known energy of the emitted γ-ray. Each inverse β-decay reaction has an energy threshold that determines the minimum neutrino energy that can initiate it. The reaction threshold defines the sensitivity of the measurement to different solar neutrino species. For detection of pp neutrinos, for example, one needs a reaction with a threshold below the maximum energy of the pp neutrinos – 0.42 MeV.

Radiochemical experiments, as well as all neutrino detection experiments, are very difficult because of the very low neutrino cross-sections. They have to deal with very small numbers of counts. The standard unit for the solar neutrino interaction rate is SNU (solar neutrino unit), 10^{-36} neutrino captures per target atom per second. One needs a very large number of target atoms and long exposure to be able to capture any number of solar neutrinos. Imagine a detector consisting of 1 ton of suitable target with mass number $A = 20$. The number of atoms will be $10^6 N_A/A \simeq 3 \times 10^{28}$ and the detector will have less than one neutrino interaction per year for a neutrino flux of 1 SNU.

The strategy is then to have a large tank containing that target nucleus $A(Z, N)$ and to store it in a radioactively clean environment for one to two half lifetimes of the unstable isotope A'. Longer exposures do not yield higher signals because most of the A' nuclei produced early in the exposure decay. The radioactive nuclei A' produced are extracted from the tank at the end of the exposure in a complicated chemical process and subjected to counting, that is going to last more than 10 nuclear decay half lifetimes. Such a long counting period is necessary for the precise determination of the background that is due to natural radioactivity in the environment, including the counters themselves. These are miniature counters, made of low radioactivity materials, and have incidental rates less than one count in 10 days. Signal counts have to have the specific characteristics of A' decay. The time dependence of the signal counts has also to be consistent with the half life of the A' isotope. Because of the small statistics this cannot be achieved in a single exposure.

The first solar neutrino experiment was realized in the Homestake gold mine in South Dakota by Ray Davis Jr. and collaborators [32]. It is based on the reaction $\nu_e + {}^{37}\text{Cl} \longrightarrow {}^{37}\text{Ar} + e^-$ that has a neutrino threshold energy 0.814 MeV. It can detect only the high energy solar neutrino channels, mainly one of the *Be* lines and the *B* neutrinos. The tank contains 615 tons of the cleaning liquid C_2Cl_4. The Homestake detector has worked reliably for 30 years. It detected first the solar neutrinos, although at a rate lower than the SSM predictions by a factor of three.

Two other experiments, GALEX [33] and SAGE [34], are based on the reaction $\nu_e + {}^{71}\text{Ga} \longrightarrow {}^{71}\text{Ge} + e^-$. This reaction has an energy threshold of 0.232 MeV, which allows the detection of the neutrinos from the *pp* branch. This is very important because *pp* neutrinos are generated in the basic reaction of the solar energetics, while *B* neutrinos are produced in a very rare reaction branch. Thus *pp* neutrinos can be linked directly to the solar luminosity. In addition the interaction cross-section on ^{71}Ga is higher than that on chlorine and the gallium experiment requires less material, which however is much more expensive than the cleaning liquid of Homestake. The half life of ^{71}Ge is 11.4 days, which allows for a shorter exposure and counting periods. GALLEX exposed 30 tons of gallium and SAGE - 57 tons. The basic difference between the two experiments is the extraction procedure, GALLEX's chemistry being simpler. These experiments also detected smaller neutrino fluxes than predicted by the SSM. Their detection rate is about 70 SNU (SSM predicts about 130 SNU), which is close to the predicted contribution from the *pp* reaction. One can thus state that the gallium experiments have experimentally proven that the *pp* chain is the main source of solar energy.

The neutrino scattering reaction $\nu_e + e^- \longrightarrow \nu_e + e^-$ provides a totally different method of neutrino detection. The scattered electron carries a large fraction of the neutrino energy and moves in a direction close to the initial direction of the neutrino. The reaction cross-section is 9.3×10^{-45} cm^2 $\times$ (E_ν/MeV)^2. The first detector capable of detecting solar neutrinos was the water-Cherenkov detector in Kamioka, Japan [35], which contained 3,000 tons of purified water. The walls of the big water tank are covered with photomultiplier tubes, which detect the Cherenkov light emitted by the scattered electron. The timing and the amplitude of the detected signals allow for the reconstruction of the direction and the energy of the electron. There is an excess of events coming from the direction of the Sun. The detection threshold was first set to 7.5 MeV, i.e. only *B* neutrinos were detected. Kamiokande took the first, somewhat fuzzy picture of the Sun in neutrinos, but detected only about one half of the SSM prediction.

The research at Kamioka was continued with the second generation Super-Kamiokande detector. It uses the same detection principle and is bigger by a factor of 10, containing a total of 50,000 tons of purified water. The fiducial volume of the detector, in the middle of the body of water, contains 21.5 Kt. In addition the photomultiplier density is substantially increased, and 40% of the wall area is covered with photo cathode. The threshold energy used for solar neutrino measurements was gradually lowered to 6.5 MeV. The daily solar neutrino statistics with this threshold is about 13 events. Figure 3.1 shows how solar neutrinos are identified in Super-Kamiokande. The most recent solar neutrino results from Super-Kamiokande are described in Refs. [36, 37].

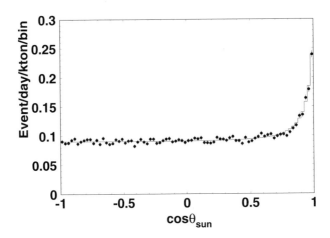

Fig. 3.1. Angular distribution of Super-Kamiokande neutrino events with respect to the direction of the Sun at the time of detection.

Because the scattered electrons move mostly in the initial direction of the neutrino the electron angular distribution peaks strongly in the direction of the Sun (actually the direction away from the Sun) at the time of the event. The flat part of the distribution away from the Sun consists only of background events, which are not related to the solar neutrinos. This background does not have any angular dependence and can be thus used to calculate the solar neutrino flux by subtraction of the background from the total event number. One can see in Fig. 3.1 a background level of about 0.09 events per day per kiloton per bin. The total number of bins is 80, i.e. the total background level is 7.2 events per day per kiloton of detector mass. The eight bins close to $\cos \theta = 1$ contain the signal events. They have a total excess of 0.48 events, which for the fiducial detector volume of 22.5 Kt convert to a daily rate of 10.8 events. The real analysis is of course done much more carefully, with estimates of all uncertainties, but follows the same principle.

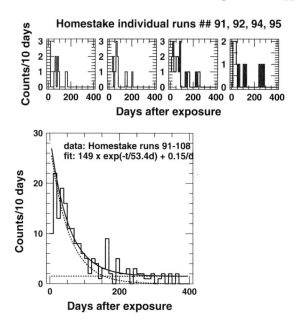

Fig. 3.2. The upper panel shows the results from four individual counting periods of the Homestake tank: ## 91, 92, 94 and 95. The lower panel shows the time dependence of the counting rate for 14 runs.

All solar neutrino detectors observe significantly smaller number of neutrinos than is predicted by all versions of the standard solar model. This is known as the solar neutrino puzzle. The puzzle has two possible sets of solutions. The first one includes possible inaccuracies of the standard solar model, such as a lower central temperature of the Sun or errors in the scaling of the fusion cross-sections to the conditions of the solar interior. The second set is related to the properties of the neutrinos. It assumes that the electron neutrinos generated in the Sun oscillate on the way to the Earth into neutrinos of different flavor, which cannot be observed by the solar neutrino detectors. This solution will be discussed later, in Sect. 7.3, together with other experimental results that suggest neutrino oscillations.

Solar neutrino results

Figure 3.1 reveals some of the difficulties in the measurement of the solar neutrino flux in water-Cherenkov detectors. The main problem is that the signal, although perfectly visible in the angular distribution graph, is only several percent of the total background in the detector. The radio-chemical method, however, has to overcome even bigger problems. Figure 3.2 shows one of them – the very low experimental statistics.

*While the upper panel of Fig. 3.2 does not reveal the lifetime of the decay-
ing isotope, the combined graph for 14 runs, shown in the lower panel, lets
you to calculate it. A simple fit with a constant background and exponential
decay gives a decay time of 53.4 days, close to the real value for ^{37}Ar. The
background is about 0.1 count per day.*

*In the 1980s there was a widely spread suspicion of the existence of a sec-
ond solar neutrino puzzle – an anticorrelation of the detection rate of solar
neutrinos with solar activity. A mild effect of that type has been observed in
the early 1970's but with the improvement of the stability of the Homestake
detector, both in the extraction technique and counting, the correlation after
1980 appeared to become stronger [40]. The further increase of the Homestake statistics and the emergence of other detectors did not confirm the effect,
which is most likely due to the difficulties in estimating the statistical significance in cases with very low statistics. Actually through the decades of its
operation the Homestake event rate has been found to correlate with almost
everything, from the market value of some famous stocks to the fraction of
republican senators in the US Senate.*

*The basic neutrino puzzle is, however, very stable. Table 3.1 gives the
predicted and detected solar neutrino fluxes for several experiments. Note
that the predictions are given in different units for the two types of detectors.
The predictions for radiochemical detectors are given in SNU, which also
account for the neutrino interaction cross-sections. For the neutrino–electron
scattering detectors the predictions are in units of neutrino fluxes above the
detection threshold, i.e. per cm^2 per second. The table also gives the results of
the SNO (Sudbury Neutrino Observatory) detector [41], which is also based on
the Cherenkov light technique. SNO is significantly smaller than Super–K, but
it also contains 1 Kt of heavy water suspended in a clear plastic container in
H_2O. Neutrinos interact not only with electrons, but also with the deuterium,
generating two protons and an electron. The two numbers in Table 3.1 refer to
these 'charge current' (CC) interactions and to neutrino–electron scattering.*

Table 3.1. Predicted and detected fluxes of solar neutrinos. The first detection
line gives the results of Homestake, Gallex and Kamiokande and the other entries
are labeled. All results are from the latest publications of all groups by the end of
2001. The numbers for the radio-chemical detectors are in solar neutrino units, and
those for water-Cherenkov dectors are in $cm^{-2}s^{-1}$.

Predicted	Cl	Gallium	Water
BPB2001 [38]	7.6 ± 1.1	128 ± 8	5.05×10^6
Brun et al [39]	7.55	129	5.10×10^6
detected	2.56 ± 0.23	74.1 ± 7	2.80 ± 0.4
SAGE/SKAM		75.4 ± 8	2.40 ± 0.1
SNO			$1.75 - 2.39 \times 10^6$

The biggest puzzle in these numbers is in the difference between the results of Homestake and the water-Cherenkov detectors. The latter detect only B neutrinos, while Homestake detects also some Be and CNO neutrinos. See Fig. 3.3 for the rates predicted in different reactions and detected by the three types of experiments. Super–K, for example, sees almost one-half of the predictions, while in the Homestake data that ratio is about one-third. On the other hand, ^7Be is produced in the nuclear reaction 3b (see the list of reactions in 3.1 and ^8B is generated from ^7Be in reaction 4b. So, if the production cross-section for ^7Be in the calculations is incorrect, it would also suppress the production of boron, and correspondingly of Be neutrinos.

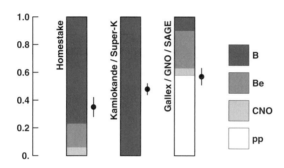

Fig. 3.3. Predictions of the contribution of different reactions to the neutrino rates in different types of detectors are compared to the observed rates.

One of the future detectors, Borexino in the Gran Sasso laboratory in Italy, is mainly sensitive to neutrinos in the medium energy range where Be neutrinos are. Borexino consists of 300 t of liquid scintillator in a sphere surrounded by organic and water shielding. Neutrinos interact quasi-elastically on the electrons of the target liquid. According to the SSM Borexino should see more than 50 solar neutrinos per day, but the real rate depends on the process that causes the solar neutrino puzzle. In the case that medium energy neutrinos are not produced by the Sun, or are absorbed on propagation to Earth, the detection rate could be very small. Borexino is scheduled to start detection in 2004.

3.1.2 Stellar evolution

The understanding of stellar evolution starts with the observation of many stars and their classification in terms of several important parameters such as mass, luminosity and surface temperature. None of these parameters can be always precisely measured. The much more precise data on the Sun help in the interpretation of other astronomical observations.

The mass of a stellar object can be derived only when the stars are in a binary system. The derivation of luminosity depends on the knowledge of the distance to the object (because of the inverse square dependence), which is often determined relatively to the better known distance to other astrophysical objects. The luminosity is expressed in the logarithmic measure of apparent magnitude, which is proportional to $-2.5 \log L$, where the stellar luminosity L is measured in a range of frequencies that cover the whole emission spectrum. The absolute magnitude of the stellar luminosity is what the apparent magnitude would have been if the star were at a distance of 10 pc. The bolometric luminosity M_{bol} is

$$M_{bol} = 4.75 - 2.5 \log \left(\frac{L}{L_\odot} \right) , \tag{3.5}$$

where 4.75 is the bolometric luminosity of the Sun. The surface temperature is usually measured as the logarithm of the ratio of the stellar magnitudes in different frequency (wavelength) intervals, such as visual (V), blue (B) and ultraviolet (U). The values (U − B) and (B − V) are called color indices.

When stars observed in a survey are plotted on the mass versus luminosity or luminosity versus surface temperature plane, they produce somewhat ordered plots, demonstrating the basic relations of stellar evolution process. From theoretical considerations one expects the luminosity to be proportional to some power of the mass – $L \propto M^n$, with $n \simeq 3$ for stars with more than one solar mass. The relation becomes more complicated for lighter stars, where the mass dependence becomes stronger.

Even more revealing is the luminosity versus surface temperature plot – the Herzsprung–Russel (HR) diagram. A cartoon of a HR diagram is shown in Fig. 3.4. The shaded area extending from the lower right to the upper left corner covers the main sequence of the stellar evolution. It is populated by stars that are supported by the proton cycle of nuclear reactions. A star is born in the contraction of a dust cloud. The contracted dust heats up and the proto-star rises upwards and leftwards of the lower right corner of the HR diagram. When the central density and temperature of the proto-star reach the threshold for the *pp* cycle, the burning of hydrogen into helium starts, and the star moves almost vertically down to settle on the main sequence.

The exact position of the star on the main sequence depends on the stellar mass. Stars stay on the main sequence until about 10% of their mass is exhausted in the fusion of helium. The total energy available to the star during its main sequence stage is thus proportional to its mass M. The stellar luminosity, on the other hand, supported by the rate of nuclear burning, depends much more heavily on the mass, see (3.5). Consequently more massive stars spend much less time on the main sequence. The duration of that period of stellar evolution is proportional to M^{-3}. Stars like the Sun spend 10 billion years on the main sequence. Stars with $20 M_\odot$ have only about one million years to pass through the main sequence.

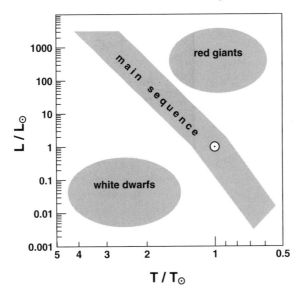

Fig. 3.4. This is a cartoon of a HR diagram, which usually consists of a scatter plot of measured stellar parameters. Note the position of the Sun at unit surface temperature and luminosity. Note also that in a classical astronomical fashion the temperature increases from right to left.

Once one-tenth of the stellar hydrogen is burnt, the further evolution depends on the stellar mass. In stars with less than about $1.5 M_\odot$ the burning of hydrogen starts at the center and slowly moves outwards. A helium core is generated which grows and expands as the hydrogen is fused. The temperature is, however, not high enough for any other nuclear reaction to start. When the size of the helium core becomes too big, the hydrogen fusion cannot support it any more and the star contracts to a white dwarf. The critical enclosed core mass for this contraction is the Chandrasekhar mass $M_{Ch} = 5.8 \times Y_e^2 M_\odot, \simeq 1.4 M_\odot$, where $Y_e \simeq Z/A$, i.e. the charge to mass ratio of the stellar material. White dwarfs are very common stars that populate the lower left corner of the HR diagram. They have surface temperatures of about 10,000 degrees and a radius of only 3,000 km. Their luminosities are low because of the small surface area.

Massive stars have a different evolution. In a $20 M_\odot$ star, for example, the fusion continues in a thick shell surrounding the contracted helium core, because the core is massive enough to reach temperatures of more than 100 million degrees. The energy is then generated in a different set of nuclear reactions in which He nuclei are fused into carbon and oxygen. When the helium fuel is exhausted another contraction raises the density and increases the temperature to a billion degrees.

The C and O core then starts burning into neon, magnesium, silicon and other heavy nuclei. These consequent fusion cycles take much less time than the *pp* cycle. During that time the stars move away from the main sequence and are on the red giant track that leads them to the upper right corner of the HR diagram. Details of the evolution process can also alter the position and bring some massive stars to the unusual blue giant location in the upper left corner of the HR diagram.

3.1.3 Supernova explosions

At the time when the core of a massive star consists exclusively of heavy nuclei and the central temperature exceeds one billion degrees, a new radiation process starts to develop – neutrino emission. Throughout the later stages of stellar evolution, stellar photons are sufficiently energetic to generate electron–positron pairs when they collide with the matter of the star. Initially the positron and the electron are not very energetic and they annihilate into a pair of photons. When the temperature starts exceeding a billion degrees the probability of annihilation into a neutrino–antineutrino pair increases. Neutrinos have a much smaller interaction cross-section than γ-rays and they leave the star unscattered. The fraction of the energy that goes into neutrinos is lost and does not participate in the maintenance of the hydrostatic equilibrium of the star. At these latter stages of stellar evolution giant stars lose much more energy through neutrino emission than they do through their photon luminosity.

The evolution continues to speed up and becomes very complicated. The temperature continues to grow and when it exceeds 3 billion degrees a fraction of the heavy elements starts to disintegrate and fuse with other residual nuclei to generate nuclei of the iron group. Further fusion is not possible because Fe nuclei are very tightly bound. The core is supported only by the electrons from the fully ionized matter which is not sufficient to prevent further contraction of the inner M_{Ch} (about $1.4M_\odot$) of the star. Two additional processes aid this last contraction.

Some of the heavy elements disintegrate. Disintegration is the reverse of fusion and it consumes energy, thus decreasing the temperature and the pressure in the central regions of the star. In addition, the electrons start interacting with the protons of the heavy nuclei converting them to neutrons in the deleptonization process $e^- + p = n + \nu_e$. The disappearance of the electrons further lowers the pressure. The decrease in the pressure does not, however, slow down these two processes, because the core continues to contract under the force of gravitation which keeps the pressure in the innermost part of the core approximately constant. This chain of events leads to a rapid collapse.

The radius of the collapsing region of the star was several thousand kilometers. In less than a second it decreases to about 50 km. The outer layers of the star at this point are still relatively undisturbed. As a result of the collapse the central density of the star increases so much that it exceeds the nuclear density of about 10^{14} g/cm^3. Further collapse is impossible and the star has to 're-adjust' to reach the normal nuclear density.

The inner core does that: it bounces back, meets the rest of the core that is still falling in, and creates a shock that now propagates outwards. The shock traverses the whole collapsing core, that carries 10^{51} ergs of energy, and then heats and ejects the outer layers of the star with very high velocity. This matter starts its expansion in space, the supernova explosion, and leaves behind the newly born proto-neutron star.

3.1.4 Supernova neutrinos

At this time the core is still hot and rich of electrons – about 40% of the original electrons exist and the deleptonization can continue. But it is not the only source of neutrinos. The star has to radiate away its binding energy, which is between 2 and 4 times 10^{53} ergs.

The core temperature exceeds 2×10^{12} degrees, or about 200 MeV. Pairs of all particles that have mass less than this temperature are produced. Most of them immediately re-interact and deposit their energy back in the core. Neutrinos have a very small interaction cross-section, but they still suffer multiple collisions before they can diffuse away from the very dense core. Electron neutrinos and antineutrinos scatter on the average about 10,000 times.

The cooling thus proceeds through the emission of $\nu\bar{\nu}$ pairs of all flavors, created in thermal equilibrium with the core temperature T_c. Electron neutrinos and antineutrinos have the highest interaction cross-section in this energy range and they suffer more interactions than the other flavors. As a result they cool down to 20–30 MeV when they manage to escape from the proto-neutron star. Muon and tau neutrinos have smaller cross-sections (they are all called muon neutrinos, ν_μ in the supernova explosion jargon), scatter much less, and escape with slightly higher energy. Because of the multiple scattering the energy of the escaping neutrinos depends very slowly on the central temperature – as $T_c^{1/5}$ [46]. Most of the binding energy of the star is radiated away in neutrinos. Neutrino emission is very strong during the first 10 seconds. The neutron star has formed when less than 5% of the original electrons remain [47, 48].

Because of the contribution of the deleptonization process, electron neutrinos are emitted in slightly greater numbers. The energy carried by neutrinos is distributed as

$$E_{\nu_e} \; : \; E_{\bar{\nu}_e} \; : \; E_{\nu_\mu} \; = \; 1.2 : 1 : 4 \,, \tag{3.6}$$

where ν_μ stands for muon and tau neutrinos and antineutrinos. The deleptonization neutrinos are emitted first in a sharp spike lasting few tenths of a second. All other species, that carry away the binding energy of the core, produce a burst of duration 10 seconds or more.

Supernova neutrinos were observed by at least three neutrino detectors during the explosion of SN1987a on 23 February 1987. (The standard designation for an optical supernova is the year plus a letter in order of the observation during that year. This was the first supernova observed in 1987.) It exploded in the Large Magellanic Clouds, a small neighbor galaxy only 50 Kpc away. One can easily estimate the order of magnitude of the supernova neutrino flux. For binding energy of the star $E_{bind} = 2\text{--}4 \times 10^{53}$ ergs and for monoenergetic 15 MeV neutrinos the flux at Earth should contain

$$F_\nu = \frac{E_b}{15\text{MeV}} \frac{1}{4\pi(50\text{kpc})^2} \simeq 3 - 6 \times 10^{10} \text{ cm}^{-2} \tag{3.7}$$

neutrinos of all flavors during about 10 seconds. The number is not dissimilar to those of the solar neutrinos. Supernova neutrinos are, however, much more energetic, have correspondingly higher interaction cross-section and are more easily detected.

The very detection of the neutrino burst of SN1987a, the measured neutrino flux duration and energy, were a strong confirmation of the theory of stellar collapse that is briefly described above. Combined with the observations in the optical, X-ray and γ-ray frequencies, this observation gave a strong boost to the theoretical development of much more realistic models of stellar collapse and supernova explosions.

The neutrinos from SN1987a

There are four published accounts for the observation of the neutrino signal from SN1987a:

- *The Kamiokande detector [42] using 2.1 Kt of water target as a fiducial volume, reported on the detection of 12 neutrinos events in as many seconds on 07:36:55 UT, 23 February 1987. The chance probability of such event is one in 10^5 years. The detection occurred about 3 hours before the optical detection of the supernova by Ian Shelton. These events are shown with circles on the energy–time plot of Fig. 3.5. Nine of the 12 events are seen in less than 2 seconds. A gap of 7 seconds is followed by three more events. Some subsequent analyses challenge one of the events, which has low energy and could be a part of the background.*
- *The IMB detector [43], with a total volume of 10 Kt, but with a significantly higher energy threshold of 20 MeV, observed simultaneously seven neutrino events. Part of the reason for the small number of events is that only a fraction of the detector was active because of power failure. We are lucky the whole detector was not switched off. Both Kamiokande and IMB were built and run for detection of proton decay and did not maintain exact synchronized clocks. The time coincidence shown in Fig. 3.5 is only good within a minute or so. Still the chance probability for any of these detections is so low that nobody doubts the real time coincidence.*

- *The Baksan detector [44], consisting of tanks of scintillating liquid (200 Kt available for SN detection) reported five events of energy between 12 and 33 MeV. The exact time of observation is difficult to assess, but it is believed to be within a minute of the Kamiokande and the IMB. This observation is sometimes challenged on the basis that the neutrino rate is much higher than the one observed by the water-Cherenkov detectors. The detector at Baksan contains 2×10^{31} free protons compared to 1.4×10^{32} in Kamiokande. One expects to see a signal smaller by a factor of 7, i.e. less than 2 events. Generally the observations of the water-Cherenkov detectors are considered more reliable. They consist of amplitude and timing data from many photo tubes, which allow the reconstruction of the energy. In contrast, the scintillator tanks give only one number – the amplitude of the event which distinguishes between signal and background.*
- *The Large Scintillator Detector in the Mont Blanc tunnel reported an outburst of five events in five seconds [45] about 4.7 hours before Kamiokande and IMB. The chance probability is very small, but the result is still controversial because it does not fit in with the other observations, or with the standard collapse theory. No neutrinos were detected at the time of the water-Cherenkov events, but not more than one event is expected. There are reports of coinciding signals also in gravity waves.*

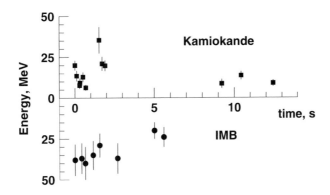

Fig. 3.5. Energy–time plot of the neutrinos detected by Kamiokande (squares) and IMB (filled circles). Note the higher energy threshold of the IMB detector. The first neutrinos detected by each detector are assumed to arrive simultaneously.

In the further discussion we shall concentrate on the results of Kamiokande and IMB, which use the same detection technique and give the energy of each event.

In water neutrinos interact with electrons, free protons and O nuclei. ν_e interactions (the same used for the detection of solar neutrinos) have cross-sections between 10^{-44} and 10^{-45} cm^2 for the different neutrino flavors. The *electrons are scattered forward, i.e. approximately track the neutrino direction. The most efficient process is the scattering of electron antineutrinos on protons ($\bar{\nu}_e + p \longrightarrow n + e^+$) with a cross-section $\sigma = 7.5 \times 10^{-44} (E/MeV)$ cm^2. The angular distribution of the emitted positrons is isotropic and does not indicate the direction of the parent neutrino. In ν_e ^{16}O scattering an electron is emitted backwards, and the cross-section is a factor of 10 lower. Because of the cross-section differences the signal in water-Cherenkov detectors is dominated by $\bar{\nu}_e$ scattering*

$$S_{\nu_e} : S_{\bar{\nu}_e} : S_{\nu_\mu} = 2 :: 40 : 1 , \qquad (3.8)$$

as charge current interaction of the higher neutrino flavors does not proceed in this energy range – the neutrino energy is less than the masses of the the muon and the tau lepton. The signals in Kamiokande and IMB show an approximately isotropic angular distribution.

Even with a small number of observed events one can analyze the data in terms of neutrino flux and energy. Figure 3.6 shows my own reconstruction of the supernova neutrino energy spectrum. All events are assumed to be $\bar{\nu}_e p$ interactions. Every event is scaled with the cross-section and the corresponding detector efficiency. The individual neutrinos, reconstructed in this way, were grouped in energy bins and compared to Planck spectra of different temperatures. Figure 3.6 shows how difficult this is. I was only able to fit the observations with a two-temperature spectrum containing $T_1 = 1.5$ MeV and $T_2 = 5$ MeV.

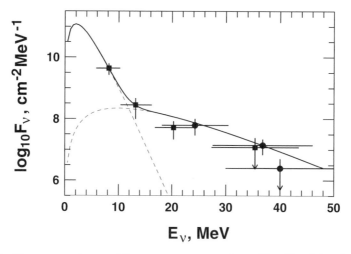

Fig. 3.6. Energy spectrum of the supernova neutrinos. The Kamiokande data are plotted with squares and IMB – with dots. The two dashed lines represent black body spectra with temperatures 1.5 and 5 MeV and the solid line is their sum.

Much more detailed analysis was performed by Bahcall and collabora-tors [49], who used Monte Carlo simulations of detection of neutrinos with different spectra and determined regions in the T–E_ν plot that are allowed by each of the observations. The regions allowed by Kamiokande and IMB intersect at $T = 4.1$ MeV, which yields neutrino energy $E_\nu = 4.5 \times 10^{52}$ ergs. Accounting for the energy in non-interacting neutrinos the total neu-trino energy from SN1987a is about 3×10^{53} ergs, in good agreement with the theoretical expectations.

The detection of the supernova neutrino burst also gave important in-formation on the basic neutrino properties. The survival of the neutrinos of relatively narrow energy range in propagation on 50 kpc allowed setting of bet-ter limits on the neutrino mass, lifetime, electric charge, magnetic moment, etc.

The mass of the neutrino can be estimated from its time delay on the propagation [50]. The time delay of a neutrino with mass m_ν emitted by SN1987a is

$$\Delta T = 2.57 \text{ s} \left(\frac{D}{50 \text{ kpc}} \right) \left(\frac{10 \text{ MeV}}{E_\nu} \right)^2 \left(\frac{m_\nu}{10 \text{ eV}} \right)^2 \qquad (3.9)$$

One would then have time delay in a reverse proportion to E_ν^2 which is a measured quantity. The actual determination of the mass is much more un-certain because it depends on the stellar collapse theory and is affected by the small statistics. The duration of the neutrino burst is in excess of 10 sec-onds and the neutrino flux and energy are not constant during that time. The more conservative estimates give limits of about 15 eV, comparable with the contemporary laboratory limits.

The neutrino lifetime may be similarly estimated to be longer than 5×10^5 seconds for m_ν of 1 eV. The electric charge, which can delay neutrinos be-cause of interactions with the galactic magnetic field, can be related to the neutrino mass. Many other neutrino and particle physics limits were set af-ter the detection of the neutrino burst of SN1987a.

3.1.5 Supernova remnants

SN1987a was a type II supernova, the evolution of which was described in section 3.1.2 in relation to the red giant branch, although not a very typical one. The precursor of SN1987a has passed through a red giant phase and through a blue giant phase. Type I supernovae result from the explosion of carbon–oxygen white dwarfs. Types I and II are distinguished by the absence (type I) or presence (type II) of a hydrogen line in their optical emission. Supernova explosions of very massive stars also do not show hydrogen lines. These are classified as type Ia, Ib, etc.

SN1987a, although not in our own Galaxy, was only the sixth nearby supernova explosion in the last millennium. The other five are galactic super-novae observed with the much less powerful instruments of the past or with the naked eye. The first known supernova is SN1006, by now almost 1000 years old. The next one, whose remnant is now known as the Crab nebula, was described by Chinese and Japanese observers in 1054. Tycho's supernova was observed by him in 1572 and Kepler observed another one in 1604. The youngest known supernova remnant is Cas A, which must have exploded about 300 years ago. It is the brightest radio source in the sky.

We have not seen a galactic supernova since, but the supernova rate must be significantly higher. Most supernovae happen in the galactic disk, which contains large amounts of interstellar dust. The optical light of the supernova may be easily obscured. Most of the information comes from observations of other galaxies, similar to ours. The supernova explosion frequency in the Galaxy is estimated to one to three per century.

Much was learned about the evolution of young supernova remnants from the observations of the remnant of SN1987a. Figure 3.7 shows a sketch of its optical light curve during the first three years after the explosion.

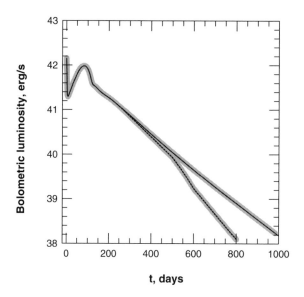

Fig. 3.7. The bolometric luminosity of SN1987a during the first 1,000 days after the explosion. The lower branch after ~ 400 days shows the optical luminosity and the upper branch adds the optical, X-ray and the γ-ray luminosities.

During the first several days the brightness of the supernova remnant decreases as the ejecta, that was heated by the shock, quickly cools down. The temperature of the ejecta after the shock passes through it is expected to reach 300,000 K. One day after the collapse it is expected to cool to about 15,000 K, and several days after that to stabilize at about 5,000 K.

After about 10 days there is a noticeable increase in the bolometric luminosity. This is due to the decreased opacity of the expanding shell. Initially all shell material is ionized by the shock wave. With the decrease of the temperature the hydrogen starts to recombine in neutral atoms. Neutral hydrogen has much lower opacity than ionized hydrogen. The recombination starts from the outer layers of the remnant and moves inwards. There are now two competing processes – the shell as a whole expands and the photo sphere moves inwards. It is not obvious a priori which one will prevail. The expansion wins during the first months after the explosion, and the bolometric luminosity slowly increases. This increase cannot last very long, because eventually all the hydrogen recombines and the photosphere reaches the base of the shell. At this time the luminosity should start to decrease.

The remnant of SN1987a went through that point three months after the collapse. The luminosity decreased to one-third of the maximum and then the time dependence changed – a new energy component started to appear. This new component decreased exponentially with time, which is characteristic for radioactive decays. After months of observations the decaying nucleus was identified as ^{56}Co (half life of 77 days), confirming the pre-supernova predictions. Further observations also established the total mass of the radioactive element – 0.07 $M_\odot$. That amount of Co was generated in the form of ^{56}Ni during the shock propagation through the ejecta. ^{56}Ni decayed into ^{56}Co with a half life of 6.1 days, and the ^{56}Co started decaying to the common and stable ^{56}Fe. Gamma-ray lines from the ^{56}Co and X-rays from the γ-ray reprocessing in the ejecta were also observed several months later.

Traces of other radioactive nuclei with longer half lives that were also produced in the explosion in smaller quantities were also revealed in later observations. Pulsar activity was, however, not detected. The radiation of a rapidly spinning pulsar in the center of the ejecta would have contributed to the luminosity of the remnant and would have changed its behavior. The change of the optical light curve at around 400 days (the lower branch in Fig. 3.7) is related to the emergence of X-rays and γ-ray lines. When they are added to the optical luminosity the ^{56}Co half life is recovered.

The supernova shell expands with very high velocity, which is proportional to the radial distance to the center of the explosion. The outer layers reach one-tenth of the velocity of light c, while the inner layers expand more slowly. The exact density and pressure distributions depend on the structure and dynamics of the precursor star. The expansion velocity is supersonic and a shock forms in front of the expanding remnant. The temperature of the remnant falls steeply with increasing radial distance, while the shocked interstellar gas is heated to high temperature, which remains constant during the free (constant velocity) expansion of the remnant.

During the expansion the shock acts as a snow plow that 'collects' interstellar matter. When the mass of that swept-up material becomes significant compared to the mass of the ejected supernova shell the remnant starts to decelerate. The evolution of the remnant enters a new phase, known as the Taylor–Sedov phase. The supernova expansion during this phase is determined by the energy of the supernova explosion which is transferred to the shell E_{shell} and by the density ρ of the supernova environment that provides the swept-up mass. The radius of the remnant $R_{SNR} \propto E_{shell}^{1/5} \, t^{2/5}$, where t is the age of the remnant. Supernova remnants enter the Taylor–Sedov phase when they are more than 1,000 years old and have a radius of more than 1 pc when they expand in the ordinary interstellar matter.

Later in the evolution of the remnant its internal temperature continues to decrease. The remnant sweeps more and more mass and slows down significantly. When the expansion velocity becomes subsonic the remnant practically ceases to exist. There is no shock, no interaction with the environment. The density of the remnant is so low that it just blends into the interstellar medium.

All the young galactic supernova remnants listed at the beginning of this section, except for the Crab nebula, are in the stage when the swept-up mass is of the order of the mass of the ejecta. The Crab nebula follows a different evolution scenario because it is energized by the Crab pulsar which constantly pumps energy into it.

3.2 Acceleration of cosmic rays

Supernova remnants in the Taylor–Sedov phase are believed to be efficient accelerators of cosmic rays as a part of the kinetic energy of the remnant that slows down is converted to cosmic rays. The idea was justified by Ginzburg and Syrovatskii [2] through simple and powerful arguments based on the energetics of supernova remnants. If we assume that a volume around the galactic disk v_{GD} is uniformly populated by cosmic rays that are contained in this volume for a characteristic time t_{GD} we can estimate the power needed to accelerate these particles. The volume of the region, v_{GD}, a disk of radius about 15 kpc and height about 500 pc, is $\pi \, (15 \text{ kpc})^2 (500 \text{ pc}) \sim 10^{67} \text{ cm}^3$. The cosmic rays energy density ρ_E is about 0.5 eV/cm^3. Let's take $t_{GD} = 10^7$ years (see Sect. 4.3 for the derivation of the age of cosmic rays in the Galaxy). The power required to replenish the cosmic rays that leak out of the galactic disk is

$$L_{CR} = \frac{v_{GD} \times \rho_E}{t_{GD}} \simeq 3 \times 10^{40} \text{ erg/s.} \qquad (3.10)$$

Three supernova remnants of mass $10 M_\odot$ expanding with velocity of 5×10^8 cm/s per century would produce 3×10^{42} erg/s. Thus an acceleration efficiency of only 1% would supply the energy content of the cosmic rays in the galactic disk.

Supernova remnants are attractive candidates for cosmic ray acceleration because they have higher magnetic fields than the average interstellar medium. They are also large and live long enough to carry the acceleration process to high energy. The acceleration mechanism is believed to be stochastic acceleration at supernova blast shocks.

3.2.1 Stochastic acceleration of charged particles

The idea of stochastic particle acceleration was first developed by E. Fermi. Fermi [51] proposed to use the charged particle interactions with interstellar clouds to accelerate cosmic rays. A simplified one-dimensional version of his scenario is shown in Fig. 3.8. Assume that a particle of energy E_0 encounters a massive cloud containing turbulent magnetic field. In the Lab system the particle and the cloud are moving towards each other. For simplicity let us assume that the particle is already relativistic, and its mass could be neglected, i.e. $E_0 \simeq p_0 c$. The cloud has infinite mass and its velocity is v_{cl}. Let the particle enter the cloud, scatter many times in the magnetic turbulence and eventually come out of the cloud moving in a direction collinear and opposite to its initial direction, as shown with track A in Fig. 3.8.

The particle energy in the coordinate system of the cloud is

$$E_0^* = \gamma_{cl}(E_0 + \beta_{cl} p_0) , \tag{3.11}$$

where $\beta_{cl} = v_{cl}/c$ and $\gamma_{cl} = (1. - \beta_{cl}^2)^{-1/2}$. The interactions of the particle with the magnetic field inside the cloud will be completely elastic, i.e. the particle energy and momentum will not change. Its direction, however, is changed, reversed in this simple example. The energy of the particle E_1 at the time it exits the cloud will be

$$E_1 = \gamma_{cl}(E_0^* + \beta_{cl} p_0^*) = E_0 \times \gamma_{cl}^2 (1 + \beta_{cl})^2 \tag{3.12}$$

The particle has gained energy ΔE. The relative gain

$$\frac{\Delta E}{E} = \frac{E_1 - E_0}{E_0} = \gamma_{cl}^2 (1 + \beta_{cl})^2 - 1 \equiv \xi \tag{3.13}$$

is proportional to the square of the velocity of the magnetic cloud. Note that the energy gain was derived for a particle which exits the cloud in the direction exactly opposite to that of its entry. The energy gain depends very strongly on the relation between the angles of the entry and the exit of the cloud relative to the cloud velocity vector. In the same one-dimensional picture, for example, the particle would not gain any energy if it left the cloud through its far side and continued moving in its initial direction as in track B in Fig. 3.8. There are also configurations in which the particle loses energy, e.g. when the particle enters the cloud along the cloud velocity.

For particles that enter and leave the cloud at angles different from zero and π radians the transformations in equations (3.11) and (3.12) include the term $\cos\theta\beta_{cl}$ where θ is the angle between the particle and the cloud directions. One can calculate the average values of $\cos\theta$ for the particles that

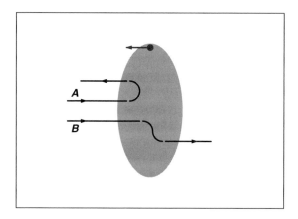

Fig. 3.8. Schematic representation of Fermi's idea of particle acceleration by scattering in magnetized clouds. Only the case when the particle trajectories are collinear with the cloud velocity are shown.

enter and leave the cloud. Since the particle direction inside the cloud is fully isotropized through multiple scatterings the exit angle is random and $\langle \cos\theta_2 \rangle = 0$. The entry angle depends on the cloud velocity and $\langle \cos\theta_1 \rangle = -\beta/3$. The average energy gain per cloud encounter then becomes $\xi \simeq 4/3\,\beta_{cl}^2$.

It is important to remember that the fractional energy gain in the process of stochastic acceleration is constant. After n encounters of magnetic clouds (with the same β_{cl} for simplicity) the particle energy will be

$$E_n = E_0(1 + \xi)^n \tag{3.14}$$

and the number of encounters needed to reach energy E_n is respectively

$$n = \ln\left(\frac{E_n}{E_0}\right) / \ln(1 + \xi). \tag{3.15}$$

At every encounter the particle can escape from the region that is occupied by magnetic clouds with some probability P_{esc}. Once it escapes its energy does not increase any more. The probability that the particle has reached energy E_n, i.e. that the particle remained in the acceleration region long enough to encounter n magnetic clouds is $(1 - P_{esc})^n$. The number of particles that are accelerated to energies higher than E_n is proportional to the number of particles that remain in the acceleration region for more than n cloud encounters. Using the definition of n in (3.15).

$$N(> E_n) = N_0 \sum_{n}^{\infty} (1 - P_{esc})^m \propto A\left(\frac{E_n}{E_0}\right)^{-\gamma}, \tag{3.16}$$

with

$$\gamma \simeq P_{esc}/\xi. \tag{3.17}$$

Stochastic acceleration thus generates power law energy spectra. In the particular case of Fermi acceleration the value of the power law index depends of the square of β_{cl}.

The energy gain per unit time depends on the frequency of encounters ν_{enc} and is

$$\frac{dE}{dt} = \nu_{enc}\Delta E = \frac{c}{\lambda_{enc}}\xi E = \frac{\xi E}{T_{enc}} \tag{3.18}$$

where λ_{enc} is the mean free path between encountering magnetic clouds and T_{enc} is the characteristic time per encounter. The acceleration time is thus proportional to the energy and reaching higher energy requires respectively longer time. As a whole Fermi's beautiful idea does not work because the acceleration is very slow. The energy gain per cloud encounter is proportional to β_{cl}^2, a number of the order of 10^{-7} or less and λ_{enc} is not shorter than 1 pc. This mechanism reproduces the power law energy spectrum observed in cosmic rays, but it does not have much to say about its index, which also depends on the velocity of interstellar clouds.

3.2.2 Particle acceleration at astrophysical shocks

The shock ahead of the expanding supernova remnant is formed because the expansion velocity of the remnant v_R is much higher than the sound velocity of the interstellar medium. The shock runs ahead of the expanding remnant with velocity v_S, which depends on v_R and the ratio of the specific heats of the shocked and unshocked media. If the interstellar medium at the shock is ionized, the shock velocity $v_S \simeq 4/3v_R$. The strength of the shock is characterized by the compression ratio $R \simeq \frac{v_S/v_R}{v_S/v_R - 1}$ and $R = 4$ in this case.

If the radial dimensions of the shock are much larger than the particle gyroradius r_g

$$r_g = pc/ZeB \simeq 3.2 \times 10^6 \text{cm} \times (E/\text{GeV})/(B/\text{G}), \tag{3.19}$$

where Z is the charge of the accelerated particle and B is the average magnetic field in the shock region, the shock could be represented as a plane for the purposes of particle acceleration, as shown in Fig. 3.9. This figure shows the expanding remnant (a slab of matter in this approximation) and the plane shock moving with velocity v_S. The interstellar matter ahead of the shock has its own velocity v_{ISM}. The shock velocity and the velocity of the interstellar matter downstream of the shock v_d are related by the requirement for continuous mass flow through the shock $v_S\rho_S = v_d\rho_d$, which gives a ratio $v_S/v_d = 4$, the same as the compression ratio R. In the frame of the shock (the bottom panel) the upstream (unshocked) stellar medium flows into the shock with velocity $u_1 = v_S$ and the downstream shocked matter flows away from the shock with a smaller velocity u_2. In the Lab system the gas behind the shock (downstream) moves away from the shock with velocity $u_1 - u_2$ and the matter upstream moves towards the shock with velocity u_1.

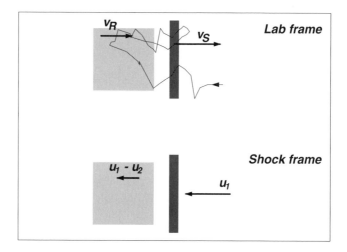

Fig. 3.9. Schematic representation of particle acceleration at astrophysical shocks. The upper panel shows the velocities and the motion of a test particle in the Lab frame. The lower panel shows the velocities in the shock frame.

The cosmic ray gas contained in the interstellar medium ahead of the shock is isotropic. As the shock approaches some of the gas particles cross the shock and move downstream. The downstream particles are isotropized. Some of them are convected away from the shock and others scatter back and cross the shock again in the upstream direction. Note that the particles always gain energy by crossing the shock because the elastic collisions are head-on in the corresponding frames. At the second crossing the sequence of encounters is the opposite of the first one, but so is the particle velocity vector. The acceleration process continues until the particles diffuse or are convected away from the shock. If one sums over all possible angles for entering and leaving the shock, the energy gain per crossing $\xi \sim 4/3\,\beta_S$, where β_S is the relative velocity of the plasma flow $\beta_S = (u_1 - u_2)/c$.

Shock acceleration is much faster than the original Fermi acceleration mechanism. The energy gain is proportional to β (first-order acceleration) rather than to β^2 (second-order (Fermi) acceleration). In addition, the supernova shock velocity is much higher than the average velocity of molecular clouds. As a result shock acceleration is orders of magnitude more efficient, and correspondingly much faster. The shock acceleration scenario was suggested in the late 1970s [52, 53, 54] and is under continuous development [55, 56, 57]. The treatment of shock acceleration in this chapter follows Gaisser's interpretation [6] of Bell's theory.

Shock acceleration gives a definite prediction for the spectral index γ of the power law spectrum of the accelerated particles. For a large plane shock the rate of shock encounters is the projection of the isotropic cosmic ray flux of density ρ_{CR} onto the plane front of the shock, which is $c\rho_{CR}/4$. The rate of escaping the shock through convection downstream away (which is the only way of leaving a plane shock of infinite length) is the product of the same cosmic ray density times the convection velocity u_2. The escape probability is the ratio of the escape rate to the encounter rate

$$P_{esc} = \frac{\rho_{CR}u_2}{c\rho_{CR}/4} = \frac{4u_2}{c}. \tag{3.20}$$

From (3.17)

$$\gamma = \frac{P_{esc}}{\xi} = \frac{4u_2}{c} \times \frac{3c}{4(u_1 - u_2)} = \frac{3}{u_1/u_2 - 1} \sim 1 \tag{3.21}$$

for strong shocks. Strong shocks are shocks with plasma flow velocity much higher than the sound speed. For supernova blast shocks with matter flow velocity of the order of 10^9 cm/s and sound speed of the order of 10^6 cm/s this condition is fulfilled during a large fraction of the supernova remnant evolution.

We can now derive the maximum energy that a charged particle can achieve in acceleration at the shock. Equation (3.18) defines the energy gain as the energy times the acceleration efficiency divided by the characteristic time of encounters, which in this case is the shock crossing time T_c. The shock crossing time could be estimated as the ratio of the mean free path for magnetic scattering λ_S and the shock velocity u_1. The limit on the particle energy comes from the requirement that the stochastic process can only continue if the mean free path for magnetic scattering, λ_S, is larger than the particle gyroradius, r_g. The energy gain is then limited by

$$\frac{dE}{dt} \leq \frac{\xi E u_1}{r_g} = \frac{u_1}{c} ZeBu_1 . \tag{3.22}$$

The maximum energy that a charged particle could achieve is then expressed as a function of the shock velocity and extension and the value of the average magnetic field as

$$E_{max} = \frac{u_1}{c} ZeB(u_1 t) = \frac{u_1}{c} ZeBr_S. \tag{3.23}$$

for relativistic particles when the energy could be approximated with the particle momentum ($E = pc$). Equation (3.23) is valid during the period of the free expansion of the supernova remnant when the radius r_s is proportional to the time t since the explosion. During the Taylor–Sedov phase the maximum energy would decrease as the radius is only proportional to the time to the power of 2/5. An important component of the expression for E_{max} is its dependence on the particle charge Z. It means that a fully ionized heavy nucleus of charge Z could achieve much higher total energy than a proton with $Z = 1$.

The maximum energy achievable at supernova remnants is estimated by Lagage & Cesarsky [58]. They show that most of particle acceleration happens before the shock has swept up mass equal to the mass of the ejecta, i.e. when the average density times the volume of the remnant equals the mass of the ejecta.

$$\frac{4}{3}\pi(u_1 t)^3 \rho_{IS} = M_{SR}. \tag{3.24}$$

For $M_{SR} = 10 M_\odot$ and $u_1 = 10^9$ cm/s and interstellar medium with density 1 nucleon per cm^3 the maximum energy is

$$E_{max} = Z \times 2.4 \times 10^5 \text{ GeV}. \tag{3.25}$$

The original number of Lagage & Cesarsky is 3×10^4 GeV since they use an expansion velocity of 5×10^8 cm/s. A higher value is derived for a lower value of the interstellar density. Some more detailed recent calculations [59] derive values close to 5×10^5 GeV.

Some additional considerations

All formulae in Sect. 3.2.2 are derived on the assumption that the magnetic field direction is normal to the shock front. In the shock acceleration jargon such shock is called a parallel shock because the direction of the magnetic field relative to the shock normal is used for the identification of the shock geometry. The acceleration process in a perpendicular shock is aided by the electric field which is created in the rest frame of the shock by the inflow of the magnetized plasma from the upstream region. The magnitude of the electric field is $E = u_1 \times B/c$. Charged particles gain energy very fast when they move along the plane shock. The energy gain is proportional to the distance along the shock and to the electric field value – $\Delta E = ZeEl$. Jokipii [56] has developed a model for acceleration in quasi-perpendicular shocks, combining the effects of stochastic acceleration with the energy gain in the drift along the shock, which is very efficient. The energy gain could be orders of magnitude higher than in parallel shocks, depending on the values of the magnetic field and of the diffusion coefficients in the upstream and downstream regions. A more recent Monte Carlo study of the acceleration on quasi-perpendicular shocks [60] confirms the high acceleration rate, but argues that the injection efficiency in such shocks is much lower than in parallel shocks. Thus to take full advantage of the high energy gain one should either somehow pre-accelerate the interstellar matter before it is injected in the shock, or arrange the shocks in a

*specific way, where the particles accelerated in parallel regions of the super-
nova shock are then injected in regions where the shock is perpendicular. In
fact, any real astrophysical shock has to be much different from the plane that
we use as an approximation. Real shocks are three-dimensional and the magnetic turbulence at the shock is so high that different fractions of the shock
must cover all possible magnetic field orientations.*

*Another important consideration is the possibility that the supernova rem-
nant may evolve not in the typical interstellar medium, but rather in the hot
and highly magnetized environment created by the stellar winds of the progenitor star [61]. Red giants, the typical supernova type II progenitors, have
very heavy mass loss during the later stages of their evolution, possibly exceeding 10^{-7} $M_\odot$ per year. The surface magnetic field of the progenitor star
is frozen into the stellar wind plasma. The medium in which at least some
supernova remnants expand is much denser and has stronger magnetic fields
than the typical interstellar medium. Völk & Biermann [61] use the example
of the extreme Wolf–Rayet stars, where the maximum acceleration energy for
protons could exceed 10^{17} eV.*

*The standard theory of shock acceleration is developed for nonrelativis-
tic shocks, i.e. $u_1 << c$. The only observational confirmations of the basic
principles of stochastic shock acceleration come from studies of heliospheric
processes. The characteristic velocities of such shocks is of order of the solar wind velocity, several hundred kilometers per second. Such shocks are not
relativistic. The theory of relativistic shocks is more complicated and is not
well studied. Kirk & Schneider [62] have worked out the main acceleration
parameters for some specific cases of relativistic shocks. One of the big differences with the nonrelativistic case is the possibility of accelerating particles
on flat power law spectra with $\gamma < 1$. The energy carried by the highest energy
particles dominates such flat spectra and thus the problem of the maximum
energy at acceleration becomes even more important.*

*The problem with the spectral shape is much more complicated than it is
presented in Sect. 3.2.2. There are several reasons for which the power laws
derived above are only approximate. One of them, the time dependence of the
maximum energy, was worked out in Ref. [63]. In this work the acceleration
spectra are power laws at any given time of the evolution of the remnant. The
time average spectrum of the cosmic rays accelerated at a remnant is the sum
of the power law spectra achieved at different stages weighted with the time
the remnant has spent at a given evolution epoch. This is a slightly concave
spectrum that can be approximated with a power law only in limited energy
ranges.*

*Another reason for concavity comes out of recent detailed treatment of the
acceleration process which also account for the effect of the accelerated protons
on the dynamics of the shock itself. The classical approach presented above
follows only the fate of the particles that are accelerated, i.e. it is a test particle approach. When the dynamics of the shock is also considered it becomes*

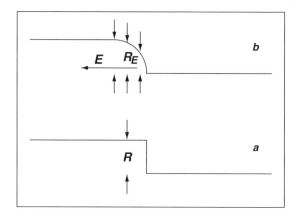

Fig. 3.10. Schematic representation of the shock front in the case of the test particles approach (a) and taking account of the shock dynamics (b). In this latter case the effective shock compression ratio R becomes energy dependent which generates a non-power law acceleration spectrum.

obvious that the shock is smoothed by the pressure of the accelerated particles and it is no longer simply a discontinuity. This configuration is illustrated in Fig. 3.10, where panel (a) corresponds to the test particle approach and panel (b) shows a shock affected by the pressure of the accelerated particles. The low energy particles cannot scatter far enough from the shock to feel the full compression ratio ($R = u_1/u_2$) and only experience a part of it. The effective compression ratio becomes energy dependent. Low energy particles would be accelerated to a steeper power law than the highest energy particles that can 'see' the total strength of the shock and which will have the flattest possible spectrum. The resulting spectrum will not be a power law. It is a concave spectrum which, if approximated with power law spectra over different energy ranges, is the steepest at energies just above the injection energy. The spectrum becomes flatter at higher energy with the flattest part coming just before the maximum energy. For a more exact description of this effect one should read the original papers by Ellison and collaborators [57] and by Berezhko and collaborators [64].

A very simple simulation procedure and useful for many applications, except for studies of the shock acceleration process itself, has been developed by R.J. Protheroe [65, 66]. It is based on the leaky box picture (see section 4.3), and implements the basic ideas of the diffusive shock acceleration described above. Particles are injected into the box with injection energy E_0. While they are inside the box, they experience acceleration with a constant energy gain $dE/dt = a$. The time for acceleration to energy E is then $t_{acc} = E/a$. Particles freely diffuse inside the box, reflecting off its walls, but can also escape *with an escape probability P_{esc} The escape time $t_{esc} = t_{acc}$ because both acceleration and escape depend on the magnetic field at the shock in the same way. One injects particles with energy E_0 into the leaky box, calculates the energy gain after a time step Δt as $\Delta E = a \times \Delta t$ and lets them escape from the leaky box with probability $P_{esc} = E/a$. The energy spectrum of the escaping particles is exactly a power law with spectral index $\gamma = 1$. The model does not have E_{max} in its basic form because it assumes that the scattering length λ is always much longer than the particle gyroradius r_g. E_{max} could be introduced in the same way as in the analytic formulae by assuming that the diffusion approximation breaks up at E_{max}. The form this is introduced in the Monte Carlo procedure is through a new definition if $t_{esc} = E/a + E_{max}/a$, which results in an exponential cutoff $\exp(-E/E_{max})$ of the spectrum of the escaping particles. This method can be very useful for studies of particle acceleration in the presence of energy loss. A very good analysis and an extension of the method is given in Ref. [67].*

3.2.3 Acceleration with energy loss

The maximum energy at acceleration E_{max} discussed in Sect. 3.2.2 is caused by the escape of the accelerated particle from the acceleration region. This is generally true for protons that are accelerated at supernova remnants with acceleration time of less than $\sim 10^5$ years. The situation, however, is very different for the acceleration of electrons, which lose energy by synchrotron radiation in the magnetic fields. The energy loss rate is proportional to the square of the electron energy and to the square of the magnetic fields $- -dE/dt = AE^2 B^2$. The energy loss time $t_{loss} = E/(dE/dt) = (AE)^{-1} B^{-2}$. With the increasing electron energy the energy loss time decreases and at some limiting energy becomes equal to the acceleration time

$$t_{acc} = t_{loss} = E/a. \tag{3.26}$$

The electron loses as much energy as it gains and cannot be accelerated to higher energy. The equality of the acceleration time and the energy loss time gives the maximum energy that the particle can reach, i.e.

$$E_{max}/a = (A \times E_{max})^{-1}/B^2, \tag{3.27}$$

which results in a limit $E_{max} = \sqrt{(a/A)}/B$. Since the acceleration efficiency a is also proportional to the magnetic field B the maximum energy of electrons is actually proportional to $B^{-1/2}$. After a substitution of the relevant shock parameters Gaisser [6] obtains the maximum energy for electron acceleration in the presence of synchrotron loss as

$$E_e^{synch} = 2.3 \times 10^4 \, \text{GeV} \frac{u_1}{c} \times (B/\text{G})^{-1/2}. \tag{3.28}$$

Synchrotron energy loss is not the only energy loss process that can interfere with particle acceleration. It is, however, a very good example because the acceleration of charged particles assumes the existence of a strong and turbulent magnetic field, which has to also cause synchrotron energy loss during the acceleration of electrons. Other common process for electron energy loss is the inverse Compton effect on the universal microwave (2.7 K) background and other ambient photon fields. Protons may lose energy in hadronic interactions with matter or with photon backgrounds. In all cases where the energy loss interferes with the particle acceleration the maximum energy is determined according to the general principle provided by (3.26) – by equating the acceleration and energy loss times.

4 Cosmic rays in the Galaxy

Once accelerated at supernova shocks, cosmic rays have to propagate through the interstellar medium before we detect them. The interstellar medium contains matter, magnetic fields and radiation fields, all of which are targets for cosmic ray interactions. Cosmic ray protons scatter in the magnetic fields and slowly diffuse away from their sources. By the time they reach the solar system cosmic ray nuclei have no memory for the position of their sources. Observations show that cosmic rays at Earth are isotropic to a very large degree, except perhaps the cosmic rays of the highest energies. Cosmic ray nuclei interact mostly with the interstellar matter and produce all kinds of secondary particles.

Electrons interact with the magnetic and radiation fields, as well as with the interstellar matter. In the magnetic fields they generate synchrotron radiation, and in radiation fields they boost gamma rays with the inverse Compton effect.

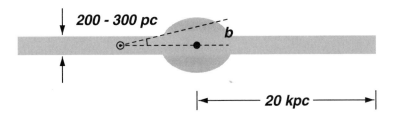

Fig. 4.1. View of the Galaxy from its side (not to scale). The black dot indicates the galactic center. The solar system is 8.5 kpc from the galactic center. In reality the galactic disk is much less regular than shown here.

The luminous matter of the Galaxy is organized in spiral arms that join in the inner Galaxy to form the galactic bulge. Viewed from the side the matter is distributed in a disk with height h about 100–150 pc (total thickness of $2h$) in the vicinity of the solar system, about 8.5 kpc from the galactic center. It is quite possible that the the height of the disk is somewhat bigger outwards

of the solar circle. The radius of the galactic disk is about 20 kpc. Figure 4.1
gives an idea of the shape of the Galaxy viewed from its side.

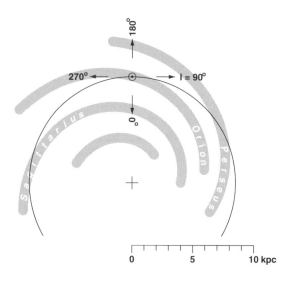

Fig. 4.2. View of the Galaxy from its North Pole. The galactic center is indicated
with a cross. The galactic longitude ℓ is shown with arrows. The thin circle shows
the solar circle.

The arrangement of luminous matter in the galactic plane is shown in
Fig. 4.2. Directions in which concentration of luminous matter are observed
are shaded. Galactic arms cannot be traced behind the galactic center region,
although the assumption is that all structures are continuous. Figures 4.1
and 4.2 also demonstrate how astrophysical objects are plotted in galactic
coordinates ℓ and b. Galactic latitude b is the angle at which an object is
seen above or below the galactic plane. The galactic longitude ℓ is measured
counterclockwise from the direction of the galactic center and the latitude b
is measured above and below the galactic plane. The galactic center is thus
at $\ell = 0°$, $b = 0°$ and the galactic anticenter is at $\ell = 0°$ and $b = 0°$. The
thin circle in the figure indicates the solar circle at galactocentric distance of
8.5 kpc.

4.1 Interstellar matter and magnetic field

Most of the interstellar diffuse matter consists of hydrogen (only about 10%
is helium and heavier nuclei) in the form of atomic neutral hydrogen (HI)
and molecular hydrogen (H_2). The atomic hydrogen is detected by its 21 cm
emission line at radio frequencies. Atomic hydrogen is present in the galactic

arms at an average density of about 1 atom/cm^3 and has a scale height of 100–150 pc. The shape of the HI distribution in the outer Galaxy is somewhat irregular and its scale height may increase. As defined by the density of the atomic hydrogen the galactic plane is warped at large galactocentric distances. The HI density decreases by factors of 2 and 3 in the space between the arms. These observations are mostly local, in the vicinity of the solar system.

Molecular hydrogen is concentrated within the solar circle and especially in the region of the galactic center. The distribution of molecular hydrogen is observed by the spatial distribution of the 2.3 mm line emitted by carbon monoxide (CO), whose density is proportional to that of H$_2$ and acts as a tracer for it. The exact relation between the CO line and the molecular hydrogen density is uncertain. H$_2$ is contained within dense molecular clouds that trace approximately the galactic arms and has somewhat smaller scale height about the galactic plane. The total mass of molecular hydrogen inside the solar circle is about 10^9 $M_\odot$. The average density H$_2$ within the solar circle is then

$$n(H_2) = \frac{10^9 M_\odot N_A}{V_{sc}} \simeq 1 \text{ cm}^{-3}$$

for a full height of the galactic disk of 200 pc and a volume $V_{sc} = 1.3 \times 10^{66}$ cm^3.

This number is uncertain as the observations of the central region of the Galaxy are very difficult because of the high matter density, and large optical depth in this direction. There is also ambiguity in the distances of objects in the direction of the galactic center, that could be either between us and the center or behind it. The density of the giant molecular clouds is 10^2–10^5 per cm^3. The hydrogen inside such clouds can be in molecular form because the large column depth of the cloud shields it from the energetic galactic photons.

A much smaller fraction of the matter is in the form of ionized hydrogen of density 0.03 per cm^3. The ionized hydrogen has been detected extending well above the galactic disk with a height of about 700 pc. Although observations of ionized hydrogen in the galactic disk are difficult, it is reasonable to assume that the disk contains a mixture of neutral and ionized hydrogen. A good round number for the interstellar matter density is 1 nucleon/cm^3. The actual matter distribution is very complicated.

Figure 4.3 gives an idea about the column density of matter in different directions. The column density is an integral from the position of the solar system to the edge of the Galaxy over the matter density in different forms. The matter density model used in Fig. 4.3 assumes cylindrical symmetry of the matter density in the Galaxy [70]. The actual matter density is not symmetric and one can detect somewhat increased column densities in the direction of the local galactic arm and a corresponding decrease towards the inter arm space. A very detailed survey of the matter distribution and maps of different forms of interstellar matter is presented in Ref. [71]. The

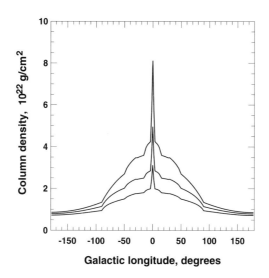

Fig. 4.3. Sketch of the column density of the Galaxy as a function of galactic longitude. The three lines are for longitudes within $2°$, $5°$ and $10°$ from the galactic plane.

total mass of HI in the Galaxy is derived to be 5×10^9 $M_\odot$ and the mass of molecular hydrogen is 0.9–1.4×10^9 $M_\odot$. The local density of the two components is respectively 0.45 and 0.2 cm^{-3}. The exact structure of the galactic magnetic field is also not very well known. Much of the knowledge comes from observations of other galaxies where the general shape of the magnetic field is better visible. Especially helpful are galaxies with disks perpendicular to the line of sight. The magnetic field values are obtained from interpretations of the Faraday rotation of linearly polarized radio signals from radio pulsars. The rotation measure (rad/m^2) gives the integral of the product of the parallel component of the magnetic field and the electron density along the line of sight to the source, i.e.

$$RM = \int_0^d B_\parallel n_e \, dr .$$

The electron density N_e has to be estimated by the delay of the arrival times of the radio signals as a function of the frequency – the dispersion measure of pulsars.

The physics and the principles of the measurements of the Faraday rotation and of the dispersion measure are given by Longair [7]. The experimental techniques and recent results are discussed in some detail in the review papers [72, 73].

The magnetic field in the vicinity of the solar system has a strength of $\sim$2 μG in direction of galactic longitude $\ell = 90°$. The pitch angle of the spiral

p is close to $-10°$ [74]. The field has a reversal at a distance of 500 pc in the direction of the galactic center $\ell = 0$ and possibly a second reversal at a distance of ~ 3 kpc [74, 75] in the opposite direction.

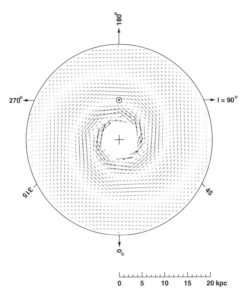

Fig. 4.4. Magnetic field strength (length of arrows) and direction in the galactic plane for the BSS model [76]. The field reversals can be best seen close to the galactic center where the field values are higher. The field is not plotted within 4 kpc of the galactic center because of the very high uncertainty in this region. The positions of the galactic center and the solar system are indicated.

The ideas of the large-scale structure are that the regular magnetic field follows the distribution of the matter, i.e. it has spiral form with either 2π (axisymmetric (ASS)) or π (bisymmetric (BSS)) symmetry. The bisymmetric model is currently favored, although axisymmetric models cannot be excluded. In bisymmetric models the field strength at a point (r, ϕ) in the galactic plane could be expressed (in polar coordinates) as [76]

$$B(r, \phi) = B_0(r) \cos\left(\phi - \beta \ln \frac{r}{r_0}\right) , \qquad (4.1)$$

where r_0 is the galactocentric distance of the position of the maximum field strength in the Orion arm, here assumed to be 10.55 kpc, and $\beta = 1/\tan p$. $B_0(r)$ could be taken as 2 μG at the position of the Sun and inversely proportional to the galactocentric distance, at least for $r > 4$ kpc [77]. Closer to the galactic center the field is higher, but its value is highly uncertain. The magnetic field strength and direction in the BSS model are shown in Fig. 4.4.

Such a model does not have a magnetic field component that is normal to the galactic disk. The r and θ components of the field should decrease exponentially with the height on both sides of the disk with a scale height of ~1 kpc. Ref. [76] gives a two scale dependence of the field strength as a function of the height above the galactic plane z:

$$|B(r, \phi, z)| \qquad = |B(r, \phi)| \times \exp(-z) \qquad \text{at } |z| \le 0.5 \text{ kpc} \qquad (4.2)$$
$$|B(r, \phi, z)| = |B(r, \phi)| \times \exp(3/8) \exp(-z/4) \text{ at } |z| > 0.5 \text{ kpc}$$

It is also not obvious whether the field is of odd or even parity, i.e. whether the field direction changes when the galactic plane is crossed.

Recent estimates of the galactic magnetic field strength give much higher values for the average field in the Galaxy as a whole, as well as locally – the field strength is estimated as 5–6 μG [78]. This requires the introduction of a different field component, which can be considered a halo field. One of the currently favorite models for the halo is the A0 dipole field [79].

This dipole field is toroidal and its strength decreases with the galactocentric distance as r^{-3}. In spherical coordinates (r, ϕ, θ) the (x, y, z) components of the A0 halo field (the solar system is at $x = 0$ kpc, $y = 8.5$ kpc) are

$$B_x \ = \ 3_{\mu G} \sin \theta \cos \theta \cos \phi / r^3 \qquad (4.3)$$
$$B_y \ = \ 3_{\mu G} \sin \theta \cos \theta \sin \phi / r^3$$
$$B_z \ = \ \mu_G \times \left(1 - 3 \sin^2 \theta\right) / r^3 \ ,$$

where $\mu_G \sim 184 \ \mu$G.kpc^3 is the magnetic moment of the galactic dipole. The halo field is very strong in the vicinity of the galactic center and is about 0.3 μG locally, where it is directed towards the North galactic pole.

The addition of the halo component does not help to solve the problem of the very high local field estimates. These could be helped by the addition of a strong random field – the current estimates give random field strengths of 0.5–2B_{reg} [78]. The characteristic cell size of the random field is of the order of 50–100 pc. It is possible that the regular and random components of the galactic magnetic field have different coupling to the matter distribution – the ratio $B_{\mathrm{ran}}/B_{\mathrm{reg}}$ may increase with the matter density. It is logical to assume that B_{ran} increases in the arms because of the magnetic fields of individual stars and molecular clouds inside them and B_{reg} is stronger in the inter-arm space.

4.2 Basic principles of the propagation

Although it is obvious from the brief description of the previous section that the interstellar conditions depend on the position in the Galaxy and there are strong fluctuations in the strength of the magnetic field and in the density of matter, there are general laws that could be used to study the propagation of charged particles in the Galaxy. The ionized gas and the magnetic field carried by it form a magnetohydrodynamic (MHD) fluid. It supports waves that travel with the Alfvén velocity v_A. Cosmic rays scatter on these waves in their propagation.

The energy in MHD waves equals the energy density of the magnetic field, i.e.

$$\frac{\rho v_A}{2} = \frac{B^2}{8\pi} .$$

(4.4)

For an average field of 3 μG the energy density of the magnetic field is 4×10^{-13} erg/cm^3 $\simeq 0.25$ eV/cm^3 and is therefore somewhat smaller than the energy density of 0.5–1 eV/cm^3 carried by cosmic rays. Cosmic rays propagating in the interstellar medium must also induce Alfvén waves which in turn act as scattering centers.

Ginzburg & Syrovatskii [2] wrote the equation of cosmic rays transport in a general form. In their approach the production of cosmic rays in the Galaxy is described by the source term $Q_j(E,t)$. The source term $Q_j(E,t)$ is defined as the number of particles of type j produced (accelerated) per cm^3 at time t with energy between E and $E + \delta E$ in a given location in the Galaxy. These particles diffuse in the Galaxy and their number changes with time. The time evolution of the density $N_j(E,t)$ of cosmic rays of given type j and with energy E at a given location in the Galaxy is a function of the following five processes:

- Cosmic ray diffusion, characterized by the diffusion coefficient $\mathcal{K} = \beta c \lambda/3$ where λ is the diffusion mean free path and $v = \beta c$ is the particle velocity.
- Cosmic ray convection, characterized by the convection velocity v_c.
- The rate of change of the particle energy dE/dt. The energy change could be positive or negative. Negative dE/dt is provided by all forms of energy loss, and mostly by synchrotron radiation for the electrons or ionization loss for protons and heavier nuclei. Energy gain could be realized in 're-acceleration' processes, additional forms of acceleration during propagation in the galactic magnetic fields away from the original acceleration site.
- Particle loss term. Because of interactions or decays, particles of type j have turned into particles of type k and their number has to be subtracted from the density $N_j(E,t)$. The loss term is $p_j N_j(E,t)$, where $p_j = v\rho/\lambda_j + 1/\gamma_j \tau_j$ could be expressed as a function of the particle velocity v, interaction length λ_j and the target density ρ for the case of loss due to interactions and by its Lorentz dilated lifetime $\gamma_j \tau_j$ in the case of decay.
- Particle gain term. Because of interactions particles of type i have turned into particles of type j and have to be added to the density. This term is a weighted sum of all interactions and decays that create particles j.

All these processes affect the source term $Q_j(E,t)$ and give the particle density $N_j(E,t)$ as a function of time and energy.

4.2.1 Particle diffusion

Diffusion is the description of particles propagating through random walk. The problem of the random walk is studied in a classical and mathematically challenging article by Chandrasekhar [80]. The article starts with the simplest problem of one-dimensional random walk, that is illustrated in Fig. 4.5. The particle is initially at position [0] and at each time step it can scatter with the same probability in positive and negative direction by one unit, i.e. after the first step it could be in position [1] or [−1]. After the second scattering the particle could be found in one of the five positions [−2, −1, 0, 1, 2]. The question Chandrasekhar asks is: what is the probability that the particle would occupy position m after N steps?

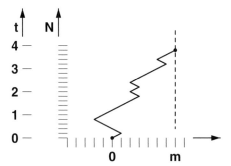

Fig. 4.5. One-dimensional diffusion problem. The particle scatters either left or right at every time step.

In order to reach the positive position m in N steps the particle should make $(N + m)/2$ positive steps and $(N − m)/2$ negative steps. Since each step is of unit length this places the restriction that N and m should both be even or odd and the probability becomes

$$W(m, N) = \frac{N!}{[(N + m)/2]![(N − m)/2]!} \left(\frac{1}{2}\right)^2 . \tag{4.5}$$

The r.m.s. displacement of the particle after N steps is $\sqrt{N}$. Chandrasekhar then uses Sterling's formula for the factorials and expands the logarithms for a small m/N ratio to obtain

$$W(m, N) = \sqrt{(2/\pi N)} \exp\left(−m^2/2N\right) . \tag{4.6}$$

The next step in the derivation is to introduce displacement x which is related to the position m through the step size ℓ, i. e. $x = m\ell$ and to assume that the particle scatters n times per unit time ($n = 5$ in Fig. 4.5). The probability that the particle will be found at x after time t becomes

$$W(x,t) = \frac{1}{2\sqrt{\pi Dt}} \exp\left(-x^2/4Dt\right), \qquad (4.7)$$

where $D = n\ell^2/2$ is the one dimensional diffusion coefficient. D measures the speed of diffusion and is measured in units of $[\text{cm}^2/\text{s}]$. The bigger D, the smaller is the probability that the particle will reach position x in time t. Chandrasekhar then develops the solution for the important cases of reflecting and absorbing boundaries.

In the case of three dimensional scattering the probability of finding the particle in a volume $[r, r + \delta r]$ is

$$W(r)dr = \frac{1}{\sqrt{4\pi Dt}} \exp(-|r|^2/4Dt), \qquad (4.8)$$

where $D = n\langle r^2 \rangle/6$ is the three-dimensional diffusion coefficient and $\langle r^2 \rangle$ is the r.m.s. scattering length.

A good example of contemporary treatment of diffusion is the treatment of relativistic particles in one and three dimensions of Achterberg et al. [81]. The paper treats the average distance, angle and time delay of charged particles of energy E after diffusion in random magnetic fields of coherence length l_{coh}. In the case of three-dimensional diffusion the r.m.s. distance of a particle to its source increases as

$$\langle r^2 \rangle \simeq 2\mathcal{K}t, \qquad (4.9)$$

where t is the diffusion time and $\mathcal{K}$ is the diffusion coefficient ($\mathcal{K} = (3c/2) \times (r_g^2/l_{coh})$), i.e. for a relativistic particle moving with velocity c the diffusion length $\lambda = r_g^2/2l_{coh}$. Since l_{coh} depends only on the random field, the diffusion length is a strong function of the particle energy.

A special case of the fastest possible diffusion is the *Bohm diffusion* where the mean free path (l_{coh}) approaches the particle gyroradius r_g. The diffusion coefficient for *Bohm diffusion* is $\mathcal{K}_B = cr_g/3$. One can use $\mathcal{K}_B$ to set a limit for the distance particles of energy E can diffuse for time t. For random galactic fields it follows from (4.9) that for protons of energy E GeV

$$\sqrt{\langle r^2 \rangle} \simeq 5\,\text{pc} \times \left[\frac{3 \times 10^{13} \times E}{\mu G} \frac{\tau_G}{10^9 \text{ yrs}}\right]^{\frac{1}{2}}, \qquad (4.10)$$

where μG is the average strength of the random galactic fields and τ_G is the age of the Galaxy. Equation (4.10) gives the average distance to which protons of energy E GeV diffuse. Since *Bohm diffusion* is an upper limit for $\mathcal{K}$ the real absolute limiting diffusion distance is larger.

4.3 Formation of the chemical composition

Understanding the chemical composition of cosmic rays starts with the experimental observation that some elements have abundances that are very different from the chemical abundances measured on Earth and in the solar system. While the majority of elements have approximately the same relative abundance, there are groups of elements that are overabundant in the cosmic rays by many orders of magnitude. The first one is the group of light elements Li, Be and B. The sub-iron elements Sc, Ti, V, Cr and Mn are also overabundant by about three orders of magnitude. The reason for the overabundance are the interactions of accelerated cosmic ray nuclei with the interstellar matter, which produce the overabundant elements as spallation products. These elements are known as secondary elements in cosmic ray composition. The most abundant nuclei in the Universe, such as H, He, C, O, and Fe, are considered primary cosmic ray nuclei. Knowing the spallation cross-sections one can estimate what column density of interstellar matter should be traversed by the cosmic rays to generate the observed overabundance. Figure 4.6 shows a collection of data on the ratio of boron to carbon cosmic ray nuclei.

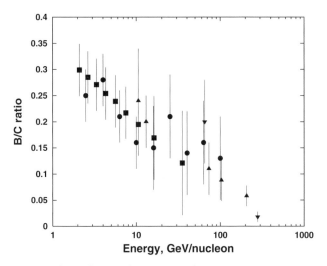

Fig. 4.6. Ratio of the fluxes of secondary boron to primary carbon nuclei.

The main processes in the propagation process are easy to describe in the leaky box approximation, which is a well accepted approximation to particle diffusion in the Galaxy. In Chandrasekhar's language the leaky box consists of a volume, where particles freely propagate, and a three-dimensional absorbing wall around it. In terms of cosmic ray propagation the leaky box model states that cosmic rays propagate in the Galaxy that contains their sources but at every step they can escape from it with certain probability P_{esc}. The important parameter is λ_{esc}, the mean amount of matter traversed by the cosmic rays in the Galaxy before they escape from it. By definition $\lambda_{esc} \equiv \rho_{ISM}\beta c\tau_{esc}$, where ρ_{ISM} is the average matter density and τ_{esc} is the lifetime of cosmic rays in the Galaxy. Neglecting the energy loss and gain of the cosmic rays in propagation and assuming an equilibrium cosmic ray density one could write a simplified transport equation for stable cosmic ray nuclei

$$\frac{N_j(E)}{\tau_j(E)} = Q_j(E) - \frac{\beta c \rho_{ISM}}{\lambda_j(E)} N_j(E) + \frac{\beta c \rho_{ISM}}{m} \sum_{i>j} \sigma_{i \to j} N_i(E) \,. \qquad (4.11)$$

The negative term on the right-hand side of the equation describes the number of nuclei of type j lost in propagation because of fragmentation. The positive term sums over all higher mass nuclei that produce j in spallation processes. The observed cosmic ray composition can be understood in terms of the general elemental abundance and the fragmentation cross-sections if all cosmic ray nuclei have the same propagation history [83, 84] and have on the average traversed 5 to 10 g/cm^2 of matter. For ρ_{ISM} of one nucleon per cm^3 this corresponds to escape time $\tau_{esc} = N_A \lambda_{esc}/c \simeq$ 3–6 $\times 10^6$ years.

Since for a given charge the diffusion process depends on the particle velocity, i.e. energy, one does not expect to have the same primary to secondary ratio at all energies. Higher energy nuclei will have shorter escape times and the fluxes of secondary elements should decrease with energy. Figure 4.6 shows the ratio of the fluxes of cosmic ray boron and carbon nuclei measured by several experimental groups.

The study of the energy dependence of the secondary to primary ratio establishes the energy dependence of the containment time for cosmic rays in the Galaxy. Using measurements like the one shown in Fig. 4.6 one could fit the energy dependence of λ_{esc} as [82]

$$\lambda_{esc} = 10.8 \, \beta \times \left(\frac{4}{R}\right)^\delta \text{g/cm}^2 \,, \qquad (4.12)$$

where R is the particle rigidity in GV and $\delta \simeq 0.6$ shows the rigidity dependence of the escape length. The formula is valid for rigidities above 4 GV. At lower rigidities the escape length is almost constant at $\lambda_{esc} = 10.8\beta$ g/cm^2.

If we make one more simplification and assume that no cosmic ray nuclei are created in propagation, i.e. only account for the loss of particles, the leaky box model gives the shape of the energy spectrum of a primary nucleus j after propagation as

$$N_j(E) = Q_j(E) \times \left(\frac{1}{\tau_{esc}^j(E)} + \frac{\beta c \rho_{ISM}}{\lambda_{int}^j}\right)^{-1} \,, \qquad (4.13)$$

where $Q_j(E)$ is the source spectrum for the nucleus j. While the escape length $\lambda_{esc}(E)$ is the same for all nuclei with the same rigidity R, λ_{int} depends on the mass of the nucleus. In the case of protons λ_{int} is the cross-section for inelastic interactions – 50.8 g/cm^2 at low energy. For heavier nuclei, interactions also include fragmentations. λ_{int} is 6.4 g/cm^2 for carbon and only 2.6 g/cm^2 for iron. Equation (4.13) suggests that the energy spectra of different nuclei will be different at low energies and will tend to become asymptotically parallel

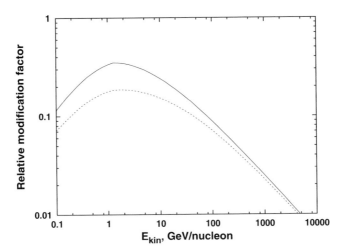

Fig. 4.7. Modification of the shape of the energy spectrum of carbon (solid line) and iron (dashed line) nuclei after propagation on the assumption of no particle gain. Only the relative shape of the spectrum of the two nuclei is correct, because the normalization depends on the average interstellar density during propagation.

to each other at high energy if they were accelerated to the same spectral index at source. Figure 4.7 shows the modification of the source spectra of carbon and iron as a function of their energy and with escape length given by (4.12). Note that the smaller λ_{int} is, the bigger the modification of the source spectrum will be. In the case of protons, λ_{esc} is always smaller than λ_{int} and the modification is simply a steepening of the acceleration spectrum from $E^{-\alpha}$ at acceleration to $E^{-(\alpha+\delta)}$ after propagation. For $\delta = 0.6$ this would suggest an acceleration spectrum with $\alpha = 2.1$ to fit the $E^{-2.7}$ observed energy spectrum of cosmic rays.

While the derivation of λ_{esc} depends only on the knowledge of the spallation cross-sections, the estimate of the escape time is uncertain because it is influenced by assumptions about the containment volume for cosmic rays in the Galaxy. Cosmic rays could have either diffused in the galactic disk (as assumed above) where the matter density is the highest or they may have spent a much longer time away from the disk, in the galactic halo, where the density is much lower. These two assumptions lead to different relations between the escape time τ_{esc} and the escape length λ_{esc}.

In principle the problem of the containment volume could be solved by observations of non-stable secondaries that decay with a half life comparable to the escape time of cosmic rays. To do this one has to put back the decay length in the negative term of (4.11), which becomes

$$\left(\frac{\beta c \rho_{ISM}}{\lambda_j(E)} + \frac{1}{\gamma \tau_j} \right) \times N_j(E) .$$

A very suitable isotope is ^{10}Be with a half life of 1.6×10^6 years. The flux of ^{10}Be can be compared with the stable isotopes ^{9}Be and ^{7}Be. The production of the three isotopes depends on the partial production cross-sections and on λ_{esc}. ^{10}Be would decay and its measured flux depends also directly on τ_{esc}. The actual estimate involves folding of the production with the decay of the isotope during propagation in a particular propagation model. The measurements are also very difficult and the results from their analysis are not fully conclusive.

Measurements of the ^{10}Be/Be ratio were performed with the IMP 7/8, SEE 3 spacecraft [85] as well as with the Voyager 1 and 2 [86] and Ulysses [87] missions. The kinetic energy of the observed Be nuclei is between 30 and 200 MeV/nucleon. Significant corrections have to be made for solar modulation effects. (See Sect. 5.2 for solar modulation effects.) The four measurements give values for τ_{esc} between 8 and 30 million years with rather large error bars. Combined with an escape length of 10 g/cm^2 this gives an average matter density derived from all three experiments of $\rho_{ISM} \sim 0.2$–0.3 cm^{-3} nucleons, i.e. less than the matter density in the galactic disk. The fraction of surviving ^{10}Be can also be used to determine what the allowed thickness of the halo is, which comes to 2.8±1 kpc if the cosmic rays are not convected away from the galactic plane. Convection velocities above 20 km/s are not allowed by such analyses.

The Ulysses data also include measurements of other radioactive isotopes: ^{26}Al and ^{36}Cl. The ^{26}Al/Al ratio shows [88] a cosmic rays containment time in the Galaxy of 16±3 million years and ρ_{ISM} of 0.28±0.05 cm^{-3}. The ratio of ^{36}Cl/Cl determines a lifetime of 11 ± 4 million years and gives a less reliable estimate of the density of $\rho_{ISM} = 0.39$±0.15 cm^{-3} [89]. An average overall measurements gives a value of 15 ± 2 million years for the average cosmic ray lifetime and of 0.26 ± 0.3 for the average density traversed by the cosmic rays. Taken together with the more recent data of the ACE satellite [90] all measurements agree with a confinement (escape) time of 17 ± 4 million years for the sub-GeV cosmic rays.

4.4 Diffuse galactic gamma rays

Protons and electrons diffusing through the Galaxy interact on the galactic matter and radiation fields and produce γ-rays. These γ-rays trace the amount of matter and the cosmic ray density as a function of the direction of the γ-rays arriving at Earth. This radiation is called diffuse galactic radiation and is mostly concentrated around the galactic plane. It has been detected by all GeV γ-ray observing satellite detectors: SAS-2 [91], COS-B [92] and most recently EGRET [93].

Gamma rays are produced by electrons in bremsstrahlung interactions, and by inverse Compton interactions with the cosmic microwave background, and with the diffuse galactic infrared/optical radiation. Protons produce gamma rays through the decay of neutral pions generated in inelastic interactions of the cosmic rays with interstellar matter.

4.4.1 Relative importance of γ-ray production processes

First we briefly discuss the general features of the expected π^0, bremsstrahlung and inverse Compton gamma-ray spectra, and make order of magnitude estimates of their relative contributions to the diffuse gamma ray flux. For this estimate, following Ref. [94] we assume that protons and electrons in the Galaxy have power law momentum spectra with identical spectral indices up to some high momentum p_{max}, much higher than the energy of the observed γ-rays. Generally the high energy γ-rays produced by π^0 decay and bremsstrahlung will have the same power law spectrum as the protons and electrons. On the other hand, the inverse Compton spectrum from an $E^{-\alpha}$ electron spectrum will be $\sim E^{-(\alpha+1)/2}$, which is much flatter than the π^0 and bremsstrahlung gamma-ray spectra. To discuss the relative importance of the various processes to the gamma-ray flux we use order of magnitude estimates. We will then give exact γ-ray yields obtained using exact formulae and Monte Carlo calculations.

For this estimate we approximate the total proton spectrum by $N(E_p) \equiv dN/dE_p = a_p E_p^{-\alpha}$ protons GeV^{-1} cm^{-3}. For $E_\gamma \gg m_\pi c^2/2$, the γ-ray source function from π^0 production is

$$Q_{\pi^0}(E_\gamma) \approx n \left(\sigma_{pp}^{inel} \frac{2Z_{N\pi^0}}{\alpha} \right) \times a_p\, E_\gamma^{-\alpha}, \tag{4.14}$$

where σ_{pp}^{inel} is the inelastic proton–proton cross section, n is the average matter density per cm^3 and $Z_{N\pi}$ is a spectrum-weighted moment of the momentum distribution of pions produced in proton–proton collisions [6]. The spectrum weighted moments give the yield of the process ab from a power law cosmic ray spectrum of integral spectral index γ

$$Z_{ab} \equiv \int_0^1 x_L^\gamma F(x_L) dx_L \,, \tag{4.15}$$

where x_L is the ratio of the energy of the secondary particle b to the primary energy in the Lab system. For $\alpha = 2.0$, 2.4, and 2.7 respectively, $Z_{N\pi^0} \approx 0.16, 0.066$ and 0.035. Thus, for $\alpha = 2.7$, close to the locally measured proton spectrum

$$Q_{\pi^0}(E_\gamma) \approx 2.5 \times 10^{-26}\, a_p\, n\, E_\gamma^{-2.7} \quad \text{photons GeV}^{-1}\,\text{s}^{-1}\,\text{cm}^{-3}, \tag{4.16}$$

where E_γ is in GeV and n is in cm^{-3}.

Similarly, we approximate the total electron spectrum by $N(E_e) \equiv dN/dE_e = a_e E_e^{-\alpha}$ electrons GeV^{-1} cm^{-3}. To obtain the bremsstrahlung source function, we assume that after an electron of energy E_e has traveled one radiation length, X_0, it is converted into a photon of energy $E_\gamma = E_e$. Hence,

$$Q_{\text{br}}(E_\gamma) \approx N(E_\gamma)n/X_0. \tag{4.17}$$

Thus, for $\alpha = 2.7$,

$$Q_{\text{br}}(E_\gamma) \approx 1.2 \times 10^{-25} a_e n E_\gamma^{-2.7} \quad \text{photons GeV}^{-1} \text{ s}^{-1} \text{ cm}^{-3}. \tag{4.18}$$

For inverse Compton scattering, we approximate the photon energy after scattering by an electron of energy $E_e = \gamma m_e c^2$ by $\gamma^2 \bar{\varepsilon}$ where $\bar{\varepsilon}$ is the mean photon energy of the radiation field under consideration. Provided the Compton scattering is in the Thomson regime ($\gamma\bar{\varepsilon} \ll m_e c^2$) this gives an inverse Compton source function

$$Q_{\text{IC}}(E_\gamma) \approx N(\gamma)n_{\text{ph}}\sigma_T/2\gamma\bar{\varepsilon}, \tag{4.19}$$

where $N(\gamma)d\gamma = N(E_e)\,dE_e$, and we obtain

$$Q_{\text{IC}}(E_\gamma) \approx a_e \frac{(\bar{\varepsilon})^{1/2}}{E_\gamma^{(\alpha-1)/2}} \frac{n_{\text{ph}}\sigma_T}{m_e c^2}. \tag{4.20}$$

For scattering on the microwave background, we use $n_{\text{ph}} = 400$ cm^{-3}, $\bar{\varepsilon} = 6.25 \times 10^{-4}$ eV, and obtain for $\alpha = 2.7$

$$Q_{\text{IC}}(E_\gamma) \approx 2.1 \times 10^{-24} a_e E_\gamma^{-1.85} \quad \text{photons GeV}^{-1} \text{ s}^{-1} \text{ cm}^{-3}, \tag{4.21}$$

where E_γ is in GeV. We note that, for an assumed matter density of 1 cm^{-3}, at 1 GeV the inverse Compton scattering contribution is an order of magnitude larger than the bremsstrahlung contribution, and the relative importance of the inverse Compton scattering contribution increases with energy. The bremsstrahlung contribution is higher than that of π^0 by about a factor of five. The π^0 γ-rays can only be important if there are many more protons than there are electrons.

4.4.2 More exact γ-ray yields

The main inaccuracy of the previous section is the assumption that galactic protons and electrons have the same energy spectrum. In fact, electrons suffer energy losses on synchrotron radiation and have a steepening spectrum with an increasing spectral index. Bertsch et al. [71] suggest the following spectrum of the galactic electrons (in cm^{-2} s^{-1} sr^{-1} GeV^{-1})

$$\begin{aligned}
dN/dE_e &= 0.019 E_e^{-2.35} & \text{for } E_e < 5 \text{ GeV} \\
&= 0.149 E_e^{-3.30} & \text{for } E_e \geq 5 \text{ GeV}
\end{aligned} \tag{4.22}$$

Figure 4.8 shows the source functions of the three processes using the electron spectrum given by (4.23) and a proton spectrum $dN/dE_p = 3.06 E_p^{-2.70}$.

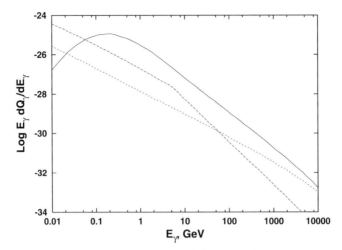

Fig. 4.8. Yields of the γ-ray production by π^0 (solid line), bremsstrahlung (dashes) and inverse Compton (dots) for the proton and electron spectra described in the text.

The matter density for π^0 and bremsstrahlung is 1 (atom/cm^3) and the microwave background is used as target for the inverse Compton effect. The π^0 yield peaks at $E_\gamma = m_{\pi^0}/2 \simeq 70$ MeV but in Fig. 4.8 the peak appears at higher energy because it is shifted by the E_γ factor. The bremsstrahlung spectrum dominates at lower energies and clearly follows the break in the electron spectrum. The inverse Compton spectrum is indeed very flat ($\sim E_\gamma^{(\alpha+1)/2}$) and approaches the π^0 contribution at $E_\gamma = 10^4$ GeV in spite of the much steeper electron spectrum.

4.4.3 Energy spectrum of γ-rays from the central Galaxy

The EGRET collaboration published the results on the diffuse γ-ray emission of our Galaxy [93] as a function of the position in the Galaxy. The data covers the whole galactic plane and latitudes between -10 and 10 degrees. We discuss here the γ-ray emission from a wide region ($300° < l < 60°$) centered on the galactic center. Instead of modeling the observed spectra for a set of assumptions we use the γ-ray production yields to extract the parameters of the astrophysical environment from data. This includes the power law indices for protons and electrons and the matter density in the observed part of the Galaxy.

We assume the momentum spectrum of particles is of the form

$$\frac{dn}{dp} = ac \left(\frac{p}{1\,\text{GeV}/c} \right)^{-\alpha} \quad (\text{GeV}/c)^{-1}\,\text{cm}^{-3} , \qquad (4.23)$$

where $a \equiv a_p$ for protons and $a \equiv a_e$ for electrons. We define Q_{π^0}, Q_{br}, and Q_{IC} to be the gamma-ray emissivities (cm^{-3} s^{-1} GeV^{-1}) for pion production, bremsstrahlung, and inverse Compton scattering calculated for $a_p = 1 \text{ GeV}^{-1}$ cm^{-3}, $R_e \equiv a_e/a_p = 1$, and $n = 1 \text{ cm}^{-3}$ as shown in Fig. 4.8. The gamma-ray intensity observed at Earth for each production mechanism is the integral of the corresponding gamma ray emissivity along the line of sight per unit of solid angle. Making the approximation that the emissivities are constant over the inner galaxy along the line of sight L (cm) we can write

$$I_\gamma(E_\gamma, \alpha) \approx \frac{La_p n_1}{4\pi} \left[Q_\pi(E_\gamma, \alpha) + R_e Q_{\text{br}}(E_\gamma, \alpha) + \frac{R_e}{n_1} Q_{\text{IC}}(E_\gamma, \alpha) \right] , \quad (4.24)$$

where $n_1 \equiv n/(1 \text{ cm}^{-3})$, n is the average nucleon number density along the line of sight and R_e is the average electron to proton ratio.

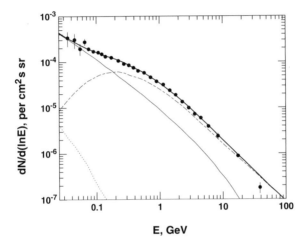

Fig. 4.9. The best fit to the EGRET diffuse γ-ray background from the central region of the Galaxy [93]. The thin solid line shows γ-rays from electron bremsstrahlung, the dash–dash line is for π^0 γ-rays and the dotted line is for IC scattering. Note that $E \, dN/dE$ is plotted to emphasize the features of the spectral shape.

Figure 4.9 shows the best χ^2 fit to the EGRET data for the inner galaxy [95]. The fit is performed with the following free parameters: electron to proton ratio R_e, particle spectral index $(\alpha_p = \alpha_e)$, maximum energy for the electrons E_e^{max} (assuming that the proton spectrum extends to 100 TeV), R_e/n_1, and the overall normalization factor. The cutoffs of the electron (at E_e^{max}) and proton (at 100 TeV) spectra are assumed exponential. The best fit (reduced $\chi^2 = 1.84$) is for $\alpha = 2.20$, $R_e = 0.12$, and $E_e^{max} = 25$ GeV. The values of all these parameters are, however, quite uncertain. The χ^2 space is very flat and could accommodate a variety of n values, all of order 1. A small change in the initial parameter values may increase the contribution of IC scattering and decrease that of π^0 γ-rays. The π^0 peak is, however, prominent in the data and every fit requires a strong π^0 contribution at GeV energies.

One surprising feature is that the spectral index α is always very flat – from 2.2 as shown in Fig. 4.9 to 2.4 when the IC contribution is stronger. The shape of the electron spectrum that comes out of the fit is not very different from the model value of Bertsch et al. [71] which is used by the EGRET group for the analysis of the matter and cosmic ray density as a function of the galactic coordinates. The detected γ-ray intensity is in very good general agreement with this model, with the exception of fluxes above 1 GeV, which are observed to be higher than the expectations from the locally observed $E^{-2.7}$ cosmic ray spectrum. The energy flux of the diffuse γ-ray emission of the inner Galaxy are 1.6×10^{-4} GeV cm^{-2} s^{-1} sr^{-1} for the total γ-ray emission and 1.2×10^{-4} GeV cm^{-2} s^{-1} sr^{-1} for its π^0 component. One might speculate that a substantial fraction of the gamma-ray emission from the inner galaxy may be due to unresolved supernova remnants. Given that supernova remnants form a large fraction of the interstellar volume, this notion is not unreasonable, given a broad definition of supernova remnants. However, we anticipate that the galactic cosmic rays are accelerated primarily at relatively young supernova shocks, and it is likely that these high intensities reflect instead the greater concentration of cosmic ray accelerators (young supernova shocks) in the inner galaxy. This is postulated in the model of Ref. [71], where the cosmic ray density is proportional to the matter density in the galactic plane.

5 Cosmic rays at the top of the atmosphere

The atmosphere of the Earth provides more than ten interaction lengths for protons going straight down. If the observations were made at sea level such a proton would retain on the average less than 0.001 of its energy. The energy loss fluctuates from event to event and the energy spectrum of cosmic ray protons would be difficult to reconstruct. Heavier cosmic ray nuclei have significantly shorter interaction lengths and lose energy much faster. For these simple reasons the observations of the cosmic rays are much easier outside the atmosphere.

Ideally, the observations of galactic cosmic rays would be done from a spacecraft well outside of the solar system. This would avoid many complications that are discussed in this chapter. Since such observations are obviously unrealistic, most of the experiments are performed either at Earth orbiting satellites or with high-altitude balloons. Balloons have the disadvantage of flying under several g/cm^2 of atmosphere. The detected cosmic ray fluxes have to be corrected for the particles that are lost or created in that thin atmospheric layer. On the other hand balloons are so much less expensive than satellites that they are currently the most widely used carriers of cosmic ray detectors.

Contemporary balloons float for durations up to several days at altitudes of 40 km. They can lift payload weights up to 4 tons, i.e. they can carry to the top of the atmosphere very large detection systems. The expense of launching such systems in space would be very great. Usually balloon flights attempt to take advantage of the changing winds in the upper atmosphere. The idea is to launch the balloon when the prevailing high-altitude winds blow west a couple of days before a change to eastward winds is expected. The ideal balloon flight would be when the high-altitude winds change direction in the middle of the scheduled flight and carry the balloon back to the vicinity of the launching pad for an easy recovery.

There are also successful attempts to launch the balloon in one continent and to recover it in another. Flights from Australia to South America have lasted for more than a week. The longest durations can be achieved in circumpolar flights. A balloon launched at the coast of Antarctica circles around the continent to come back over the site of the launch in about two weeks. If the condition of the balloon and the detectors allow it, the flight could

continue for another two weeks. Antarctic flights also have the advantage of
flying at a constant altitude, because so close to the pole there are no daily
variations of the temperature and the pressure of the atmosphere. In other
long-duration flights the altitude changes significantly at day and night time.
This should be corrected by decreasing the weight of the payload at night,
which of course requires carrying a lot of ballast and decreases the possible
weight of the experimental arrangement.

5.1 Cosmic ray detectors

All kinds of particle detectors have been used in cosmic rays research. Many
detectors were invented for that purpose. The first detection of cosmic rays
by Hess used a gold leaf electroscope that measures the level of ionization.
It is schematically shown in Fig. 5.1. The gold leaves are charged until they

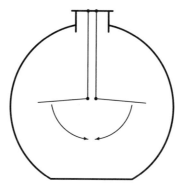

Fig. 5.1. Schematic drawing of an electroscope such as that used by Hess in the
discovery of cosmic rays.

come apart from each other. If the ionization level is low they can stay in
this position for a long time. With increased ionization level the gold leaves
discharge and come back close together. Hess discovered that the ionization
level increases with altitude and thus proved that cosmic rays indeed come
from outside the atmosphere.

The typical contemporary cosmic ray detector is quite different. A sketch
of such a detector is shown in Fig. 5.2. There are two layers of detectors, A
and B, that can determine simultaneously the time and position of a cosmic
ray crossing that layer. These could be proportional counters, scintillators,
spark chambers, or combinations of different particle detectors. A different
detector C measures the particle charge. This could be done by measuring
the amplitude of the signal in the scintillation counter or by using a gas-
Cherenkov detector. In the middle of the arrangement there is a powerful

magnet. The charged particle trajectory is bent in the magnetic field. The particle then crosses the second layer on the bottom of the arrangement, that measures the displacement from the crossing of the top layer.

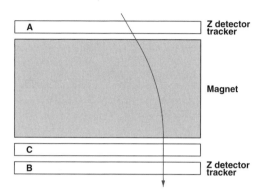

Fig. 5.2. Schematic drawing of a magnetic spectrometer. In some versions there is a calorimeter at the bottom of the arrangement.

Detectors like the one shown in Fig. 5.2 measure the bending of the charged particle in the magnetic field. The bending is proportional to the product of the magnetic field and the particle rigidity. The time of flight between the top and the bottom of the detector, Δt, measures directly the velocity, i.e. the ratio of particle energy and mass. When the charge of the particle is also known, the combination of the two measurements describes fully the cosmic ray particle. Occasionally there is a stack of counters layered with target material on the bottom of the arrangement. The particles interact in this calorimeter and deposit their energy in it. This provides an independent measurement of the particle energy.

Detectors using magnetic field, usually called magnetic spectrometers, give a very good measurement of the cosmic ray particles. The only disadvantage is their limited energy range. High energy particles have large gyroradii, r_g, and are not deflected much by the magnetic field. With the particle energy the measurement becomes increasingly inaccurate and then impossible, depending on the resolution of the layers A and B. The energy range is also limited from below as the probability that a low energy particle with small r_g would be able to cross both layers A and B decreases.

A very good example for a high-resolution magnetic spectroscope is the BESS detector [96] developed for precise measurement of cosmic ray fluxes with very good discrimination between different species. BESS employs a superconducting magnet of strength 1 T, which is designed not to cause too many interactions of the cosmic ray particles. BESS is mounted inside

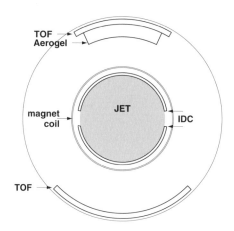

Fig. 5.3. The BESS detector (not to scale).

a cylinder of diameter 1.7 m and consists of the following components, as shown in Fig. 5.3:

– top and bottom time of flight (TOF) scintillators that also measure the particle energy loss.
– aerogel Cherenkov counter mounted under the top TOF.
– two inner drift chambers (IDC) inside the magnetic field space.
– central tracking device in the magnetic field region which is realized as a jet type drift chamber (JET) and measures up to 28 particle positions.

All these detectors provide for multiple measurements of all parameters necessary for the reconstruction of the particle energy, mass and charge, which have to be consistent with each other for a successful reconstruction. The existence of multiple detectors also increases the energy range of the whole detector as different elements have different threshold and saturation energies.

In principle the magnetic spectrometers can be a combination of the two major classes of cosmic ray detectors: nondestructive detectors and calorimeters. Nondestructive detectors generally measure the charge and the momentum of the cosmic ray particle that does not interact in the detector. A good earlier example for a nondestructive detector is the University of Chicago detector carried by the *Challenger* space shuttle in 1985 [97]. It measures the charge of the particle in two gas-Cherenkov counters on the top and the bottom of the arrangement. The particle Lorentz factor $\gamma = E/m$ is measured with transition radiation detectors. Transition radiation detectors measure the X-rays generated when the charged particles traverse radiators of varying dielectric constant. The biggest advantage of the nondestructive detection technique is that the detectors are relatively light and a balloon (or satellite) could carry a large sensitive area detector. The disadvantages are that any type of nondestructive detector has a limited energy and charge range. In the case of the University of Chicago detector the gas-Cherenkov counter is not sensitive enough for low charge nuclei and thus cannot measure protons. The transition radiation detector, on the other hand, saturates at high Lorentz factors and thus cannot measure very high energy nuclei.

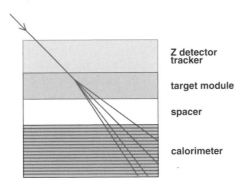

Fig. 5.4. Schematic drawing of a JACEE calorimeter detector.

The other type of detectors, particle calorimeters, are usually used for detection of high energy cosmic rays. A calorimeter consists of layers of particle detectors and target material. The particle detectors could be of any type, from scintillation counters to nuclear emulsion. A very heavy electronic calorimeter was launched in the late 1960's in the Proton-4 satellite [98]. A modern version is the thin calorimeter employed by the Japanese–American balloon experiment (JACEE) [99], different versions of which have successfully flown more than 10 times. The JACEE calorimeter, shown schematically in Fig. 5.4 consists of three different parts. The top of the calorimeter measures the particle charge by monitoring its ionization. Then there is target material in which the particle interacts. After a little space that allows the tracks of the secondary particles to spread, there is a calorimeter consisting of layers of lead, X-ray films and nuclear emulsion. Similar calorimeters are also used by the Russian–Japanese collaboration RUNJOB [100] These calorimeters measure the amount of energy carried by the neutral secondaries created in the inelastic particle interaction. The biggest advantage of calorimeters as a cosmic ray detectors is that they have similar sensitivity to all cosmic ray nuclei.

The main problem with the calorimeters is that they are heavy. Balloons can carry not more than 2 m^2 area of a thin calorimeter and thus obtaining reasonable statistics requires the long duration of a circumpolar flight. The charge resolution of a JACEE calorimeter is worse than that of a nondestructive detector. The energy resolution on an event by event basis is limited by the fact that only the energy in neutral secondary particles, which is subject of large fluctuations, is measured. Cosmic ray spectra can only be reconstructed on a statistical basis from comparatively large data sets. With increasing particle energy more and more of the energy leaks out of the calorimeter. This requires energy-dependent corrections to the measured fluxes. Such corrections are model-dependent and lead to systematic errors.

5.2 Solar modulation

Charged cosmic ray nuclei entering the solar system have to overcome the magnetic field that is frozen in and carried by the solar wind. The lower the rigidity of a cosmic ray particle is, the lower is the probability of its penetration through the heliosphere to Earth. The physics of the particle interactions with the solar wind is discussed by Longair [7]. The study of the properties of the solar wind and the heliosphere is a major topic of research in space and plasma physics. Experimentally it is conducted through spacecraft experiments, some of which have already explored not only the heliospheric environment of the Earth but also a large fraction of the solar system. We will just touch the subject here and demonstrate the existence of the solar modulation and the way in which cosmic ray fluxes are corrected for it.

Solar wind is the outflow of material from the surface of the Sun. The existence of solar wind was predicted by Parker in 1958 [101] and was experimentally confirmed soon after that. The solar wind originates in the solar corona which has a temperature of about 10^6 K, a factor of hundred higher than the photosphere of the Sun. The magnetic field is frozen in the ionized material and is dragged outwards from the Sun. The field is attached to the rotating Sun and the expansion leads to the creation of an Archimedes spiral which is the large-scale field structure. This structure was named the *Parker spiral*. The radial and azimuthal components of the magnetic fields are

$$B_r = B_\odot R_\odot^2 \frac{1}{r^2} \quad B_\phi = B_\odot R_\odot \frac{\sin\theta}{r} , \tag{5.1}$$

where $B_\odot$ is the magnetic field on the surface of the Sun, $R_\odot$ is the solar radius and θ is the zenith angle measured from the center of the Sun. At sufficiently large distance from the Sun the radial component vanishes and the azimuthal component, which can be approximated as circular, dominates.

The solar wind carries along the magnetic fields characteristic for the hot coronal regions. In the vicinity of Earth the solar wind particles, mostly protons, have velocities of 300 to 600 km/s, which corresponds to an average kinetic energy of 500 eV. The solar wind flux is 1.2×10^8 cm^{-2} s^{-1} and its energy density is about 2.5 KeV cm^{-3}. The magnetic field strength is about 5×10^{-5} G which translates into energy density 40 times lower than that of the solar wind particles.

The modulation of the galactic cosmic rays in the solar system was first observed as an anticorrelation of the neutron monitor counts with the sunspot number. The number of sunspots reflects the number of active regions on the sun and correspondingly the epoch of solar activity. The sunspot number (and many other measures of the solar activity) have roughly an 11 year cycle, one half of the magnetic cycle of the Sun. Neutron monitors are ground based detectors that effectively measure the total amount of cosmic ray energy that reaches the observation level. Figure 5.5 shows the monthly average number of counts for the last thirty years of the Swarthmore/Newark neutron monitor [102], which is located at a moderate latitude.

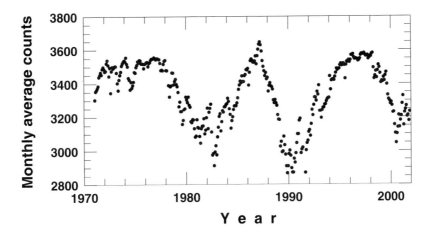

Fig. 5.5. Monthly averages of the counting rate of the Swarthmore/Newark neutron monitor [102]. Note that this is a suppressed zero plot and the variation is of the order of 20%.

The maximum of the neutron monitor counts follows the minimum of the solar activity and vice versa with a delay of about 1 to 2 years. Note that consecutive solar cycles are very different in the strength of the cosmic ray flux and in the shape of its time dependence. The same is true for the degree of solar activity. The maxima and minima of two solar cycles could be very different. Although experiments attempt to measure the cosmic ray fluxes at the same epoch of solar activity, there is always the need for corrections of the measured flux for the exact magnitude of the solar activity. As an example I will follow the arguments presented in the paper of the LEAP [103] experiment that describes the proton and helium fluxes detected during the 1987 solar minimum. Such a procedure has become standard for later experiments.

The LEAP collaboration uses the standard three-dimensional spherically symmetric model of solar modulation first developed by Gleeson & Axford [104]. The model accounts for the following three processes:

- cosmic ray diffusion through the magnetic field carried by the solar wind,
- the convection by the outward motion of the solar wind, and
- the adiabatic deceleration of the cosmic rays in this flow.

The first two processes lead to a rigidity dependent decrease of the particle flux. The third one leads to a decrease of the energy of the particles that penetrate the heliosphere.

LEAP has normalized the cosmic ray spectra of protons and helium to their measured high energy shape. At rigidities exceeding 20 GV cosmic rays are not affected by the solar wind. The diffusion coefficient used $\kappa = C_0 \beta R$ is proportional to the particle rigidity where the coefficient C_0 is adjusted to match the detected flux at high energy. The solar wind speed v is taken to be 400 km/s. The data is best fit by a solar modulation parameter $\phi = 500 \pm 75$ MV. The solar modulation parameter is the integral

$$\phi = \frac{1}{3} \int_{r_1}^{r_{hs}} \frac{v}{\kappa} \, dr, \qquad (5.2)$$

where r_1 is the heliospheric radius of the Earth (1 AU) and r_{hs} is the boundary of the heliosphere assumed here to be 50 AU. Current data suggest that the boundary is further away, not any closer than 80 AU.

In the force field approximation [105] the effect of solar modulation is expressed in terms of the single modulation parameter ϕ. A particle that has total energy E_{IS} in interstellar space would reach the Earth with energy $E = E_{IS} - |Z|\phi$, where Z is its charge. The flux of particles of that type at Earth Φ is related to the interstellar flux Φ_{IS} as

$$\Phi(E) = \frac{(E^2 - m^2)}{(E_{IS}^2 - m^2)} \times \Phi_{IS}(E_{IS}) , \qquad (5.3)$$

where m is the particle mass. The first term in (5.3) accounts for the loss of flux and the second one accounts for the particle energy loss.

Figure 5.6 shows the LEAP proton flux compared to the interstellar flux fit of the data above 20 GeV with different ϕ values. Note that the force field approximation works well only in a limited, relatively high rigidity range. Our crude fit is somewhat different from that of the original paper [103], which is made with a solar modulation code. It still demonstrates the main effect of solar modulation and shows how the proton flux varies during the solar cycle. It is much easier to perform this procedure when cosmic ray particles with different rigidity for the same energy per nucleon are detected. The classic case is when the flux of protons and helium nuclei are measured by the same experiment. The helium nuclei have higher rigidity by a factor of 2 and are thus modulated less than the protons.

There are indications [106] that positive and negative (electrons or antiprotons) cosmic ray particles are modulated differently as a function of the solar polarity. It is qualitatively explained by the different directions of cosmic ray drifts. When the solar polarity is positive, i.e. magnetic field points

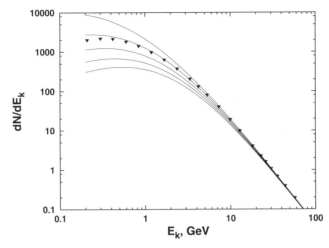

Fig. 5.6. Comparison of the LEAP proton flux to a fit of the measurements above 20 GeV with solar modulation using modulation parameters $\phi = 200, 400, 600, 800,$ and 1,000 MV from top to bottom. The modulation is performed under the force field approximation (5.3).

at the North pole of the Sun, in the northern heliosphere positive particles drift into the solar system radially inwards along the magnetic sheet and then out of it along the magnetic field lines. Negative particles drift of course in the opposite direction. The directions of the drift cause different degrees of solar modulation.

5.3 Geomagnetic field effects

The last hurdle for the galactic cosmic rays on their way to the atmosphere is the geomagnetic field. It bends the trajectories of the charged particles, preventing the low rigidity particles from reaching the atmosphere. The cosmic ray flux on top of the atmosphere is therefore not isotropic and depends on the detector position x, $\Phi(R, x_d, \Omega) = \Phi_0(R) \times \epsilon_B(r, x_d, \Omega)$, where $\Phi_0(r)$ is the flux at distances more than several earth radii, already corrected for the solar modulation. The penetration probability, ϵ_B, can take only the discrete values 0 or 1, i.e. a particle can or cannot reach the position x as a function of its rigidity R and the angle of its motion in the geomagnetic field frame.

Stoermer [107] had solved analytically the equation of motion for the case of a dipole field and neglecting the shadow of the Earth even before the discovery of cosmic rays. The solution expresses the particle motion in units of Stoermer radius $r_S = \sqrt{(\mu_0 M/4\pi R)}$, where M is the magnetic dipole moment of the Earth ($M \simeq 8.1 \times 10^{25}$ G cm^3). For particles that penetrate vertically towards the center of the magnetic dipole the minimum rigidity required for penetrating to distance r from the center of the magnetic dipole is

$$R_S \geq 59.4\,\mathrm{GV} \times \left(\frac{r_\oplus}{r}\right) \cos^4 \lambda_B / 4 \,, \qquad (5.4)$$

where λ_B is the magnetic latitude and 59.4 GV $\simeq M/(2r_\oplus^2)$ is the rigidity of a particle in a circular orbit of radius $r_\oplus$ in the equatorial plane of the dipole field. The minimum rigidity for a particle that penetrates to the surface of the Earth at the magnetic equator is correspondingly $\sim$14.9 GV (total energy of 14.9 GeV for protons or about 7.5 GeV/nucleon for He nuclei). At magnetic latitude of $\pm60°$ the minimum rigidity is 0.93 GV which translates to energies of 1.32 and 1.05 GeV/nucleon. The vertical cutoffs change slightly with altitude throughout the atmosphere.

The complete formula for the Stoermer rigidity cutoff, R_S, is

$$R_S(r, \lambda_B, \theta, \varphi_B) = \left(\frac{M}{2r^2}\right) \left\{ \frac{\cos^4 \lambda_B}{[1 + (1 - \cos^3 \lambda_B \sin\theta \sin\varphi_B)^{1/2}]^2} \right\} , \qquad (5.5)$$

where θ is the particle zenith angle and φ_B is the azimuthal angle measured clockwise from the direction of the magnetic south. The dependence on φ_B contains the well known east–west effect: for positively charged particles at the same zenith angle the cutoff is higher from the east direction and vice versa for negatively charged particles. The expression $\cos^3 \lambda_B \sin\theta \sin\varphi_B$ has a maximum for $\sin\varphi_B = -1$, pointing roughly at the geographical East direction, 270° clockwise from the geographic south.

Stoermer's formula gives a good idea of the magnitude of the geomagnetic cutoffs, but has limited accuracy because the geomagnetic field is only approximately an offset dipole (with a geomagnetic North pole at latitude 81°, longitude 84.7°W). The formula also generally underestimates the cutoffs because it neglects the shadow of the Earth, i.e. allows the penetration of charged particles with trajectories that would have intersected the surface of the Earth. More exact calculations could be done using the backtracking technique [108] and more realistic models of the geomagnetic field. A detailed, time dependent, geomagnetic field model is developed and maintained in the International Geomagnetic Reference Field model [109].

The backtracking technique consists of the integration of the equation of motion of a particle with the opposite charge starting at a position with λ_B and angles θ and φ_B at an altitude $(r - r_\oplus)$ above the surface of the Earth. If a backtracked antiproton reaches a certain large distance r_{free} from the Earth it is assumed that a proton can penetrate to the initial position from interplanetary space. If the backtracked particle is trapped in the geomagnetic field closer than r_{free} for a pathlength longer than l_{trap} or if its trajectory intersects the surface ($r < r_\oplus$) the trajectory is considered forbidden. Such a calculation was performed in Ref. [110] with $r_{free} = 30r_\oplus$ and $l_{trap} = 500r_\oplus$. Figure 5.7 shows the penetration probability ϵ_B for particles of different energies backtracked east and west with zenith angle of 30° from an altitude of 20 km at the location of Kamioka, Japan (latitude 36.4°, longitude 137.3°E).

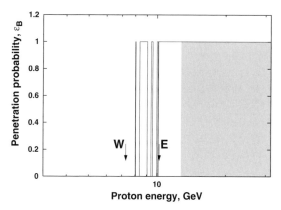

Fig. 5.7. Penetration probability for protons of different energy and with zenith angle of 30 degrees coming from east (shaded area) and from west. The arrows show the Stoermer cutoffs.

In the magnetic dipole case particles with rigidity above R_S are always allowed from the same direction. It the case of a realistic field the situation is quite different: in the vicinity of R_S particle trajectories change rapidly with the rigidity and the sharp cutoff is replaced with a series of allowed and disallowed rigidities (the penumbra region), as shown with the histogram for the western direction in Fig. 5.7. The arrows show the Stoermer cutoffs for the same angles.

Figure 5.7 also shows the difference in ϵ_B for protons moving from east and from west. The cutoff rigidity for West is the lowest and has the most complicated shape. The cutoff for east is the highest and has a relatively simple form – a box in this case. For all practical purposes one measures averages in small energy and angular bins. The averaging replaces the integer values of ϵ_B with the probability p_B for particles in given rigidity interval to penetrate to given altitude in the atmosphere from some angular direction $\Delta\Omega$. Figure 5.8 shows p_B for protons of several different energies averaged over the azimuthal angle and in 0.1 bins in $\cos\theta$ at the location of Kamioka. This figure was generated for a calculation of neutrino fluxes and extends to negative values of $\cos\theta$, i.e. includes particles that propagate to the atmosphere on the other side of the Earth.

The vertical Stoermer cutoff for the location of Kamioka is 13.7 GV. In fact there are no protons of energy 5 GeV (lowest curve in Fig. 5.8) that penetrate to altitude of 20 km above the horizon. The penetration probability p_B increases with the proton energy, although even for 31.6 GeV protons there is a range of zenith angles with p_B less than 1. This is the shadow of the Earth, which is better visible in p_B graphs for locations closer to the magnetic poles.

The magnetic field in the vicinity of the Earth has also a contribution from electric currents in the solar system. This part of the field introduces

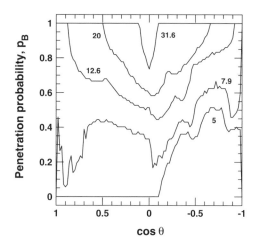

Fig. 5.8. Penetration probability for protons of several different energy as a function of $\cos\theta$. p_B is averaged over the azimuthal angle and in 0.1 bins over $\cos\theta$. The proton energy is indicated at the respective lines.

a time dependence of the geomagnetic cutoffs. Periodicities of 1 year and 24 hours are present, as well as more complex time variations related to the level of solar activity. These variations of the 'external' magnetic field can additionally modify the cutoffs and have to be taken into account for short-duration experiments.

5.4 Cosmic ray spectra and composition

Cosmic ray spectra are steep and most of the particles and the energy carried by cosmic rays are concentrated in the lower energy end of the spectrum, in particles with kinetic energy less than 1 GeV/nucleon. The cosmic ray spectra in this energy range are measured well for all nuclei up to Ni and for many of their isotopes. Most of the results come from long-term spacecraft missions that carry sophisticated small detectors with very good charge and energy resolution. The composition of cosmic rays measured in these experiments, corrected for solar modulation and for propagation effects, reveals the composition of the material that was injected for acceleration in supernova shocks. Knowledge of that composition identifies the environment of the supernova shock and potentially contributes to the identification of specific cosmic ray sources.

Figure 5.9 shows the abundance of different elements in cosmic rays normalized to the general galactic abundance for elements from hydrogen to iron. The data is taken from Ref. [111]. Plotted on a logarithmic scale the cosmic ray abundances are very similar to the general abundances of the elements in the universe with two notable exceptions: hydrogen and helium. All other nuclei can be divided into two general groups: one having the same abundances as the general ones - in the other they have abundances lower by a factor of ~3-4. The classical interpretation involves [112] the difference in the first ionization potential (FIP) of these elements. FIP is approximately the amount of energy that has to be deposited in the atom to kick an electron out of its outermost energy level and to singly ionize it. Once ionized, the charged atom could be injected for acceleration. Elements such as S, O and Ar that have lower cosmic ray abundances all have FIP greater than 10 eV, while Ca, Mg and Fe have FIP lower than 10 eV and cosmic ray abundances equal to the general ones. The division at 10 eV is partially suggested by the experimental data shown in Fig. 5.9 and also by the fact that 10 eV (or ~1.1×10⁶K is the typical temperature of stellar coronae, including that of the Sun. Observations of the solar wind and of solar flare particles show abundances very similar to those of the cosmic rays. The division of abundances at FIP of ~10 eV suggests that atoms are injected for acceleration from stellar coronae. Shapiro [113] argues that the very common flare stars are injectors for cosmic ray acceleration.

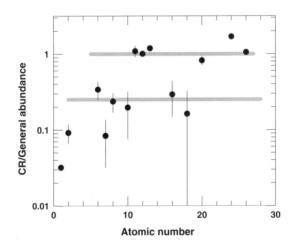

Fig. 5.9. The relative abundance of cosmic ray elements as a function of the atomic number of the element. The two shaded strips show the average abundance for high and low FIP elements. Note that H and He are notable exceptions with very low abundances.

A different picture was more recently developed by Meyer, Drury & Ellison [114]. They confirm the approximate ordering of cosmic ray abundance by FIP, but note that FIP is related to other quantities and search for exceptions in the FIP–abundance correlation. The study also involves new data for elements heavier than Ni. As a result the cosmic ray abundance is better ordered by volatility, elements that form grains more easily are more abundant than those that do not. The measure is the condensation temperature T_c that replaces FIP in the abundance ratios. Highly volatile elements, with T_c less than 400K are preferentially abundant as a function of their mass. This probably reflects the increasing mass to charge ratio which leads to growing acceleration (injection) efficiency. The suggestion is that elements with higher T_c form dust grains, which are fed into the supernova shock, are accelerated very efficiently to energies of the order of 0.1 MeV per nucleon. At these energies grains are destroyed and the the participating ions are further accelerated to cosmic ray energies. There are still several notable exceptions to the rule, because H, C, O and a few other elements do not fit well within this pattern. These are special cases that have to be treated separately in this scenario. Carbon, for example, is very abundant in the pre-supernova winds of very heavy stars which may supply the injection material for acceleration. Oxygen, on the other hand, is partially locked in grain silicates in the interstellar medium.

The under-abundance of the most common elements, hydrogen and helium, by almost one order of magnitude is not fully understood, although some solutions exist. Silberberg & Tsao [115] suggest that the lightest elements are suppressed because their rigidities are the highest, especially in the case in the case of partial ionization. If a cutoff (of about 1 MeV/nucleon) exists at the boundaries of the astrospheres, similarly to the heliosphere, smaller fraction of H and He would be able to penetrate through that boundary compared to much heavier elements. This mechanism would also suppress the abundances of other cosmic ray nuclei which have to be restored by the introduction of additional processes. In the Meyer et al. [114] scenario hydrogen is under abundant among the highly volatile elements because of its mass to charge ratio of 1.

A possibility is that a large fraction of the cosmic rays are accelerated material from winds of very massive Wolf–Rayet stars with their particular chemical composition which is very different from the winds of Sun-like stars. Although independently developed, such an idea [116] fits well with the model of Völk & Biermann [61].

At higher energy the measurements are much more difficult, both because of the decreased cosmic ray flux and the need of much bigger detectors to measure precisely the cosmic ray energy. On the other hand, as explained in Sect. 4.3, λ_{esc} becomes much smaller than λ_{int} and the spectra of all nuclei approach their source acceleration spectrum corrected at the same rigidity for λ_{esc}.

Figure 5.10 shows the cosmic ray spectra above kinetic energy of 2 GeV/nucleon in five different mass ranges: H, He, CNO, Ne–Si and $Z > 17$. We present only the spectra for energy above 2 GeV because we are mostly interested in the high energy behavior of the spectra and want them to be approximately free of solar modulation effects, which are, however, still visible in the hydrogen flux.

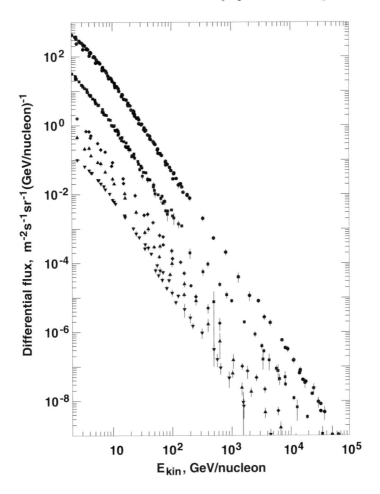

Fig. 5.10. Energy spectrum of cosmic rays in five mass groups:H, He, CNO, Ne–Si, and $Z > 17$.

Multiplying the differential cosmic ray spectrum by a a power of the energy is often used to flatten the spectrum and make spectral changes readily visible. On the other hand this technique is somewhat risky because it exaggerates small differences in the energy normalization in different experiments, which show in the overall normalization of the detected spectrum. A difference of 10% in the energy normalization of two experiments introduces an error of about 20% in the normalization of their results and we prefer to plot the differential fluxes in Fig. 5.10.

The experimental data are diverse. Some of the differences are significantly bigger than the combined statistical and systematic errors derived by the experimental groups. In the 2–20 GeV region there may be some small effect of the solar modulation (for which the data is not fully corrected) because few measurements are made in exactly the same epoch of the solar cycle. Above 20 GeV/nucleon solar modulation is insignificant.

The first impression from Fig. 5.10 is that the spectra of all heavier nuclei are somewhat flatter than the proton spectrum. To the extent that acceleration and propagation of cosmic rays depend on collisionless processes, such as acceleration at supernova blast waves and diffusion in turbulent plasmas, the particle spectra at high energy should depend only on gyroradius, or equivalently on magnetic rigidity. Thus a difference between the spectra in rigidity would require different acceleration sources and/or propagation histories.

H and He are two interesting cosmic ray components because both have large λ_{int} and the dominance of the escape over interaction should be achieved earlier than for the heavier nuclei. It is obvious from Fig. 5.10 that these two components dominate the flux of cosmic ray nucleons and are thus extremely important for the production of secondary particles in the atmosphere which we will discuss in the following chapters. During the 1990s there were several measurements of the H and He flux with contemporary instruments that are discussed in some detail in the next subsection. The data sets used in Fig. 5.10 are also identified in the next subsection.

In the range below 100 GeV/nucleon there are no significant differences between the spectral indices of H and He. This was noticed by Webber et al. [122] who concluded that below 100 GV the rigidity spectra of hydrogen and helium are consistent with each other within experimental errors. The natural conclusion from such observations is that these two species have already reached their energy independent spectral range and that they have the same source and propagation histories.

A different picture emerges when the highest energy data are included in the comparison. All four experiments that have measured the spectra above 10 TeV have the He spectrum flatter than the H spectrum by $\Delta\gamma \sim 0.1$. This has brought in the suggestion that protons and all heavier nuclei have a different acceleration history [117].

In Fig. 5.10 the spectra of heavier nuclei are divided into three groups: C–O, Ne–Si and Fe. The grouping is necessary for the extension of the spectra to high energy because of the decreasing charge resolution and the low fluxes, and correspondingly the low experimental statistics at high energy. The C–O data of HEAO [118] and CRN [119] are the sums of carbon and oxygen, while the JACEE statistics [120] contain some nitrogen nuclei. For heavier nuclei the HEAO and CRN points are the sums of the Ne, Mg and Si fluxes, while JACEE data are for all nuclei between Ne and S. JACEE data may contain nuclei of the sub-iron group.

The spectra of all heavy nuclei indicate a flattening of the energy spectrum with energy. It is once again difficult to draw definite conclusions about the reason for that, because the very high energy end of the spectra comes mostly from JACEE and RUNJOB data that contain a mixture of primary and secondary nuclei. In addition to that JACEE, as well as all other calorimetric detectors, measures only the energy of the neutral secondaries produced by the primary cosmic ray and the energy of individual cosmic ray nuclei is subject of strong experimental fluctuations. The highest energy data points that contain only a small number of events have thus very large errors. It is tempting, however, to conclude that protons and heavy nuclei are accelerated at a different type of supernova shocks [117]. Protons are accelerated at blast shocks expanding in the ISM, while heavy nuclei are accelerated at shocks expanding in the pre-supernova winds. This suggestion has also some relevance to the under-abundance of H and He. Hydrogen in such a scenario is under-abundant because it is accelerated in a small fraction of all cosmic ray accelerators. He is under-abundant because the winds of Wolf-Rayet stars are are depleted of He [121].

5.4.1 Energy spectra of different cosmic ray components

In this subsection we give a quantitative description of the fluxes of different components that have been measured in the most recent experiments. The experimental data are presented in graphs that plot the differential fluxes as a function of the kinetic energy of the cosmic ray nuclei. The differential fluxes are multiplied by $E_k^{2.5}$ to make it possible to plot them on a linear scale. Since we aim at a good quantitative description of the cosmic ray fluxes we do not mind the corresponding increase of the errors – we want to emphasize the errors in order to obtain a better feeling for the uncertainty in the fits of the energy spectra of different nuclei. All nuclei heavier than hydrogen and helium are grouped in the same way as in Fig. 5.10.

The cosmic ray fluxes of the different groups of nuclei are fit with the expression

$$F(E_k) = K \left[E_k + B \exp \left(-\frac{C}{\sqrt{E_k}} \right) \right]^{-\alpha} , \qquad (5.6)$$

where the coefficients B, C express the kinematic change of the shape of the cosmic ray spectrum expressed in kinetic energy.

Hydrogen and helium fluxes

About ten years ago the absolute magnitude of the H and He fluxes was not well known. There were two normalizations, that of Webber et al. [122] and of the LEAP [103] collaboration, which were different by about 50% at proton energy of 10 GeV. Since then measurements have been performed by a number of new experiments: MASS [123], IMAX [124] and CAPRICE [125], which all support the lower normalization of LEAP. The breakthrough came in 2000, when the results of two independent measurements with the new detectors BESS [126] and AMS [127, 128] were published. BESS and AMS measured the fluxes of H and He almost simultaneously in June and July 1998 in a solar epoch close to solar minimum. BESS, the magnetic spectrometer shown in Fig. 5.3, was launched on a long-duration balloon flight. AMS, which is the prototype of a future Space Station experiment, flew aboard a Space Shuttle flight. The agreement between the H fluxes measured by these two experiments is indeed surprising. If one estimates the difference between them as

$$\Delta = \frac{1}{N} \sum \frac{|F_{BESS} - F_{AMS}|}{(F_{BESS} + F_{AMS})} , \qquad (5.7)$$

$\Delta = 0.009$ for the 31 data points of BESS. Since AMS has wider energy range we interpolate AMS data to obtain AMS flux at the energies for which BESS has measured the proton flux.

On the other hand the four other magnetic spectrometer data sets (referred to above) cluster at hydrogen flux values, smaller by 15–20%. The average value of $dN/dE \times E_{kin}^{2.5}$ at 20 GeV for the BESS and AMS sets is 6,590 $GeV^{1.5}\ m^{-2}\ s^{-1}\ sr^{-1}$, while it is 5,587 for the other four experiments. At 50 GeV the values are respectively 5,707 and 4,986. This gives a measure of the uncertainty of the cosmic ray flux of hydrogen.

It is not possible to obtain a good fit of all available data sets. We do not include in the current fits the old data set of Webber et al. [122], which is very different from the modern measurement. For the purposes of fitting we create several separate data sets, that also include the data of Refs. [129, 130, 131, 132, 133]. Three of the these four data sets are from emulsion chamber experiments above 1 TeV. Ryan et al. [129] is an early calorimeter experiment that measured protons between 50 and 2,000 GeV, i.e. covers the gap between the two groups. We re-normalize it down by 25%, which is fully consistent with the uncertainty of its efficiency (John Ormes, private communication). Table 5.1 shows the parameters of the different fits.

Table 5.1. Fit parameters for hydrogen. Data sets are marked as follows: 1, all 11 data sets; 2, AMS, BESS, Ryan et al. + TeV data; 3, AMS, BESS only; 4, LEAP, IMAX, CAPRICE94, MASS91 + TeV data.

Fit #	α	K	B	C	χ^2/n.d.f.
1	2.68 ± 0.04	9646 ± 1100	2.11	0.45	153.0
2	2.74 ± 0.01	14500 ± 360	2.20	0.25	1.03
3	2.74 ± 0.01	14510 ± 560	2.21	0.25	0.27
4	2.68 ± 0.05	9345 ± 1500	2.12	0.48	246.0

One could use different combinations of data sets, but this would not change the conclusion – in combination with TeV measurements AMS and BESS produce a better fit. It is amazing that 'AMS & BESS only' fit (set 3 in Table 5.1) is so close to the fit including TeV data. It is true that the errors of the TeV data points are much larger than the 1 to 100 GeV measurements, but it also means that TeV data are in line with the GeV ones and the χ^2 of the joint fit is still very good.

Figure 5.11 shows all data compared to the fits 2 and 3. Closed circles indicate AMS data; closed squares BESS; inverted closed triangles LEAP; closed triangles IMAX; closed diamonds CAPRICE94; pluses MASS91; and crosses - Ryan et al. Open symbols indicate TeV data sets: squares - Ivanenko; diamonds Zatsepin et al.; triangles Kawamura; and circles JACEE.

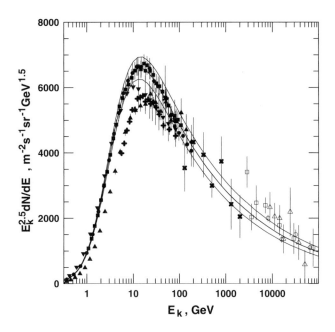

Fig. 5.11. Comparison of all hydrogen data sets to the fits 2 and 3. The thick line shows the central value of the fit, and the thin lines use the errors of K and α to show the maximum and minimum fluxes allowed by the fit.

We adopt this fit as a good representation of the proton data in a wide energy range. The differences between the individual experiments are reasonably small. To emphasize this we give in Table 5.2 the fluxes at several predetermined proton energies that come from fits of the individual data sets.

The situation with the He flux is not that good. To start with, there is a slight disagreement between the measurements of AMS [128] and BESS [126]. BESS presents data per E_k, while AMS gives a table in rigidity, but this is not the reason for the disagreement according to the experimental groups. One should note that AMS measured fluxes above the atmosphere while BESS has to correct the measured He fluxes for absorption in the atmosphere. The comparison between the two data sets as in (5.7) gives $\Delta = 0.08$ indicating a systematic excess of He nuclei in the BESS data.

Some of the difference can be traced to the different interpretation of the measurements by the experimental groups. AMS assumes 15% ^{3}He contribu-

Table 5.2. Hydrogen fluxes (in $^{-2}\,s^{-1}\,sr^{-1}\,GeV^{-1}$) measured by different experiments.

reference E_k, GeV	[126]	[127]	[103]	[125]
1	9.6×10^2	9.5×10^2	1.2×10^3	9.8×10^2
2	4.5×10^2	4.6×10^2	4.9×10^2	4.4×10^2
3	2.4×10^2	2.5×10^2	2.5×10^2	2.3×10^2
5	9.5×10^1	9.8×10^1	9.3×10^1	8.7×10^1
10	2.1×10^1	2.1×10^1	1.9×10^1	1.8×10^1
20	3.7×10^0	3.7×10^0	3.3×10^0	3.1×10^0
30	1.3×10^0	1.3×10^0	1.1×10^0	1.0×10^0
50	3.3×10^{-1}	3.2×10^{-1}	2.8×10^{-1}	2.5×10^{-1}
100	5.0×10^{-2}	4.8×10^{-2}	4.1×10^{-2}	3.6×10^{-2}
200	7.4×10^{-3}	7.2×10^{-3}	5.9×10^{-3}	5.0×10^{-3}

Table 5.3. Fit parameters for helium data sets are marked as follows: 1, all 11 data sets; 2, all excluding BESS; 3, all excluding CAPRICE94; 4) no TeV data.

fit #	α	K	B	C	χ^2/n.d.f.
1	2.60 ± 0.02	426 ± 31	1.22	0.42	7.00
2	2.61 ± 0.02	426 ± 34	1.22	0.43	7.95
3	2.61 ± 0.02	434 ± 36	1.20	0.41	7.40
4	2.63 ± 0.04	462 ± 56	1.22	0.37	7.86

Table 5.4. Helium fluxes (in $^{-2}\,s^{-1}\,sr^{-1}\,(GeV/nucleon)^{-1}$) measured by different experiments.

reference E_{kin}, GeV	[126]	[128]	[103]	[125]
1	9.0×10^1	7.8×10^1	1.0×10^2	9.2×10^1
2	3.3×10^1	3.0×10^1	3.4×10^1	3.3×10^1
3	1.6×10^1	1.5×10^1	1.5×10^1	1.5×10^1
5	5.7×10^0	5.2×10^0	5.1×10^0	4.9×10^0
10	1.2×10^0	1.0×10^0	9.7×10^{-1}	9.2×10^{-1}
20	2.1×10^{-1}	1.8×10^{-1}	1.7×10^{-1}	1.6×10^{-1}
30	7.2×10^{-2}	6.1×10^{-2}	5.9×10^{-2}	5.7×10^{-2}
50	1.8×10^{-2}	1.6×10^{-2}	1.6×10^{-2}	1.5×10^{-2}
100	2.6×10^{-3}	2.4×10^{-3}	2.6×10^{-3}	2.5×10^{-3}

tion to their measured rigidity spectrum. This increases the calculated kinetic energy by about 7% (at high rigidity) and raises the normalization through the $E_k^{2.5}$ factor. We present the He data of the two groups treated equally, i.e. converted from rigidity to kinetic energy assuming 100% ^{4}He.

The AMS helium spectrum roughly agrees with the measurements of the other four magnetic spectrometers, that all have to account for the destruction as well for the production of He nuclei in the atmosphere above the detector. This is described the best in the CAPRICE paper [125]. Both BESS and CAPRICE calculate that 90% to 93% of the helium flux on top of the atmosphere will reach the detector at float altitude. This correction is much smaller than the systematic differences between different sets.

In all we use 11 data sets, same as for hydrogen at energies 1 to 100 GeV, and also Ryan et al. (scaled by 0.75), RICH [134], Sokol, Kawamura et al. and JACEE. As for hydrogen, we attempt to create different sets for fitting that include data sets with relatively similar normalization. Table 5.3 gives the fitting parameters for the following fitting sets: 1, all data sets; 2, all data sets excluding BESS – the data sets with highest normalization; 3, all data sets without CAPRICE; 4, no TeV data.

It is indeed possible to create sets that fit together with much smaller normalized χ^2, but this would require an arbitrary exclusion of individual measurements. The way fits appear in Table 5.3 the fit parameters are consistent with each other in spite of the large χ^2 values. In all four combinations the He flux is fitted with a much flatter spectral index than the H flux. Both the normalization and the spectral index are consistent within the error bars for all four sets. A large fraction of the χ^2 is due to the difference between AMS and BESS measurements, both of which have small errors. A comparison of fit 1 with all data sets used in the fitting in shown in Fig. 5.12.

We accept the fit shown in Fig. 5.12 as the best representation of the He energy spectrum. The spectral shape is the same as in all data set combinations, but the relative error of the normalization is significant. Table 5.4 gives the flux values for He at the same set of energies as in Table 5.2.

Heavier nuclei

There are two data sets available that give information either for all charges between 6 and 26 individually [118] or in relatively narrow groups [135]. We use these two data sets in the GeV energy range plus the data of the CRN [119] detector and all available data at much higher energy. For the CNO group we include RICH and JACEE, for the Mg–Si group we include JACEE only and for $z > 17$ we include JACEE and Ichimura et al. [136].

Fitting presents even bigger problems here because of the inconsistency of the different data sets. HEAO-3 and CRN have generally much smaller normalization than the measurement of Simon et al. [135]. So in addition to fitting all available data one can construct two different fitting sets, one including Simon et al. and other 'high' fluxes and a 'low' set. The high sets include Simon et al. data and RICH data for CNO, Simon et al. for SiMg, and Simon et al. and Ichimura for $z > 17$. The low sets include HEAO-3 and CRN for all nuclei. All fitting sets include JACEE data. Table 5.5 gives the fitting parameters for all groups of nuclei with $z > 2$.

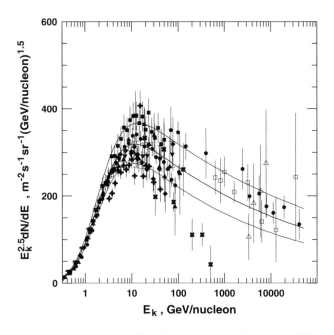

Fig. 5.12. Fit of the helium flux. Symbols are as in Fig. 5.11. The results of the RICH experiment are shown with filled pentagons.

Table 5.5. Fit parameters for heavier nuclei. For each group there are three fitting sets, that include all data sets, the 'high' data sets and the 'low' data sets.

fit #	γ	K	B	C	χ^2/n.d.f.
CNO					
all	2.68 ± 0.06	37.1 ± 6	21.2	1.67	13.9
high	2.60 ± 0.07	33.2 ± 5	0.97	0.01	2.52
low	2.71 ± 0.09	38.8 ± 9	36	1.95	19.0
SiMg					
all	2.78 ± 0.08	18.4 ± 4	3.10	0.40	2.66
high	2.79 ± 0.08	34.2 ± 6	2.14	0.01	0.74
low	2.80 ± 0.07	18.6 ± 6	5.08	0.59	1.81
z >17					
all	2.75 ± 0.04	4.70 ± 0.75	5.35	0.74	2.15
high	2.68 ± 0.01	4.45 ± 0.50	3.07	0.41	0.54
low	2.84 ± 0.04	5.71 ± 1.10	6.20	0.78	1.62

For all heavy nuclei the 'high' fitting sets generate the best χ^2 values. Note that in these particular cases the small χ^2 reflects not only the goodness of the fit, but also the size of the errors associated with each experiment. While HEAO-3 data have very small errors, Simon et al. data have generous errors and together with the uncertain JACEE data set generate unrealistically low χ^2. It is interesting that the combination of 'high' GeV data sets generally requires flatter energy spectrum, which is counterintuitive. The reason is that the fits are attracted to the 'low' HEAO-3 data, which are fitted rather well and do not account at all for the high energy JACEE data. Such fits miss the error bars of all JACEE data points. For this reason we prefer to use the fits of the 'high' fitting sets, which use the wide energy range of all experiments for the description of the cosmic ray fluxes. One unexpected result from these fits, however, are the different spectral indices for the three components, varying from 2.60 for CNO to 2.79 for SiMg and to 2.68 for $z > 17$. On the other hand all extreme spectral slopes are accompanied by reasonably high errors.

As an example for the fit quality we show in Fig. 5.13 a comparison between the 'high' fit for $z > 17$ and the full data set. This is the typical situation for all fits of $z > 2$ nuclei. One could imagine how a fit attracted to the HEAO-3 and CRN points (triangles and filled circles) could fit these two measurements very well, but not be able to represent the other data sets. The fit of the 'high' fitting set, which is shown in Fig. 5.13 with its errors, splits the difference between the high and low normalization data, which are fairly well described by the one sigma errors of the fit.

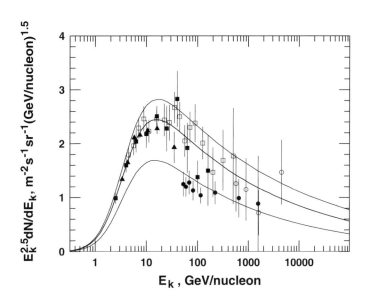

Fig. 5.13. Fit of the $z > 17$ flux. HEAO-3 data are shown with filled triangles, Simon et al. with filled squares, CRN with filled circles, JACEE with open circles and Ichimura et al. with open squares.

Nucleon fluxes

Although the fits of the energy spectrum of nuclei heavier than He appear to be very uncertain we can still attempt to generate the spectrum of the cosmic ray nucleons that hit the atmosphere. The reason is that such a spectrum is dominated by the relatively well fitted H and He spectra. If we use the tables in the previous two subsections we can calculate the quantity $E_k^{2.5} \times dN_i/dE_k$ for all groups. The rough numbers are respectively 6,500, 320, 25, 14, and 2.6 $GeV^{1.5} m^{-2} s^{-1} sr^{-1}$ for the five groups. If we multiply these numbers by the mass of each group we will have a good idea about the contribution of different components to the all nucleon flux at 20 GeV. The biggest uncertainty comes not from the fits, rather from the average mass for the last three groups where the individual elements in a group may have very different energy spectra. Still, if we assume masses of 14, 26, and 56 for the three groups of heavy nuclei, we obtain relative contributions of 75%, 15%, 4%, 4%, and 2% for all five groups. Even if the contributions of all heavy nuclei are wrong by 50% in the same direction the nucleon flux would only change by 5%. Figure 5.14 gives the relative contributions to the all nucleon flux calculated using the fits from the previous two subsections (the 'high' fits for heavy nuclei) and the average mass numbers used in the rough estimate above. We preferred to use somewhat unrealistically high mass numbers for the groups of heavy nuclei because that exaggerates their contributions to the all particle flux, and correspondingly gives the biggest possible uncertainty in the flux.

Since Fig. 5.14 assures us that the error in the all particle flux cannot be very large, we now calculate the flux and present in Fig. 5.15. The shaded area gives the uncertainty from the fits of the five nuclear components. The relative error of the all nucleon flux is shown in the lower panel of Fig. 5.15.

We remind the reader again that most of the uncertainty comes from the fact that the data sets used contain secondary as well as primary nuclei. The ratio of secondary/primary nuclei is energy-dependent, which means that the average charge and mass, especially for the MgSi and $z>17$ groups, will be energy-dependent. We somewhat exaggerate the contributions of these two groups by assuming high average masses for them (by about 10%) and increase both their contribution and the relative error of the presented spectrum. To check for the magnitude of the error due to the changing ratio of secondary to primary nuclei we used the HEAO3 [118] data at 3.5 GeV/nucleon, where the fraction of secondary nuclei should be larger than at high energy. The average charge of the $z>17$ group at this energy is 23.6. Assuming a constant mass to charge ratio, the error introduced by using $A = 56$ at this energy is only 9% for the contribution of this group to the nucleon flux, and well below 0.1% on the total nucleon flux. With increasing energy the average z value increases and the error decreases. On the other hand, the flux is quite flat, which probably keeps the error almost constant. The total uncertainty reaches a minimum of about 6% at kinetic energy below 3 GeV and a maximum of more than 25% at high energy.

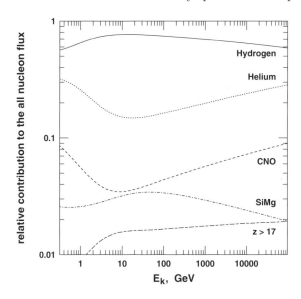

Fig. 5.14. Relative contribution of different groups of nuclei to the all nucleon flux. The 'high' fits from the previous section are used for heavy nuclei. Average mass numbers of 1, 2, 14, 26, and 56 are assumed – see text.

Figure 5.15 also shows with a dotted line the neutron flux. It is very useful to have the all nucleon fluxes separated into proton and neutron fluxes for calculations of the flavor content of the muon and neutrino fluxes. Such a separation, however, will be affected more strongly than the all nucleon flux by the uncertainties in fitting the cosmic ray mass components. There are also some additional uncertainties, due to the isotopic composition of cosmic rays.

The most important uncertainty comes from the energy-dependence of the $^3He/^4He$ ratio. Since (see Fig. 5.14) He nuclei always contribute more to the all nucleon flux than all heavier nuclei, even a small error in this ratio will noticeably affect the flux of neutrons. To estimate this ratio one can fit the low energy data on the $^3He/^4He$ ratio. The fits give ≈ 0.2 for $\beta = 1$. At higher energies one can use the E_k^δ dependence detected in other secondary/primary particle ratios with two extreme δ values of 1/3 and 0.6 and study the effect on the neutron to proton ratio. At low energy the ratio is also affected by the solar modulation, because all components heavier than H have higher rigidity. At high energies the ratio is mostly affected by the increasing contribution of the flat heavy nuclear components. In our estimates the neutron to proton ratio starts at a high value of 0.23 at 0.3 GeV/nucleon and declines to 0.14 at energy around 3 GeV. After reaching a minimum at 10 GeV the ratio slowly rises to reach 0.21 at 10^5 GeV

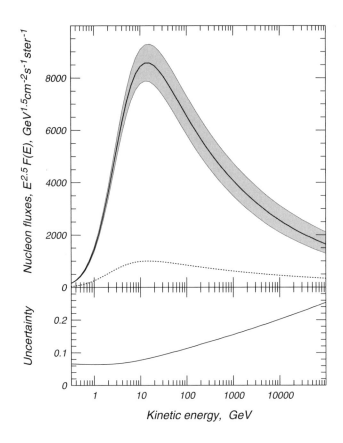

Fig. 5.15. The upper panel gives the all nucleon flux with its errors (shaded area) and the neutron flux (dotted line). The lower panel gives the relative error.

5.4.2 Electron spectrum

The flux of electrons at the top of the atmosphere has also been the subject of many measurements after the first unambiguous detection of electrons by Earl [137]. Figure 5.16 shows some of the results on the total electron flux, i.e. the sum of the fluxes of electrons and positrons. Electrons are believed to be primary particles and positrons are generated by primary electrons and nuclei on propagation in the Galaxy [138, 139]. The energy loss processes of the electrons (and positrons) are very different from those of charge nuclei. They lose energy on synchrotron radiation in the galactic magnetic field and on inverse Compton scattering on the radiation fields in addition to bremsstrahlung on the interstellar matter. The shape of the electron spectrum at Earth and the magnitude of the electron flux give thus additional, complementary information about the conditions in interstellar space.

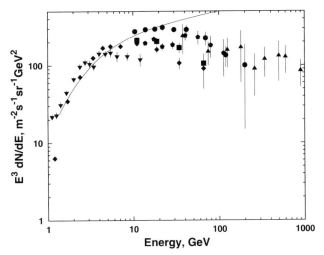

Fig. 5.16. Differential spectrum of all electrons multiplied by E_e^3 measured by several different experiments (see text). The line shows the proton spectrum divided by 100 and multiplied by E^3.

The data included in Fig. 5.16 are from Refs. [140] squares; [141] circles; [142] triangles; [143] inverted triangles; [144] diamonds; and [145] pentagons. The differential flux is multiplied by E_e^3 to make the spectral shape better visible. There is a spectral break at E_e between 5 and 20 GeV. At higher energy the electron spectrum has a power law shape with $\alpha_e = 3.3$, consistent with the assumptions of Ref. [71]. A steepening of the electron acceleration spectrum by one power of E_e is expected on propagation because of electron radiation loss. At lower energy the measured electron fluxes are subject to solar modulation, similar to this of cosmic ray nuclei. The flux of cosmic ray protons, also multiplied by the total proton energy E_{tot} and divided by 100, is shown with a line. Up to the break in the electron spectrum electrons are about 0.01 of the hydrogen nuclei, and have approximately the same spectral shape. At higher energy the proton to electron ratio increases by an energy dependent factor, presumably proportional to the differences in the spectral shape, i.e. $E^{\alpha_e - \alpha_H}$ of electrons and hydrogen nuclei.

The difference in the spectra of electrons and of positrons is also of great interest. In principle the spectrum of all electrons consists of two components: primary electrons accelerated at astrophysical shocks and secondary electrons and positrons, that are produced in approximately equal numbers in propagation. If all positrons are of secondary origin, their spectrum can be estimated quite accurately to the extent that the acceleration spectrum of the primary species is known. There is, however, the interesting possibility that also some primary positrons are generated, for example by the decay of dark matter particles. If an unknown dark matter particle decays into a $e^+ e^-$ pair, it will create a positron and electron component that peaks around one half of the mass of that particle. Since the electron spectrum is dominated by electrons this component would be more clearly visible in the positron spectrum. Earlier there were some indications that the positron spectrum flattens at energy above 10 GeV. One has to keep in mind that the measurement of the positron flux is very difficult because of the dominant positive proton background, which can easily mimic positrons. Figure 5.17 shows two of the recent measurements of the positron fraction of the electron flux.

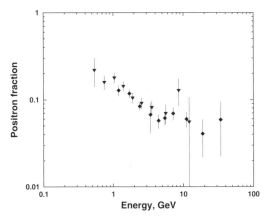

Fig. 5.17. Positron fraction of the total electron flux derived in Refs. [143] and [144]. Symbols are the same as in Fig. 5.16.

The positron fraction of the electron flux by the two experiments is 0.2 or higher at energies below 1 GeV and decreases to about 0.05 at energies exceeding 5 GeV or so. Such a behavior, after accounting for solar modulation, is fully consistent with the secondary origin of all positrons [139]. The recent measurements of the positron fraction of the electron flux give no indication for the existence of primary positrons. Because of the large experimental uncertainty above 10 GeV, however, only limits on the fraction of primary positrons can be currently set.

5.4.3 Antiprotons

The antiproton flux is interesting for the same reasons as the positron flux. Unless there is a section of our galaxy that consists of antimatter, all antiprotons should be secondary. Primary antiprotons could also be created by heavy particles (with mass greater that two proton masses) decays into $\bar{p}p$ pairs, which should in principle show up in the $\bar{p}$ spectrum. There is, however, one more complication – in the antiproton production cross-section.

In hadronic interactions antiprotons have to be produced in $\bar{p}p$ pairs. If a primary proton collides with interstellar matter and produces a $\bar{p}p$ pair, its energy should exceed the energy threshold of $7m_p$, i.e. the absolute energy threshold for $\bar{p}$ production is about 7 GeV. Above this threshold the production cross-section grows rapidly and combined with the rapidly decreasing proton flux forms a distinct interstellar spectrum of the secondary antiprotons that peaks around kinetic energy of 2 GeV. This is very different from the monotonic proton acceleration spectrum. The formation of the antiproton production spectrum is discussed in detail in the book of Gaisser [6].

In the leaky box approximation the equilibrium $\bar{p}$ spectrum is obtained with (4.13), where λ_{esc} is the same as for protons and λ_{int} includes annihilation and other inelastic interactions of antiprotons, all energy dependent. The source term of antiproton production is usually divided into two parts: the antiproton production summed over all primary cosmic ray and interstellar target nuclei, and a term that accounts for the $\bar{p}$ energy loss in inelastic interactions.

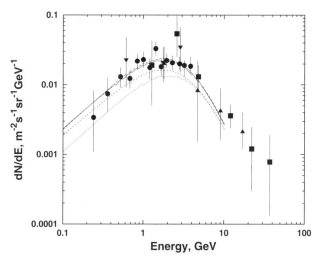

Fig. 5.18. Antiproton fluxes detected by the CAPRICE [146], squares, IMAX [147], inverted triangles, MASS [148], triangles, and the BESS [149], circles, experiments. The predictions with solar modulation for different solar epochs are from Ref. [150].

Figure 5.18 shows experimental data on the $\bar{p}$ flux compared to predictions [150] including the calculation of interstellar proton fluxes and solar modulation at different solar epochs. The solar modulation of the antiproton flux is different from this of protons for two reasons. The modulation at low energies is smaller than for protons, because the interstellar $\bar{p}$ spectrum increases up to about 2 GeV and $\Phi_{IS}(E_{IS})$ in (5.3) grows up to that energy and diminishes the modulation. In addition, antiprotons are negatively charged, and if the suggestion of Ref. [106] is correct their solar modulation would be different from this of protons. This last suggestion was made by Bieber et al. [150], who calculated a larger $\bar{p}/p$ ratio below 1 GeV for the epoch just before the change of the solar polarity in year 2000. This prediction was confirmed by the last measurement of the BESS group (which is not shown in Fig. 5.18).

The recent high statistics antiproton measurements agree well with the more sophisticated predictions for the $\bar{p}$ flux, which leaves little space sources of primary antiprotons.

6 Cosmic rays in the atmosphere

As we discussed in the previous chapter, primary cosmic rays interact with the atmosphere and produce fluxes of secondary, tertiary, etc. particles. All these particles together create a cascade, called an air shower. In principle, measurements of air shower particles can be interpreted in terms of the energy spectrum and the composition of the primary cosmic rays. The interpretation of these measurements requires one additional step – calculation of the fluxes generated by a cosmic ray nucleus of mass A, charge Z and energy E.

Such calculations would be easy if we knew exactly the properties of inelastic interactions of nucleons with medium-heavy nuclei, which provide most of the atmospheric targets. These properties are, however, not known that well. High energy physics has mostly concentrated on proton–proton collisions that reveal the basic structure of matter. Nuclear physics experiments usually probe the other extreme – central collisions of heavy nuclei that allow the study of collective effects. Here we have to deal with collisions of protons and heavier nuclei on the nuclei of the atmosphere, and then with the collisions of the secondary particles with the same atmospheric target. A small error in the energy spectrum and final state composition of the secondaries in a single interaction, which would be negligible compared to the experimental uncertainties, grows with every generation and could alter the interpretation of a cosmic ray measurement.

The other problem in the calculation of the secondary atmospheric fluxes is the limited knowledge of the primary cosmic ray flux, which provides the other major set of inputs for the calculation. Other uncertainties are related to the structure of the atmosphere and its variations, and with the method of calculation.

The measurements of the atmospheric fluxes give us a different type of information. While the direct satellite and balloon measurements of the primary cosmic rays ideally tell us the exact type and energy of the cosmic ray nucleus, here we study the slope and normalization of the cosmic ray energy spectrum and more general descriptions, such as the proton to neutron ratio and the average mass of cosmic rays in certain energy range.

We shall first discuss the atmospheric structure, then give some analytic estimates of the fluxes in the atmosphere and finally show experimental data and compare them to Monte Carlo calculations.

6.1 Atmospheric structure

The main parameter we want to know about the atmosphere is what is the amount of matter above any atmospheric layer, in which the primary cosmic ray particle has interacted, produced secondaries and started the air shower. This quantity is the atmospheric depth, X, measured in g/cm^2. The depth is the integral in altitude of the atmospheric density above the observation level h, i.e.

$$X = \int_h^\infty \rho(h_1)\, dh_1 \ . \tag{6.1}$$

The altitude dependence of the atmospheric density ρ is thus the key. Density is also of practical importance for cascade calculations since it determines the ratio between particle interactions and decays.

The ratio of the atmospheric pressure ($\equiv X$) to density in the atmosphere is proportional to the temperature. If the temperature were constant, then the relation between altitude and depth would be very simple

$$X = X_o \exp(-h/h_0) \ , \tag{6.2}$$

where X_0 is the atmospheric depth at sea level (1,030 g/cm^2) and h_0 is the scale height of the atmosphere. This is true for perfect gas with constant composition that is in hydrostatic equilibrium.

The measurements of the altitude dependence of the temperature can be approximately fit with a two temperature dependences. The temperature decreases with altitude up to the troposphere, and is then constant. Equation (6.2) can then be used for the upper atmosphere. A fit of the US Standard Atmosphere for medium latitudes and the seasons of spring and fall gives tropospheric altitude of 11 km, average surface temperature of 15°C and temperature drop rate of 6.5°C per kilometer. M. Shibata has added to it a third layer that accounts for the increase of the temperature at very large altitudes. The altitude dependence of the atmospheric depth X then becomes:

$$\ln X = 5.26 \ln\left[(44.34 - h)/11.86\right] \ for \ h < 11\,\mathrm{km} \tag{6.3}$$
$$= (45.5 - h)/6.34 \ \text{for} \ 25\,\mathrm{km} > h > 11\,\mathrm{km}$$
$$= 13.78 - 1.67\left[68.47 - 1.2 \times (48.63 - h)\right]^{\frac{1}{2}} \ \text{for} \ h > 25\,\mathrm{km}$$

All coefficients are in kilometers and the atmospheric depth X is in g/cm^2. One should not forget that these formulae are approximate. For practical use all coefficients should be much more precise to interface better the three layers. Expressed in terms of the isothermal atmosphere (6.2) this gives a scale factor h_0 of 8 km below 5 km and a scale factor of ~6.4 km above the troposphere.

All the formulae and numbers above are for the vertical direction. For zenith angles less than 60° one can scale the slant depth as $X \cos\theta$ - the flat Earth approximation. For larger angles one has to account correctly for the

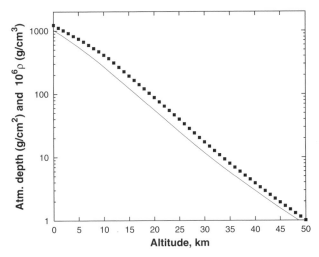

Fig. 6.1. Altitude dependence of the vertical atmospheric depth X (line), and the atmospheric density ρ (points) from the atmospheric model of (6.3)

curvature of the Earth. This is usually done in tabular form after mapping the altitude dependence of the atmospheric density and integrating in altitude numerically along the angle. The atmospheric profile of (6.3) gives total horizontal atmospheric depth of 36,000 g/cm^2.

Equation (6.3) gives an atmospheric profile, shown in Fig. 6.1, which is a good representation for the atmosphere averaged all over the Earth. The profiles at specific locations could be very different. Figure 6.2 shows measurements of the atmospheric temperature versus altitude for the austral summer and winter at the US McMurdo station in the Antarctic. These atmospheric conditions will obviously create different atmospheric profiles.

During the austral summer the troposphere level is at about 8 km. The temperature is approximately constant at −40°C between altitudes of 10 and 25 km and then increases up to −30°C. An atmospheric model like (6.3) with updated parameters can be created on the basis of these measurements. During the austral winter the tropospheric level is not detected and the atmospheric density profile has to be modeled differently.

These variations strongly suggest that a correct atmospheric profile has to be used for the interpretation of every experiment that is sensitive to the exact atmospheric conditions.

6.2 Analytic approximations

The air shower development can be modeled analytically through the solution of transport equations similar to those of Sect. 4.3. We follow Gaisser's

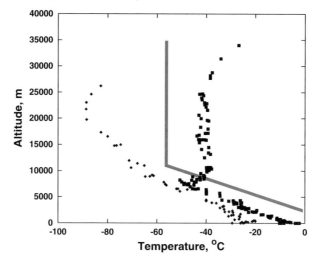

Fig. 6.2. Measurements of atmospheric temperature at McMurdo during the austral summer, squares, and winter, diamonds. The shaded line shows the assumptions of (6.3).

approach [6] in this section. The basic equation for uncorrelated[1] particle fluxes of type j at atmospheric depth X is

$$\frac{dF_j(E, X)}{dX} = -\left(\frac{1}{\lambda_j} + \frac{1}{d_j}\right) F_j(E, X) + \sum_i \int \frac{g_{ij}(E_i, E_j)}{E_i} \frac{F_i(E_i)}{\lambda_i} dE_i$$

(6.4)

where λ_j is the mean free path for inelastic interactions of a particle of type j and d_j is its decay length. The mean free path (m.f.p.) $\lambda_j \equiv A/(N_A \sigma_j)$ is the inverse of the product of the number of nuclei with mass A per gram of matter and the interaction cross-section. The average mass of an air nucleus is 14.5 and for a proton cross-section 300 mb $(3 \times 10^{-25}$ cm$^2)$ we obtain $\lambda_p(air)$ of 80 g/cm^2. The mean free path λ_j is energy dependent. The decay length d_j is also expressed in g/cm^2. For zenith angles θ below 60° (in the flat Earth approximation) it is defined as

$$\frac{1}{d_j} \equiv \frac{\epsilon_j}{EX \cos\theta},$$

(6.5)

where ϵ_j is the critical energy at which the interaction probability in the atmosphere equals the decay probability. Particles of energy $E \ll \epsilon_j$ always decay and of energy $E \gg \epsilon_j$ always interact. The second term in (6.4) is the source term, which sums over the production of secondary particles j with energy E_j by particles of type i with energy E_i.

---

[1] Uncorrelated particles are those that are not detected in groups. In practical terms it means that every cascade produces only one particle that can be detected at any level.

For stable particles, and when the source term in (6.4) can be neglected, the solution becomes simple [6]

$$F_j(E, X) = F_j(E, 0) \exp\left(-\frac{X}{\Lambda_j}\right), \tag{6.6}$$

where Λ_j is the particle absorption length and $F_j(E, 0)$ is the boundary condition, i.e. the flux of particles of type j at the top of the atmosphere. The energy spectrum of nucleons, neglecting the production of nucleon–antinucleon pairs in the atmosphere, is given by (6.6) with Λ_p defined as

$$\frac{1}{\Lambda_N} = \frac{1}{\lambda_N}[1 - Z_{NN}], \tag{6.7}$$

where λ_N is the nucleon m.f.p. and Z_{NN} is the spectrum weighted moment of the nucleon production cross section by nucleons – see (4.15). To the first approximation $\gamma = 1.7$ for atmospheric processes. Λ_N than has a value of about 120 g/cm^2. See Gaisser's book [6] for tables of spectrum weighted moments and absorption lengths.

The fluxes of atmospheric mesons are much more complicated. One has to write explicitly the coupled cascade equations given in a general form by (6.4) and use the appropriate Λ and Z values. Since Λ and Z are mildly energy dependent the solutions are useful in restricted energy range. Gaisser solves the equation for the pion fluxes in the atmosphere in the two extreme cases: for pion energy $E_\pi \gg \epsilon_\pi$ and for $E_\pi \ll \epsilon_\pi$. $\epsilon_\pi = 115$ GeV for vertical fluxes.

Neglecting decay, and with the boundary condition $\Pi(E, 0) = 0$, i.e. there are no primary pions, the vertical pion flux at depth X is

$$\Pi(E, X) = F_N(E, 0)\frac{Z_{N\pi}}{(1 - Z_{NN})}\frac{\Lambda_\pi}{\Lambda_\pi - \Lambda_N}[\exp(-X/\Lambda_\pi) - \exp(-X/\Lambda_N)], \tag{6.8}$$

where $F_N(E, 0)$ is the nucleon flux on the top of the atmosphere. This flux reaches maximum at $\sim$140 g/cm^2 and then declines. Deep in the atmosphere the flux decreases as $\exp(-X/\Lambda_\pi)$.

In the low energy case, including decay, the solution requires more work and becomes

$$\Pi(E, X) \simeq F_N(E, 0)\frac{Z_{N\pi}}{\lambda_N}\frac{XE}{\epsilon_\pi}\exp(-X/\Lambda_N). \tag{6.9}$$

Figure 6.3 shows the development of the pion flux in the atmosphere in the two approximations. The normalization of the two curves is arbitrary. Both curves have to be multiplied by the nucleon flux on the top of the atmosphere, and the dashed curve ($E_\pi \ll \epsilon_\pi$) has to be scaled by E_π/ϵ_π. Both curves reach maximum at atmospheric depth between 100 and 200 g/cm^2. This behavior is typical for the fluxes of secondaries in the atmosphere, excluding the fluxes of atmospheric neutrinos.

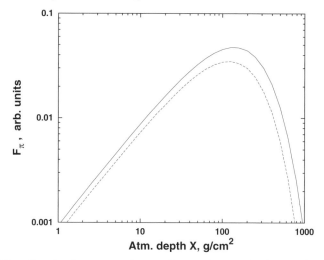

Fig. 6.3. Pion flux in the atmosphere calculated with (6.8) and (6.9) neglecting decay, solid line, and with decay, dashes. The normalization of the two curves is arbitrary, only the shape is correct.

6.2.1 Muons from meson decay

After calculating pion fluxes in the atmosphere (as well as the fluxes of other mesons by using the corresponding values of Λ and Z) Gaisser proceeds to calculate the fluxes created by pion decay. The first step is the calculation of muons and muon neutrinos from charged pion decays – $\pi^+ \longrightarrow \mu^+ + \nu_\mu$ and respectively $\pi^- \longrightarrow \mu^- + \bar{\nu}_\mu$. The production spectrum of the decay particles of energy E at depth X is a sum of all atmospheric decays of particles of type i that generate particles of type j.

$$P_j(E,X) = \sum_i \int_{E_{min}}^{E_{max}} \frac{dg_{ij}(E,E')}{dE} D_i(E',X)\, dE' , \qquad (6.10)$$

where $dg_{ij}(E,E')/dE$ is the spectrum of secondaries j from decaying particles of type i with energy E'. D_i is the spectrum of decaying mesons of energy E' at depth X, which is the flux of such particles weighted by the decay probability $\epsilon_i/(E'X\cos\theta)$. E_{min} and E_{max} are the minimum and maximum energy of the decay particles determined by (2.37). For two-body pion decay the minimum muon energy $E_{min}^\mu = E_\pi(m_\mu^2/m_\pi^2) \simeq 0.57 E_\pi$ and $E_{max}^\mu = E_\pi$.

Accounting for the two-body decays of pions and kaons and assuming a proton spectrum of $1.8E^{-2.7}$ cm^{-2} s^{-1} sr^{-1} GeV^{-1} Gaisser obtains the muon spectrum in the same units after an integration over the muon production in the whole atmosphere.

$$\frac{dN_\mu}{dE_\mu} \simeq 0.14 E_\mu^{-2.7} \left[\frac{1}{1 + \frac{1.1 E_\mu \cos\theta}{115\ \text{GeV}}} + \frac{0.054}{1 + \frac{1.1 E_\mu \cos\theta}{850\ \text{GeV}}} \right] , \qquad (6.11)$$

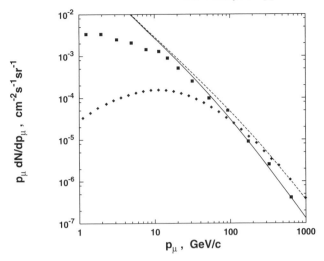

Fig. 6.4. The muon flux of (6.11) (solid line for vertical and dashed line for 75°) is compared to the vertical muon flux of Ref. [152], squares. The diamonds show the muon flux at the angle of 75° presented in Ref. [153].

where the first term in the square brackets represents the muons from pion decay and the second one represents the muons from kaon decay. 850 GeV is the value of ϵ_K and 0.054 is the weight of the kaon contribution of the muon flux coming from the kaon production cross-section and the muon branching ratio of kaon decays. Equation (6.11) does not account for the muon energy loss and will only be correct at relatively high energy when the energy loss is only a small fraction of the muon energy. This equation is developed further with an account for the energy loss by P. Lipari [151].

Equation (6.11) reveals two important features of the muon spectrum. At low energy ($E_\mu \ll \epsilon_\pi$) the muon spectrum has the shape of the primary cosmic ray spectrum, while at high energy it steepens by one power of E_μ as the thickness of the atmosphere is not big enough for pions to decay. The pion decay length $c\tau_\pi$ is 7.8 m and for $E_\pi = 1,000$ GeV it becomes 55.7 km, longer than the vertical extent of the atmosphere. Because of the $\cos\theta$ factor pions decay more easily in nonvertical showers and muons at large angles have a flatter energy spectrum. Note also that the contribution of kaons to the muon flux increases with the muon energy because of higher value of ϵ_K. At low energy the relative contribution of kaons is about 5% and it grows up to 27% asymptotically.

Figure 6.4 compares two muon measurements, that of Allkofer et al. [152] for vertical muons and that of Jokisch et al. [153] at an angle of 75° to the muon fluxes from (6.11). Flat Earth atmosphere is assumed in the estimate for the 75° flux. At low muon energy Gaisser's formula overshoots the measured fluxes by a large amount because of the neglected muon energy loss. At high energy, however, the agreement is very good. The comparison between the two

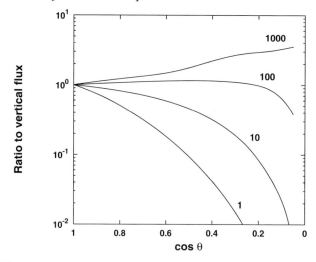

Fig. 6.5. The ratio of the inclined to the vertical muon flux as a function of the cosine of the zenith angle θ, from Ref. [154]. Muon momentum is given by each curve in GeV/c.

experimental data sets also shows the change of the muon energy spectrum with angle. At low muon energy the 70° muon flux is lower because of the increased energy loss and decay. The decay length for 1 GeV muons is about 6 km. At energies exceeding ϵ_π the inclined muon spectrum is flatter then the vertical one and the muon flux is respectively higher.

Figure 6.5 gives a quantitative impression of this effect, as presented in Ref. [154]. The calculation is a Monte Carlo type and accounts for the curvature of the Earth. 1 GeV/c muons fade fairly quickly with angle and their flux decreases by about factor of 10 at 60°. The flux of 100 GeV/c muons is relatively flat up to $\cos\theta$ of 0.2 and then quickly declines. 1, 000 GeV/c muons flux monotonically increases with the zenith angle. Especially sensitive to the zenith angle is the flux of TeV muons at the approach to the horizontal direction, which is not shown here. A small difference in $\cos\theta$ changes drastically the thickness and density profile of the atmosphere and the muon energy spectrum. For this reason the measurements of almost horizontal muons are very difficult to interpret.

6.3 Muon fluxes in the atmosphere

Muons of momenta from hundred MeV/c to several TeV/c have been measured continuously during the last 50 years. The interest in these results has changed during the years. In the beginning muons were used to study the geomagnetic effects [155]. Then muon measurements and the muon charge ratio were interpreted in terms of the primary cosmic ray spectrum and composition. Now the geomagnetic field is very well known, the interest in the primary cosmic ray spectrum continues, but the most recent measurements are more often interpreted in terms of the hadronic interactions that produce them and in the details of the development of the atmospheric cascades.

Generally muons are measured with magnetic spectrometers, starting with the same spectrometers that measure directly the primary cosmic ray spectrum and ending with the use of the LEP experiment L3 [156]. These devices measure the μ^+/μ^- ratio as well as the total flux. Measurements are made at the top of the atmosphere, at sea level, and at all intermediate altitudes. These are made at mountain level and during the ascent of balloon experiments.

All measurements are compared to calculated muon fluxes. Almost all current calculations employ the Monte Carlo technique. These consist of a hadronic collisions event generator and a cascade code that handles the propagation of individual muons through the atmosphere. The event generator has to represent correctly all features of the inelastic hadronic collisions in the relevant energy range. The cascade code has to deal with particle decays and energy loss. While decay and energy loss are not very important for high energy muons, the calculations of GeV muons are very sensitive to the correct representation of the muon energy loss. The main reason is that the energy loss changes the muon decay probability and the two processes have to be handled simultaneously.

6.3.1 Experimental results on atmospheric muons

Most of the experimental results are taken at ground level. They are usually presented binned in momentum and the flux is multiplied by a power of the muon momentum. The power depends on the energy range that the experiment intends to emphasize. If the measurement is in the GeV range, then dN/dp_μ or $P_\mu dN/dp_\mu$ is plotted. For the 10 GeV range $p_\mu^2 dN/dp_\mu$ is more appropriate and one more power of p_μ is added for higher energy muons as the muon energy spectrum continuously becomes steeper. Figures 6.6 and 6.7 show collections of muon data taken at ground level compared to calculations. The data are from Ref. [152], full circles, Ref. [157], full triangles, Ref. [158], full inverted triangles, Ref. [159], full diamonds, Ref. [160], full pentagons, and Ref. [161], full squares. Two more sets, that are still preliminary, are shown with open symbols – Ref. [162] with squares and Ref. [156] with circles (only statistical errors are given for this last measurement).

There is a definite trend for a lower normalization of the muon spectra in the new experimental data. In the 10 GeV range the difference between the data of Refs. [161, 162] and the measurements of Refs. [152, 159] is about 20–25%. The disagreement is even slightly bigger than shown in the figure because the data of [161] are taken at an altitude of 960 g/cm^2 where the muon flux should be somewhat higher. The differences cannot be explained by solar modulation or geomagnetic effects. The calculation [154] matches well the older data sets but overestimates the muon flux according to the

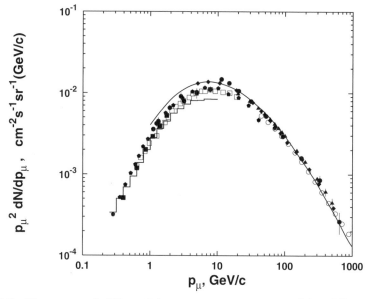

Fig. 6.6. Muon ground differential energy spectra measured by different experiments weighted by p_μ^2. See text for the references and symbols representing the data sets. The line is from Ref. [154] and the histogram, from Ref. [163].

new data. The calculation that generated the histogram will be discussed below.

Figure 6.7 emphasizes the muon fluxes at high energy. The 100 to 1,000 GeV/c data show a spectrum that is somewhat flatter than the predictions, although the very high energy data [158] seem to agree with the predictions. Table 6.1 gives the values of muon fluxes at fixed muon momenta measured by different experiments. These numbers have been interpolated (except for Ref. [159] which gives fluxes at the same momenta) from the published values accounting for the experimental errors. In case the interpolation does not fit the measured fluxes well, the value is not included in the table. Experiments that do not fit the energy range or have too few data points are not included.

Although the general features of the muon spectrum on the ground are well understood, differences up to 25% are not acceptable in modern calculations. The calculations of [154] are one-dimensional, i.e. the assumption is that all secondary particles move in the direction of the primary cosmic ray particle. At high energy (well above the nucleon mass) this approximation is good enough, because all secondaries are emitted in a narrow forward cone in the Lab system. At GeV energies, however, this approximation could break. Other reasons could be associated with the main assumptions of every calculation – the primary cosmic ray flux and the model of inelastic collisions. The primary energy responsible for the muon flux on the ground is about $10E_\mu$. For the muon fluxes high in the atmosphere the primary cosmic ray energy is somewhat lower. It is therefore important to understand the the

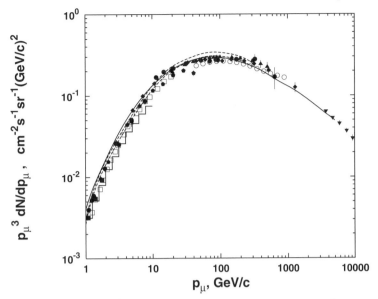

Fig. 6.7. Ground muon differential energy spectra weighted by p_μ^3. The symbols and the solid lines are the same as in Fig. 6.6. The dashed lines are the predictions of Ref. [164].

Table 6.1. Muon fluxes at ground level. Muon momentum is in GeV/c and the fluxes are in $\mathrm{cm}^{-2}\,\mathrm{s}^{-1}\,\mathrm{sr}^{-1}\,(\mathrm{GeV/c})^{-1}$.

p_μ	[152]	[157]	[159]	[160]	[162]	[156]
1.	–	–	–	3.6310^{-3}	–	–
2.	1.6610^{-3}	–	–	1.6210^{-3}	1.3910^{-3}	–
3.	8.9910^{-4}	–	–	9.3910^{-4}	8.6110^{-4}	–
5.	3.9210^{-4}	–	5.3310^{-4}	4.2810^{-4}	4.0510^{-4}	–
10.	1.3710^{-4}	–	1.3510^{-4}	1.0910^{-4}	1.0410^{-4}	–
20.	2.7110^{-5}	2.4310^{-5}	2.6610^{-5}	2.1210^{-5}	–	–
30.	9.1410^{-6}	8.8010^{-6}	8.8110^{-6}	7.9010^{-6}	–	–
50.	2.1010^{-6}	2.2010^{-6}	2.1810^{-6}	1.8010^{-6}	–	1.9210^{-6}
100.	2.7410^{-7}	2.9610^{-7}	2.7010^{-7}	–	–	2.7110^{-7}
200.	3.4110^{-8}	3.5610^{-8}	3.5210^{-8}	–	–	3.0810^{-8}
300.	9.1410^{-9}	9.7710^{-9}	1.0710^{-8}	–	–	8.1810^{-9}
500.	–	1.7910^{-9}	2.3810^{-9}	–	–	1.5410^{-9}

development of the muon flux in the the atmosphere – its growth curve for muons. Such a curve for the integral flux of muons of momentum > 1 GeV/c is shown in Fig. 6.8.

The calculations of the growth curve of muons shown in Fig. 6.8 bracket the experimental data, pretty much in the same way as they bracket the energy spectrum on ground level. One has to have in mind that the growth

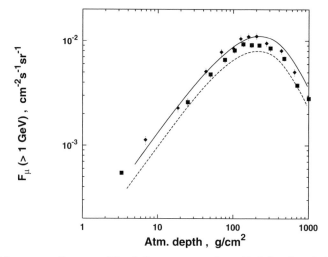

Fig. 6.8. Muon growth curve. The full squares are from Ref. [161] and the diamonds are from Ref. [165]. The predictions are from Refs. [154, 163], shown with a full and a dashed line respectively.

curve is fully determined by 1–10 GeV/c muons and the behavior of the higher energy muons does not alter it. The differences between the two predictions will be discussed in the next section.

Another piece of data that is of common interest is the muon charge ratio $R = F_{\mu^+}/F_{\mu^-}$. R is a measure of the ratio of protons to neutrons in the primary cosmic ray flux. In pp interactions the ratio of positive to negative pions in the final state is well above 1, especially among the fastest pions produced in the collision. It is related to the charge conversion probability for protons, i.e. interactions of the type $p+p \rightarrow n+p+\pi^+ +....$. The opposite is true for nn interactions. These effects are milder, but still noticeable, for interactions on air nuclei, where the target is half protons and half neutrons. The dominance of protons in the primary cosmic ray flux makes R greater than unity and the measured R values can be interpreted in terms of the average p/n ratio of the cosmic ray flux. Figure 6.9 shows several measurements of R.

The low energy data of Refs. [161, 165] are taken at high geomagnetic latitude. The ground data on the charge ratio agree well with each other within their stated errors. Since the muon spectra themselves are measured with considerable errors, the error bars on their ratio are significant. There is not a significant energy dependence in the charge ratio, although the preliminary L3 [156] data show a slightly higher R value. The 'float' data of [161] have, however, significantly higher charge ratio.

Muon data at 'float' are generated by a single interaction of the cosmic rays. Only $[1 - \exp(-3.9/\lambda_p)] \simeq 5\%$ of the primary nucleons have interacted above the 3.9 g/cm^2 observation level and the probability for secondary interactions is much smaller. The charge ratio thus reflects directly the pro-

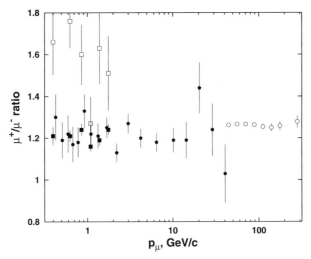

Fig. 6.9. Muon charge ratio. The full squares are from Ref. [161], the diamonds, from Ref. [165], and the circles from Ref. [156]. The open squares show the measurements of [161] at float level – average atmospheric depth of 3.9 g/cm ^{2}.

ton/neutron ratio of the cosmic ray flux the average π^+/π^- production ratio. With the primary spectrum described in Chap. 5, R of 1.6 suggests π^+/π^- ratio of about 1.4, consistent with accelerator measurements and theoretical estimates.

The GeV muons observed at 'float' never reach sea level. 3.9 g/cm^2 corresponds to an altitude of about 40 km – the decay length ℓ_d of ~7 GeV/c muons. The muons observed at sea level are produced further down in the atmosphere, within one decay length of the observation level. The contribution of the late cascade generations is much more pronounced, the memory of the charge of the primary nucleons starts to fade and R decreases.

Muon flux calculations

In the previous chapter we presented two predictions for the muon flux – those of Ref. [154] and Ref. [163]. Both are Monte Carlo style calculations, but there are several important differences between them. The first one in done in one dimension (1D), the second in three dimensions (3D). We shall use the dimensionality to label the calculations.

The 1D calculation uses the primary cosmic ray flux defined in Ref. [154], which is not very different from the flux fits shown in Sect. 5.4. 3D uses the flux of Ref. [125] which for hydrogen is about 10–20% lower than our best fit of all data. 1D is performed with the original version of the hadronic interaction code TARGET (see Ref. [154] for a description of the main features of the code). The 3D calculation uses the updated 3D event generator TARGET2.1 [166] and 3D cascade geometry. Table 6.2 compares the Z factors for charged pions and kaons (that determine the production of muons in the atmosphere in (6.9)) in these two models and gives several experimental Z values.

Table 6.2. Z factors and their effect on the muon flux. Model values are for energy of 1000 GeV

Reference	$Z_{p\pi\pm}$	$Z_{pK\pm}$	$R_{K/\pi}$	κ_μ
[167] 175 GeV	0.076	0.0097	0.13	10.54
[168] 400 GeV	0.074	0.0074	0.10	9.70
[20] 1,500 GeV	0.083	0.0100	0.12	11.34
[154] 1D (10 GeV)	0.089	0.0014	0.016	–
[154] 1D (100 GeV)	0.079	0.010	0.12	–
[154] 1D (1,000 GeV)	0.072	0.0105	0.15	10.34
[166] 3D (10 GeV)	0.076	0.003	0.039	–
[166] 3D (100 GeV)	0.077	0.010	0.13	–
[166] 3D (1,000 GeV)	0.071	0.014	0.20	9.80

The parameter κ_μ is the asymptotic ratio between the vertical muon flux to the primary cosmic ray nucleon flux. At very high energy the flux of muons $F_\mu = \kappa_\mu F_N/E_\mu$. The two models have similar κ_μ values when the 1,000 GeV Z factors are used, but the energy dependence of the Z factors in the relevant energy range is different. The 3D hadronic model has lower Z values for pions and a higher ones for kaons. The increased kaon production does not play much of a role for muons below 10 GeV/c. The pion Z factors are smaller enough to generate a lower muon/primary flux ratio below 10 GeV/c. The use of a lower primary cosmic ray flux additionally decreases the predictions and explains the difference between the fluxes of [154] and [163].

Another important factor is the 3D calculation itself. It has been noticed that 1D calculations agree pretty well with the shape of the muon spectra measured at very high altitude (the balloon 'float' altitudes are at depth of 3 to 7 g/cm^2) but overestimate the muon flux below ~2 GeV at the ground and especially at intermediate atmospheric levels. Figure 6.10 shows the data of Ref. [161] at different atmospheric depths and compares them to three calculations, all performed with the event generator TARGET2.1. The histogram is the 1D calculation of the vertical muon flux. The dashed curve is calculated in 3D and the solid curve is a 3D calculation that takes account of the geomagnetic field. The 1D calculation fits pretty well the muon spectrum at 'float' and clearly overshoots the low momentum muon flux at sea level. The 3D calculation decreases the flux below ~2 GeV/c and the account for the geomagnetic field (solid curve) makes the final correction. The shape of the muon spectrum below 5 GeV/c agrees well with data. The steeper shape at

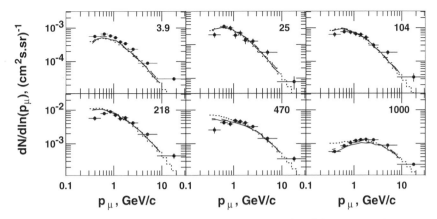

Fig. 6.10. Energy spectra of negative muons from Ref. [161] (data points) compared to predictions of 1D (histogram) and 3D calculations. See text for explanation of the curves. Note the change of scales in the top panels as the muon flux increases.

higher momenta is related to the steep cosmic ray spectrum of Ref. [125]. The normalization of the cosmic ray spectrum is also too low.

The effect of the 3D calculation is a combination of the inelastic interaction properties and the cascade geometry. Lower energy particles are produced at larger angles because the average transverse momentum is almost constant. For this reason the contribution of inclined showers to the sub-GeV/c vertical muon flux is large. At small atmospheric depths $X \ll \lambda_p$ there is a compensating effect, as inclined cosmic rays see larger column depth and interact more often above the 'float' altitude. At bigger depths there is no compensation and the low momentum muon flux decreases. There is also a corresponding increase of the muon pathlength in the atmosphere.

Pathlength is important because muons decay easier when they lose energy as shown in Fig. 6.11. Increased pathlength causes lower muon flux, especially at low momenta. Look at Fig. 6.11: at a distance of 10 km from the injection point the 2 GeV/c flux has hardly changed, the energy loss has decreased the 1 GeV/c flux by less than 2, but the 0.5 GeV muons have simply disappeared. The account for the geomagnetic field increases the pathlength because muons bend in it. This pathlength increase is the reason for the further decrease of the sub-GeV muon flux shown with solid line in Fig. 6.10.

The geomagnetic field also alters the charge ratio at sea level. Negative muons coming from the west bend in such a way that their trajectory becomes closer to the vertical. Positive muons bend in the opposite way. Because of the east–west effect the cosmic ray flux coming from western direction is higher than the one from the east and the number of μ^- arriving almost vertically to the detectors increases. This effect is stronger at low geomagnetic latitude and the BESS experiment reports a low muon charge ratio measured in Japan.

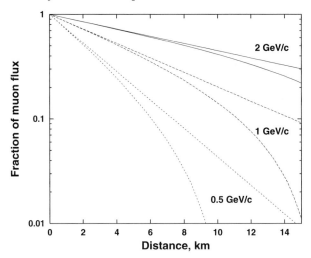

Fig. 6.11. Relation between decay probability and energy loss for muons injected vertically at altitude of 20 km with momenta of 2, 1, and 0.5 GeV/c. The straight lines show the fraction of muons with the corresponding momentum surviving as a function of the distance to the injection point. Curves are with energy loss.

At 0.6 GeV/c the charge ratio is 1.1 and slowly increases to about 1.2 at 2 GeV/c.

7 Cosmic rays underground

Primary cosmic rays almost never reach sea level. Secondary particles do. Hadrons, electrons and γ-rays interact immediately with the rock and are quickly absorbed. 10 meters of rock provide two to three times more column depth than the whole atmosphere. Only very high energy muons ($E >$ 500 GeV) can penetrate deep underground where they can be detected by underground detectors. So can neutrinos with their very small interaction cross-section.

High energy muons are produced in the decay of energetic mesons, as described in Sect. 6.2.1. The spectrum of these muons still keeps the information about the magnitude and the shape of the primary cosmic ray flux. Deep underground detectors are not subject to the time restrictions of balloon and satellite experiments. They can measure the muon flux for a very long time and provide reliable information on the primary cosmic ray spectrum at arbitrarily high energy. There are, however, other restrictions: the flux of muons declines with energy by one power more than the primary cosmic rays. At 1 TeV the muon flux decreases 1,000 times more than the cosmic rays flux. Detectors have to be bigger by the same factor. The traditional method of energy measurement does not work either – 1 TeV muons are difficult to bend, they require expensive superconducting magnets and high resolution position detectors. Calorimeters do not work either, because of the small muon interaction cross-sections.

Nature itself provides a method and we go back to the first generation of cosmic ray muon experiments. Atmospheric muons were measured by counting the particle fluxes under iron or lead absorbers of different thicknesses. We can now measure the muons under different thicknesses of rock. A single underground experiment can do this by measuring muons arriving at the detector from different zenith angles. The world data set extends the energy range as the experiments are built in deep mines and especially constructed underground laboratories. There is one more complication: in addition to a good description of the muon production spectrum we have to know very well the muon energy loss at high energy and the thickness and composition of the rock above the detector.

Underground neutrinos do not have to be of high energy. As we shall see later, low energy neutrinos penetrate easily through the whole Earth while higher energy neutrinos occasionally interact. Exactly because of their ability to penetrate through matter the detection of neutrinos presents a formidable task. As we saw in the case of solar neutrinos, one needs huge detectors. Although the interaction cross-section of GeV neutrinos is higher, the flux of atmospheric neutrinos is much lower than the flux of solar ones and the detector size requirement is still very high.

Atmospheric neutrinos were first detected [169, 170] soon after the experimental discovery of the neutrino, when its properties were not well known. For many years afterwards atmospheric neutrinos were viewed mostly as a background noise for other rare processes. In the last twenty years they have become a topic of increasing interest because the observations have suggested that neutrinos convert from one to another flavor – they oscillate. Neutrino oscillations are not a part of the standard model of particle physics, they are a new fundamental process and thus the subject of the current interest.

In this chapter we show the underground results on high energy muons and on atmospheric neutrinos. We shall then discuss neutrino oscillations as an explanation for the observed anomalies in the atmospheric and solar neutrino measurements.

7.1 High energy muons underground

For the GeV muons discussed in Chap. 6 the only essential energy loss process is ionization. The reason is that the atmospheric depth is vertically only 1,000 g/cm^2. For the muons underground, however, other processes become increasingly important. Imagine a laboratory 1 km underground. For the average density of the upper crust of the Earth of 2.65 g/cm^3 the column depth is 2.65×10^5 g/cm^2. Because the rock density for experiments at different locations varies, it is often measured in kilometers of water equivalent (1 k.w.e. = 10^5 g/cm^2). At such depths other muon energy loss processes become important. These are the radiation processes of bremsstrahlung, direct pair-production, and photoproduction. Direct pair-production is a process in which the muon emits a virtual photon and the virtual photon produces an electron–positron pair. Because of the two electromagnetic vertices the cross-section is proportional to α^2, not to α as for bremsstrahlung, and is thus smaller by a factor of 100. Photoproduction is also related to the emission of a virtual photon. In this case the photon interacts hadronically with matter and generates secondary hadrons. Since the generation of a pion requires center of mass energy at least equal to the sum of the proton and pion masses, only higher energy virtual photons are important and the cross-section at muon energies in the GeV range is very small.

Ionization energy loss depends only weakly on the muon energy and in first approximation can be considered constant at about 2 MeV per g/cm^2. Radiation processes on the other hand cause energy loss that is proportional to the muon energy, i.e. $dE_\mu/dx = -bE_\mu$. The total muon energy loss then is

$$\frac{dE_\mu}{dx} = -a - bE_\mu , \tag{7.1}$$

where $b = b_{br} + b_{pair} + b_{ph}$ is the sum of the fractional energy loss in the three radiation processes. For rock b is roughly 4×10^{-6}. The critical energy for muons is the energy ϵ at which ionization energy loss equals radiation energy loss, i.e. $\epsilon = a/b \simeq 500$ GeV. The energy loss is dominated by radiation at $E_\mu \gg \epsilon$ and by ionization at $E_\mu \ll \epsilon$. These simple formulae let us calculate the average energy E_μ of a muon with initial energy E_μ^0 after propagating through X g/cm^2 of rock

$$E_\mu = (E_\mu^0 + \epsilon) \times \exp(-bX) - \epsilon \tag{7.2}$$

and the reverse quantity

$$E_\mu^0 = (E_\mu + \epsilon) \times \exp(bX) - \epsilon . \tag{7.3}$$

The minimum energy for a muon to penetrate at depth X can be obtained from (7.3) when we set E_μ to 0, i.e.

$$E_\mu^{min} = \epsilon \left[\exp(bX) - 1 \right] . \tag{7.4}$$

At small column depths, $X \ll 1/b$ g/cm^2 muons lose energy mostly on ionization and $E_\mu^{min} \simeq aX$. The muon energy spectrum underground then reflects the surface muon spectrum with a flattening under $E_\mu \simeq aX$. At big depths, $X \gg 1/b$ g/cm^2, the spectrum has almost constant shape up to $E_\mu \simeq \epsilon$ and steepens above that energy.

The traditional measure of the muon spectrum underground is the depth–intensity relation. This is the integral flux of muons as a function of the column depth. Individual experiments measure the flux of muons under different zenith angles and then convert the fluxes to vertical direction using muon production models. In this way every experiment generates several experimental points. Figure 7.1 shows the world data set on the depth–intensity curve.

The data in Fig. 7.1 are from the compilation of M. Crouch [171], open squares, Baksan [172], open circles, LVD [173], open diamonds, MACRO [174], full diamonds, and Frejus [175], inverted full triangles. The experimental data are taken at different depths and under rock with different chemical composition. To be compared the various depths are converted to standard rock. This is an artificial material characterized by atomic mass A of 22, charge Z of 11 and density of 2.65 g/cm^3. The line that disappears from the graph just above 15 km.w.e. shows the expectations for fluxes of vertical muons. The shaded area shows the predictions for muons generated from interactions of muon neutrinos that will be discussed in Sect. 7.2.

One can qualitatively understand the shape of the depth intensity curve if one assumes a power law muon energy spectrum on the ground $E_\mu = KE_\mu^{-\alpha}$ and accepts the definition of the minimum muon energy reaching depth X as in (7.4). The depth intensity relation then becomes

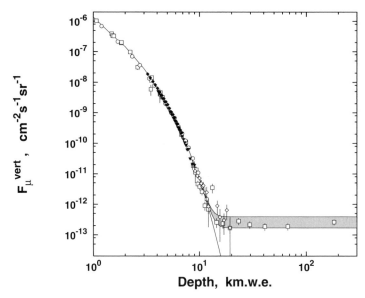

Fig. 7.1. Depth–intensity relation – the integral muon flux measured at different depths and angles and converted to vertical muon flux is compared to predictions. See text for the references to different data sets.

$$F_{\mu}^{vert} = \frac{K\,\epsilon^{-\alpha+1}}{\alpha - 1} \times \exp(-(\alpha - 1)bX) \times (1 - e^{-bX})^{-\alpha+1} \; . \qquad (7.5)$$

The first term is a constant with depth and reflects the muon energy spectrum on the ground. The third one is always larger than 1 and approaches unity at large depths. At large depths the second term determines the depth intensity curve which has the exponential shape $\exp(-X/X_0)$ with $X_0 = [b(\alpha - 1)]^{-1}$ as seen for large depths in Fig. 7.1. Since this asymptotic behavior is reached from above, the curve has a curvature at depths below ~10 km.w.e.

A better prediction for the muon fluxes underground cannot be made without more exact accounts for the muon energy loss. Neither the ionization, nor the radiation processes are energy independent. Figure 7.2 compares the relative importance of the four processes for the muon energy loss as a function of the muon energy.

Accounting for the fluctuations in the muon energy loss is also very important when one needs to calculate the number of muons that survive to certain column depth. We discuss the muon energy loss processes and the muon survival probability in detail in the following section.

Muon propagation

It is easy to calculate the depth to which a muon can penetrate in the analytic presentation of the muon energy loss of (7.1). From (7.4) we obtain the range

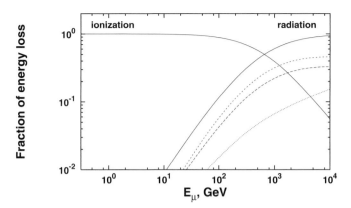

Fig. 7.2. Relative importance of different radiation processes as a function of the muon energy normalized to the total energy loss per g/cm^2. The long-dashed curve is for bremsstrahlung, the short-dashed curve for direct pair-production, and the dotted curve for photoproduction.

of a muon of energy E_μ, i.e. the underground depth that this muon will reach, as

$$R(E_\mu) = \frac{1}{b} \ln\left(\frac{E_\mu}{\epsilon} + 1\right). \tag{7.6}$$

The range of a muon of energy E_μ to be detected with certain nonzero energy can be obtained in a similar way from (7.3).

These expressions are good under the assumption of (7.1) – that the muon energy loss is continuous and that muons lose equal amount of energy in propagating through 1 g/cm^2 of matter. This is approximately true when ionization is the main energy loss process – up to 100 GeV as shown in Fig. 7.2. At higher energy the muon energy loss is not continuous - muons occasionally interact and lose a relatively large fraction of their energy. Fluctuations are inherent to the radiative processes and they replace the range $R(E_\mu)$ with a distribution of ranges. Muons that did not radiate propagate much further than $R(E_\mu)$ but the majority of muons did radiate and cannot reach this depth. One can compare the average of the range distribution $\langle R_\mu \rangle$ with R to estimate the effects of the fluctuations. Figure 7.3 shows the survival probability in standard rock for muons of energy between 1 and $10^{1.5}$ TeV.

The ratio $\langle R_\mu \rangle/R$ is energy dependent. The higher E_μ is, the more dominant are the radiation processes and the more important are the fluctuations of the energy loss. The range distribution becomes wider and $\langle R_\mu \rangle/R$ decreases. The account for the range distribution thus affects the detection rate of high energy muons and changes the slope of the derived primary cosmic ray spectrum.

The detection rate of muons at a certain depth X is

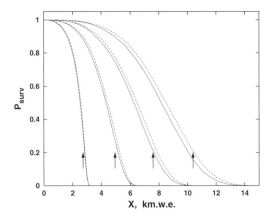

Fig. 7.3. Survival probability of muons with energy of 1., 3.16, 10., and 31.6 TeV in standard rock. The two curves for each energy indicate the uncertainties in the bremsstrahlung cross-section as stated in below. The arrows show the average depth for muon survival calculated from (7.4).

$$\int dN/dE_\mu \, P_{surv}(E_\mu, X) \, dE_\mu \,,$$

where P_{surv} is the muon survival probability at depth X. The effective muon range $R_{eff} \equiv \langle R_\mu \rangle$ is calculated as an integral over the survival probability.

$$R_{eff} \;=\; \int_0^\infty P_{surv}(E_\mu, X) \, dX \qquad (7.7)$$

and can also be defined as the range for reaching depth X with energy higher than 0 using the calculated muon energy distribution instead of (7.3).

For the steep muon energy spectrum the small probability that low energy muons survive at depth X is more than compensated for by the much higher flux of such muons. The detection rate is dominated by the tail of the survival probability P_{surv}. The account for the fluctuations will increase the rate at large X [176] and derive a steeper muon spectrum on the ground than (7.5).

Analytic treatments of the muon survival probability are only possible when the muon energy loss is approximated with simple functions that do not adequately represent these processes. More detailed calculations have to be made with the Monte Carlo method, simulating each muon interaction. The Monte Carlo procedure has to implement all cross-sections very precisely, because high energy muons suffer many interactions to reach large depths and each small error is raised to a large power. The cross-sections are given below and their implementation is discussed in Ref. [176]. The general approach to the calculation is to divide the energy loss in 'soft' and 'hard' parts, depending on the amount of relative energy loss in the interaction. Finding the dividing line between 'soft' and 'hard' is the key to a good calculation.

Muon energy loss

Muon energy loss processes are summarized in Refs. [177, 178]. These two treatments agree quite well, although there are some differences in the calculated ionization energy loss at low energy. We will give the basic energy loss formulae here and refer the reader to these two publications for the details.

The muon energy loss on ionization is calculated according to Bethe–Bloch (2.5), where $\gamma\beta$ is replaced with the same quantity for the muon and $W = (E_\mu^2 - m_\mu^2)/(m_e^2 + m_\mu^2 + 2m_e E)$. The constants for the density effect are the same as for electrons.

The bremsstrahlung cross-section for muons (2.15) is modified by the squared ratio of the electron and muon masses $(m_e/m_\mu)^2$, i.e. is smaller by a factor of more than 40,000.

The cross-section for muon bremsstrahlung becomes:

$$\frac{d\sigma_{Br}}{du} = \alpha \left(2Zr_e\frac{m_e}{m_\mu}\right)^2 \frac{1}{u}\left[\frac{4}{3}(1-u) + u^2\right]\xi(\delta) \tag{7.8}$$

where u is the fraction of the energy transferred to the photon. The screening function $\xi(\delta)$ depends on the minimum momentum transfer to the nucleus: $\delta = m_\mu^2 u/[2E(1-u)]$. The screening function has been parametrized [179] with the form:

$$\xi(\delta) = \ln\left[f_n \frac{m_\mu}{m_e} \frac{189Z^{-1/3}}{1 + (\delta/m_e)\sqrt{e}189Z^{-1/3}}\right] \tag{7.9}$$

where $e = 2.7182....$ There is some uncertainty in the value of of the nuclear form factor f_n. Reference [179] gives for $Z > 10$ and $f_n = 2/3\ Z^{-1/3}$ while Rozental [180] gives $f_n = 3/2\ Z^{-1/3}$, A measurement of the bremsstrahlung cross-section [181] fitted their data with $f_n = \exp(-0.128R_{0.5})$, where $R_{0.5} = 1.18A^{1/3} - 0.48$ fm, closer to the result of Rozental. These different form factors affect quite strongly the survival probability of muons of energy above several TeV. Figure 7.3 compares the survival probability with the form factors of Rosental (solid line) and of Ref. [179].

The pair-production by charged particles exist for all particles, but for electrons it has a much smaller cross-section because of the α^2 factor. In general form the cross-section can be written as

$$\frac{d\sigma_{pair}}{dv} = \frac{2\alpha^2 r_e^2}{3\pi}Z^2\frac{1-v}{v}\int\left[\mathcal{F}_e(r) + \frac{m_e^2}{m_\mu^2}\mathcal{F}_\mu(r)\right]dr , \tag{7.10}$$

where v is the fractional energy loss of the muon and r is $(E_{e^+} - E_{e^-})/v$. The integration limits in v are $v_{min} = 4m_e/E$ and $v_{max} = 1 - 3m_\mu\sqrt{e}Z^{1/3}/4E$. The functions $\mathcal{F}$ have the following complicated forms, as parametrized in Ref. [182]:

$$\mathcal{F}_e = \left[\ln\left(1+\frac{1}{\varsigma}\right)\left((2+r^2)(1+\zeta)+\varsigma(3+r^2)\right)+\frac{1-r^2-\zeta}{1+\varsigma}-(3+r^2)\right]\mathcal{L}_e\,,$$

$$(7.11)$$

with

$$\varsigma = \left(\frac{vm_\mu}{2m_e}\right)^2\frac{(1-r^2)}{(1-v)}\,;\quad \zeta = \frac{v^2}{2(1-v)}$$

and

$$\mathcal{L}_e = \ln\frac{189Z^{-1/3}\sqrt{(1+\varsigma)(1+\mathcal{Y}_e)}}{1+\frac{2m_e\sqrt{e}189Z^{-1/3}(1+\varsigma)(1+\mathcal{Y}_e)}{Ev(1-r^2)}} - \frac{1}{2}ln\left[1+\left(\frac{3m_e}{2m_\mu}Z^{1/3}\right)^2(1+\varsigma)(1+\mathcal{Y}_e)\right],$$

$$\mathcal{Y}_e = \frac{5-r^2+4\zeta(1+r^2)}{2(1+3\zeta)\ln\left(3+\frac{1}{\varsigma}\right)-r^2-2\zeta(2-r^2)}$$

The other $\mathcal{F}$ term is

$$\mathcal{F}_\mu = \ln(1+\varsigma)\left[(1+r^2)(1+\frac{3\zeta}{2})-\frac{1}{\varsigma}(1+2\zeta)(1+r^2)\right]\mathcal{L}_\mu$$

$$+\left[\frac{\varsigma(1-r^2-\zeta)}{1+\varsigma}-(1+2\zeta)(1-r^2)\right]\mathcal{L}_\mu$$

with

$$\mathcal{L}_\mu = \ln\left(\frac{189Z^{-2/3}\frac{2m_\mu}{3m_e}}{1+\frac{2m_e\sqrt{e}189Z^{-1/3}(1+\varsigma)(1+\mathcal{Y}_\mu)}{Ev(1-r^2)}}\right)$$

and

$$\mathcal{Y}_\mu = \frac{4+r^2+3\zeta(1+r^2)}{1+(1+r^2)(\frac{3+4\zeta}{2})\ln(3+\varsigma)-\frac{3}{2}r^2}$$

The direct pair-production cross-section for muons is significantly higher than the bremsstrahlung cross section because of the very strong decrease of the latter. The energy distribution of the virtual photon is, however, very soft and the energy losses of the two processes are of the same order.

The photoproduction cross-section for muons is given by

$$\frac{d\sigma_{ph}(E_\mu)}{dv} = \frac{A\alpha}{2\pi}\sigma_{\gamma N}(vE_\mu)\,v\times\mathcal{F}[E_\mu,v,\sigma(vE_\mu)]\,,\qquad(7.12)$$

where v is again the fractional energy loss of the muon and $\sigma_{\gamma N}$ is the interaction cross-section of real photons with nucleons. To obtain the total cross-section one integrates in v from $v_{min} = (m_\pi^2 + 2m_p m_\pi)/2m_p E_\mu$ to $v_{max} = 1 - m_\mu/E_\mu$. Bezrukov and Bugaev [183] have parametrized the the function $\mathcal{F}$ in the following way:

$$\mathcal{F} = \frac{3}{4}\mathcal{L}(x)\left[\kappa \ln\left(1 + \frac{m_1}{t}\right) - \frac{\kappa m_1^2}{m_1^2 + t} - \frac{2m_\mu^2}{t}\right]$$

$$+ \frac{1}{4}\left[\kappa \ln\left(1 + \frac{m_2}{t}\right) - \frac{2m_\mu^2}{t}\right]$$

$$+ \frac{m_\mu^2}{2t}\left[\frac{3}{4}\mathcal{L}(x)\frac{m_1^2}{m_1^2 + t} + \frac{m_2^2}{4t}\ln\left(1 + \frac{t}{m_2^2}\right)\right] .$$

The variable x is a function of the photoproduction cross-section and the medium − $x = 2.82 \times 10^{-3} A^{1/3} \sigma_{\gamma N}$ *where* $\sigma_{\gamma N}$ *is measured in* μb. *The function*

$$\mathcal{L}(x) = \frac{3}{x^3}\left(\frac{x^2}{2} + \frac{1 + x}{e^x} - 1\right)$$

and the variables κ *and* t *are functions of* v

$$\kappa = 1 - \frac{2}{v} + \frac{2}{v^2}; \quad t = \frac{m_\mu^2 v^2}{1 - v} .$$

The constants m_1^2 *and* m_2^2 *are respectively 0.54 and 1.8 GeV^2. Bezrukov and Bugaev have used used the following energy dependence for the photo production cross-section*

$$\sigma_{\gamma N}(E) = 114.3 + 1.647 \times \ln^2(2.13 \times 10^{-2}(E/\text{GeV}))\ \mu b ,$$

but other parametrizations for the cross-section can also be plugged into (7.12).

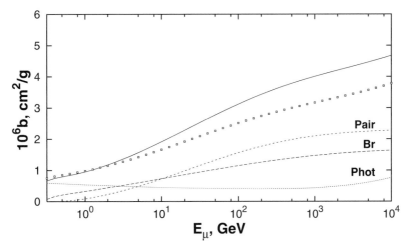

Fig. 7.4. Relative energy loss on radiation in standard rock. Solid line shows the sum of the three processes. Circles show the *b* value for clean ice.

Figure 7.4 gives the relative energy loss for the three processes in standard rock. The solid line is the sum of the three processes and the circles show the total relative energy loss for clean ice with density 0.93 g/cm³.

The process cross-sections given above can also be used for calculations of the electromagnetic cross-sections for the τ lepton. The bremsstrahlung cross-section is very strongly suppressed because of the very high mass of the τ lepton and is of no practical importance. Pair-production cross-section (and energy loss) is also suppressed but only with about the ratio of the τ and μ masses. At high energy the τ energy loss is dominated by photoproduction [184].

7.2 Atmospheric neutrinos

Atmospheric neutrinos are produced in the decay chain of secondary mesons in atmospheric cascades. GeV neutrinos come predominantly from the charged pion decay chain: $\pi^+ \to \mu^+ + \nu_\mu$; $\mu^+ \to e^+ + \nu_e + \bar{\nu}_\mu$ and the corresponding one for π^-. The contribution of kaons grows with energy. Some of the most general features of the sub-GeV atmospheric neutrinos can be deduced from their parent decay chain. The decay length for a 1 GeV muon is about 6 km, while the characteristic length for vertical atmospheric cascades is 20 km. Thus not only all mesons, but practically all muons in this energy range decay and the energy spectrum of the resulting neutrinos follows the spectrum of the cosmic ray primaries. In addition, the charged pion decay chain generates a muon neutrino/antineutrino pair $\nu_\mu\bar{\nu}_\mu$ for every electron neutrino ν_e or antineutrino $\bar{\nu}_e$. These three neutrinos take each about 1/4 of the pion energy. The ration of muon to electron neutrinos then is $\frac{\nu_\mu + \bar{\nu}_\mu}{\nu_e + \bar{\nu}_e} = 2$. This rule of thumb is confirmed by precise numerical calculations and is modified only slightly by the contribution of other meson decays. The ratio of muon to electron neutrinos is known much better than the absolute flux of neutrinos. With increasing energy, the secondary pions are more and more likely to interact because of the relativistic increase of their decay length $\gamma c\tau$. Above ϵ_π (= 115 GeV in vertical cascades) kaon decays become a much more important source of neutrinos and dominate the neutrino production by 3:1 at asymptotically high energy. The interaction/decay competition of the secondary mesons steepens the energy spectrum of the neutrinos originating in meson decays by one power of the energy, i.e. if the primary cosmic ray spectrum is $N(E) = A \times E^{-\alpha}$ the spectrum of the meson decay neutrinos is proportional to $E_\nu^{-(\alpha+1)}$. Gaisser [6] calculates the neutrino flux from pion and kaon decay using the decay probability for the mesons and the decay kinematics in the same ways as in (6.11). The result for neutrinos can be expressed as

$$\frac{dN_\nu}{dE_\nu} \simeq 0.0096 E_\nu^{-2.7} \left[\frac{1}{1 + \frac{3.7 E_\nu \cos\theta}{115 \text{ GeV}}} + \frac{0.38}{1 + \frac{1.7 E_\nu \cos\theta}{850 \text{ GeV}}} \right] . \qquad (7.13)$$

Because of the difference in the decay kinematics the contribution of kaons is higher than for muons. Combined with the increasing importance of the second term of (7.13) at high energy this makes kaons asymptotically the major contributor to the muon neutrino flux.

Another power of the energy is added for neutrinos originating in muon decays, which steepens their spectrum to $E_\nu^{-(\alpha+2)}$. For that reason the spectrum of electron neutrinos, that come mostly from muon decays, is steeper than that of muon neutrinos, which are also generated directly in meson decays. At very high energies, when the atmosphere is too thin for muons to decay, neutral kaon decays are the only source of electron neutrinos through the the decay branch $K_L^0 \to \pi^\pm e^\mp \bar{\nu}_e(\nu_e)$.

The ratio of muon to electron neutrinos at high energy depends strongly on the π/K ratio in multiparticle production interactions, which is known only with an accuracy of about 30%. This uncertainty also affects the absolute normalization of the atmospheric neutrino flux. The major factor in the absolute normalization of the atmospheric neutrinos is however the magnitude of the galactic cosmic ray flux interacting in the atmosphere.

Fig. 7.5. Isotropy of the atmospheric neutrino flux underground in the absence of geomagnetic cutoffs. The detector is much closer to the atmosphere above it but sees a much larger fraction of the atmosphere below it. The flux of downward-going and upward-going neutrinos is the same.

The Earth is totally transparent to neutrinos until their mean interaction length $\lambda_\nu = (N_A \sigma_{\nu N})^{-1}$ reaches 10^{10} g/cm^2 – the column density through the center of the Earth. This happens for neutrino energies about 10^5 GeV and does not affect the detection of atmospheric neutrinos because by that energy the atmospheric neutrino flux has decreased by many orders of magnitude. A neutrino detector thus can detect neutrinos in 4π solid angle. Neutrinos coming from below are produced in the atmosphere of the opposite hemisphere. A major feature of the atmospheric neutrino flux is its isotropy. In the absence of geomagnetic cutoffs the neutrino flux underground would be the same from above and from below as shown in Fig. 7.5. The r^{-2} effect from the distance to the production layer of atmosphere is compensated exactly by the larger fraction of the atmosphere seen in any solid angle.

Local geomagnetic effects influence mostly the 'downward-going' neutrinos, while for 'upward-going' neutrinos one has to integrate the field values over the whole area of the atmosphere. The flux of upward-going neutrinos at different locations thus varies much less than the flux of downward-going neutrinos. For high energy neutrinos coming from primary cosmic rays that are not affected by the geomagnetic field the isotropy is real and the neutrino flux is symmetric for negative and positive cosines of θ. The convention is that $\cos\theta = -1$ corresponds to vertical upward-going neutrinos and $\cos\theta = 1$ defines vertical downward-going neutrinos.

Neutrinos are detected mostly through their charged current interaction $\nu_i + N \Rightarrow \ell_i + X$ where the lepton number i is transferred to the corresponding lepton. What experiments see is an electron or muon track originating inside the detector. Assuming that the atmospheric neutrino flux is of the order of 1 cm^{-2} s^{-1} and the neutrino cross-section is of the order of 10^{-38} cm^2 one can estimate the necessary detector size. The event rate is

$$F_\nu \times \sigma_\nu \times N_A$$
$$= 1 \times 10^{-38} \times 6.10^{23} = 6.10^{-15} \text{ per gram per second}$$
$$= 1.8 \times 10^{-7} \text{ per gram per year}$$
$$= 180 \text{ per 1,000 tons (Kt) per year}$$

Detectors have to have sizes of thousands of tons and to operate for years to collect reasonable atmospheric neutrino statistics. They have to be shielded from downward-going atmospheric muons, which places them deep underground. A large fraction of the total volume of the detector (close to the walls) is not used in neutrino detection because it is often polluted by natural radioactivity. It is used as anti-coincidence counter to reject entering tracks. The inner part of the detector is its fiducial volume.

There are two types of detectors:

–   traditional calorimeters, where layers of particle detectors are interspaced by layers of target material, e.g. iron.
–   water-Cherenkov detectors: photo tubes on the walls of a large tank filled with water detect Cherenkov radiation from the charged particles created in the neutrino–nucleon interaction.

Each detector type has its advantages: Cherenkov detectors are cheaper and can reach total volumes of tens of Kt. Particle calorimeters can measure the charged particle tracks more precisely.

Lepton identification (electron or muon) is based on the different modes of energy loss for these particles. Electrons suffer bremsstrahlung with radiation length in water of 36 cm, much less than the detector dimensions. Even low energy electrons thus initiate electromagnetic (electron–photon) cascades. Muons of energy less than 100 GeV lose energy mostly through ionization ($\sim$2 MeV/cm). In traditional calorimeters muons have a straight track with possibly a little bit of a kink before stopping. Electrons, on the other hand, show a bunch of short tracks left from individual cascade electrons. In water-Cherenkov detectors this translates into well defined Cherenkov rings for the muons and 'filled in' circles for the electrons.

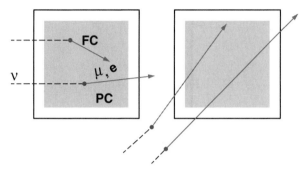

Fig. 7.6. The shaded area represents the fiducial volume of the detector. The left-hand panel shows the geometry of the fully (FC) and partially (PC) contained events. The right-hand figure shows through-going (below) and stopping neutrino-induced muons.

Experimentally neutrino events are classified generally in three groups, which are shown schematically in Fig. 7.6:

- Fully contained (FC) events for which the interaction vertex and all tracks are fully contained in the fiducial volume of the detector. Both the direction and the energy of the daughter leptons can be measured.
- Partially contained (PC) events where only the interaction vertex is contained. Only the direction can be measured and a minimum energy of the lepton can be assigned. PC events are almost exclusively muons.
- Upward-going muons, which are generated by neutrino interactions in the rock surrounding the detector. Upward-going muons can be through-going (exit the detector) or stopping inside the detector. The direction is known but the energy of the muon at production is not.

FC and PC events are observed in the full 4π solid angle, while upward-going muons may only come from below (2π acceptance). Downward-going neutrino-produced muons are swamped by the atmospheric muons. Neutrino-induced upward-going muons are a nice way to detect neutrinos because the effective volume of the detector is increased by the trajectory of the muon generated in the neutrino interaction, i.e. by its range R_μ in the rock.

Each type of experimental neutrino event is generated by an energy range of atmospheric neutrinos. The exact distribution depends on the experimental details and on the location of the experiment. Figure 7.7 shows the energy distribution of muon neutrinos responsible for the different event types in the Super-Kamiokande detector. FC events are generated by neutrinos of average energy about 1 GeV. The neutrino energy response for PC events and

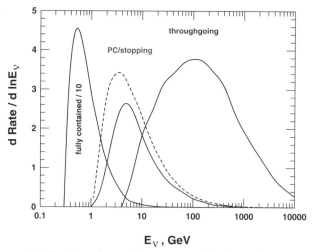

Fig. 7.7. Energy distribution of neutrinos responsible for different types of experimental events. The curves are labeled and the one for PC events is plotted with a dashed line. Note that the curve for fully contained events is scaled down by a factor of 10. The units for the rate are $(Kt.yr)^{-1}$.

stopping upward going-muons peaks at 3–5 GeV and through-going upward muons are produced by neutrinos in a wide energy range that peaks at about 100 GeV. Other detectors have slightly different energy responses, but the energy ranges are roughly the same.

The rate of contained neutrino events (FC or CC) is calculated by an integration over the neutrino spectrum and the probability that a neutrino of energy E_ν can generate a lepton of energy E_l. For leptons of energy between E_1 and E_2 the rate is

$$\text{Rate} = N_A \int_{E_1}^{E_2} dE_l \int_{E_l}^{E_\nu - m_l} \frac{dN}{dE_\nu} \left(\frac{d\sigma(E_\nu)}{dy} \right)_{y=1-E_l/E_\nu} dE_\nu \, , \quad (7.14)$$

where σ is the neutrino interaction cross-section, m_l is the mass of the lepton, and $y = 1 - E_l/E_\nu$ is the Bjorken variable. The rate for upward-going muons includes the muon range and is discussed in Sect. 7.2.1.

The neutrino interaction cross-sections in the whole energy range of atmospheric neutrinos are relatively well known. Below 1 GeV the cross-section is dominated by the quasi-elastic scattering [188] $\nu(\bar\nu) + n(p) \to l^\mp + p(n)$ and at higher energy the most important process is the deep inelastic scattering (DIS) [189] $\nu + N \to l + N + X$, where X could be one or more hadrons. The charge current process (CC), where the lepton number of the neutrino is transferred to a lepton, is the most important one for atmospheric neutrino detection. After the threshold energy range the DIS cross-section is proportional to the neutrino energy up to energy of about 1 TeV. Figure 7.8 shows the muon neutrino cross-section as a function of the neutrino energy. In the

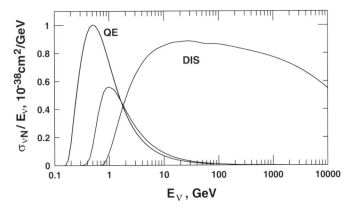

Fig. 7.8. Muon neutrino cross-section as a function of the neutrino energy. QE and DIS cross-sections are labeled. In between is the 1π cross-section.

GeV energy range the antineutrino cross-section is about 1/3 of the neutrino cross section. The cross-sections are strongly suppressed at threshold, i.e. at energies slightly above the mass of the secondary lepton.

Experimentally the neutrino cross-sections are better known at higher energy. The low energy cross-sections are so small that the collection of experimental statistics becomes very difficult. The intermediate energy (1π) cross-section [190], where the inelastic neutrino interaction is accompanied by the production of a single pion, has been studied even less.

7.2.1 Upward-going muons

Neutrino detection through upward-going muons was developed long ago, because it increases significantly the effective volume of the neutrino detector. Imagine a detector with surface area A, height h and volume $V = h \times A$ as shown in Fig. 7.9. For fully contained events the fiducial volume will be less than one half V. For upward-going muons the effective volume of the detector will be increased R_{eff}/h times which could be a very significant number. For muons of energy 1 TeV $R_{eff} = 2.7$ km.w.e. For a water-Cherenkov detector of height 27 m (taller than a eight-story house) the increase is 100 times. The first detections of atmospheric neutrinos [169, 170] were with neutrino induced muons.

The rate of neutrino-induced upward-going muons is obtained similarly to (7.14) with the evaluation of $d\sigma_\nu/dy$ replaced by the probability $P_{\nu\mu}(E_\nu, E_\mu)$ that a neutrino of energy E_ν will produce a muon of energy above E_μ at the detector [185, 186, 187], i.e.

$$\text{Rate}_{up}(> E_\mu) = \int_{E_\mu}^{\infty} dE_\mu \int_{E_\mu}^{E_\nu - m_\mu} \frac{dN}{dE_\nu} P_{\nu\mu}(E_\nu, E_\mu) \, dE_\mu \ . \qquad (7.15)$$

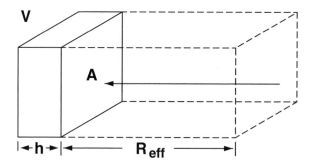

Fig. 7.9. Schematic presentation of the idea for detection of upward-going neutrino-induced muons.

The probability $P_{\nu\mu}(E_\nu, E_\mu)$ is given by an integral over the differential neutrino cross-section $d\sigma/dy$ evaluated at $y = 1 - E'_\mu/E_\nu$ folded with the effective range $R_{eff}(E'_\mu, E_\mu)$ of a muon of energy E'_μ to reach the detector with energy E_μ.

$$P_{\nu\mu}(E_\nu, E_\mu) = N_A \int_0^{E_\nu} dE'_\mu \left(\frac{d\sigma(E_\nu)}{dy}\right)_{y=1-E'_\mu/E_\nu} R_{eff}(E'_\mu, E_\mu) \quad (7.16)$$

Figure 7.10 gives the probability $P_{\nu\mu}$ for muon energy of 1 GeV and of 1 TeV. The expectations for the vertical (lower line) and horizontal (upper line) flux of neutrino-induced muons of energy above 1 GeV are plotted in Fig. 7.1.

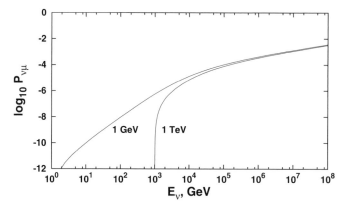

Fig. 7.10. Probability that a neutrino of energy E_ν will produce a muon of energy 1 GeV and 1 TeV at the detector in standard rock.

An experimental requirement that the muons reach the detector with energy above E_μ does not decrease the rate significantly for modest muon energies. The reason is that both the neutrino cross-section and the muon range are proportional to E, up to 1 TeV and 500 GeV respectively. Even for atmospheric neutrinos with a power law spectrum $E_\nu^{-3.7}$ the detector rate dependence on energy is only $E^{-1.7}$. This fact has made neutrino-induced muons very popular. The disadvantage is that one has to deduce not only the energy of the atmospheric neutrino, but also the energy of the detected muon at production. The interpretation of the results becomes quite complicated and is possible only on the basis of good experimental statistics.

There is only one way to judge the energy spectrum of the neutrinos that generate upward-going muons - the ratio between the rates of through-going and stopping muons. The tracks of through-going muons cross two walls of the detector on its entry and exit. Stopping muons do not have enough energy to penetrate the whole detector. There are thus two energy ranges: for stopping muons from E_{min} to E_{thru}, where E_{min} is the minimum energy for analysis of the muon track and E_{thru} is the energy needed for penetration through the whole detector. Through-going muons have energy above E_{thru}. In a detector these ranges depend on the angle of the muon track in detector coordinates. The rate for stopping muons is calculated by replacing the ∞ by E_{thru} in the integration limit of (7.15). The neutrino energy response to stopping and through going muons is shown in Fig. 7.7.

Neutrino cross-sections

This section gives the main formulae that the reader can use to calculate the quasi-elastic (Q) and the CC deep inelastic scattering (DIS) cross sections. The cross-sections are given following Ref. [188] for the QE process and Ref. [189] for DIS. Both assume an isoscalar nucleon with mass $M = (m_p + m_n)/2$ as target.

The differential quasi-elastic cross-section is given as a function of the square of the four momentum transfer Q_2 as:

$$\frac{d\sigma}{dQ^2} = \frac{G_F^2 \cos^2 \theta_c M^2}{8\pi} \frac{1}{E_\nu^2} \times \tag{7.17}$$

$$\times \left[\mathcal{A} \mp \frac{vF_0}{(1+Q^2)^2} (F_1 + F_2)\mathcal{S} + \tfrac{1}{4} \left(F_A^2 + F_1^2 + \tfrac{v}{4}F_2^2 \right) \mathcal{S}^2 \right]$$

The $-$ sign is for ν interactions and the $+$ sign is for antineutrinos, $v = Q^2/M^2$, r is the square of the ratio of the lepton and nucleon masses ($r = m_l^2/M^2$) and the term $\mathcal{A}$ is

$$\mathcal{A} = \tfrac{v+r}{4} \left[(4+v)F_A^2 - (4-v)F_1^2 + vF_2^2(1-v/4) + 4vF_1F_2 \right] \tag{7.18}$$
$$- \tfrac{r(v+r)}{4} \left[(F_1+F_2)^2 + F_A^2 + 4F_AF_p - v F_p^2 \right] .$$

The term $\mathcal{S}$ is given by

$$\mathcal{S} = \frac{4E_\nu M - Q^2 - m_l^2}{M^2} . \tag{7.19}$$

The functions F are defined as

$$F_1 = \frac{4 + v\,(1 + C_s)}{(4 + v)(1 + Q^2/Q_0^2)^2} \qquad F_2 = \frac{4 C_s}{(4 + v)(1 + Q^2/Q_0^2)^2}$$

$$F_A = \frac{1 + F_0}{(1 + Q^2/Q_0^2)^2} \qquad F_p = \frac{M^2 F_A}{m_\pi^2 + Q^2} \; .$$

The constants have the following values: $F_0 = -1.2546$; $C_s = 3.71$; $Q_0 = 0.71$ and θ_C is the Cabbibo angle. The differential cross-section as a function of other variables can be obtained from (7.18) by substitution using the Bjorken variables

$$y = 1 - E_l/E_\nu; \qquad x = \frac{|Q^2|}{2M E_\nu y} \qquad (7.20)$$

Equation (7.18) gives the differential cross-section for interaction on in-dividual isoscalar nucleon. Corrections are needed for the calculation of the differential and total cross-sections on nuclear target at low energy. The cor-rections are guided by the quantity $Q_s = (E_\nu^2 - 2E_\nu \cos\theta + P_l^2)^{1/2}$, where $\cos\theta$ is the angle between the neutrino and the lepton in the Lab system. Q^2 is replaced by $[Q_s^2 - (E_\nu - E_l)^2]$ and the differential cross-section is scaled down with the quantity

$$\frac{3}{2}\frac{Q_s}{p_F^3}\left(\frac{2M^2 E_\nu p_l}{E_l}\right).$$

There is also a logarithmic function of the lepton and Fermi (p_F momentum of the nucleus of order 1).

The deep inelastic scattering cross-section is given in terms of the Bjorken variables and the parton structure functions. The double differential cross-sections for CC interactions [191] $\nu_l + N \to l + X$ are given by

$$\frac{d^2\sigma}{dx\,dy} = \frac{G_F^2 M E_\nu}{\pi}\,\frac{1}{(1 + Q^2/m_W^2)^2}\,x\,\left(Q_1 + (1 - y)^2 Q_2\right) , \qquad (7.21)$$

with

$$Q_1 = u_v + d_v + u_s + d_s + 2s_s + 2b_s$$

for neutrino scattering and

$$Q_1 = u_s + d_s + 2c_s + 2t_s$$

for antineutrino scattering.

$$Q_2 = u_s + d_s + 2c_s + 2b_s$$

for neutrinos and

$$Q_2 = u_v + d_v + u_s + d_s + 2c_s + 2t_s$$

for antineutrinos. u, d, c, s, t and b are the the structure functions of different quark flavors and the subscripts v and s identify valence and sea quarks. The lower integration limit in x for the calculation of the total cross-section is $x_{min} = Q_0^2/(2MyE_\nu)$ and the upper limit depends on neutrino energy. The lower limit in y is calculated the same way: $y_{min} = Q_0^2/(2ME_\nu)$ and Q_0 is the minimum momentum transfer. Q_0^2 is of the same order as in the QE process.

The contribution of different quarks to the cross-section is driven by their fraction in the quark sea. Valence quarks, having flat structure functions, dominate the cross-section at low energy. While they do, the neutrino cross-section is higher than the antineutrino one because of the $(1 - y)^2$ factor. At high energy the contribution from the valence quarks decreases and the neutrino and antineutrino cross-sections become equal. b and t quarks have small contributions even at very high energy. See the original article [189] for the contribution of different flavors.

The W propagator $(1 + Q^2/m_W^2)^{-2}$ determines the energy dependence of the total cross-section. At neutrino energies much smaller than m_W the cross-section increases linearly with E_ν and at $E_\nu \gg m_W$ the increase is logarithmic. The exact shape of the neutrino cross-section at high energy depends on the parametrization of the quark structure functions. The cross-section shown in Fig. 7.8 is calculated with the GRV structure functions [192]. The neutral current cross-section $(\nu_l + N \to \nu_l + X)$ is calculated in a similar way and the W propagator is replaced by the Z_0 propagator where m_{Z_0} plays the same role.

7.2.2 Flux calculations

Early calculations of the atmospheric neutrino flux used the relations between the production of muons and neutrinos to estimate the energy spectrum of atmospheric neutrinos. Since the decay kinematics of mesons and muons is well known, one can in principle use the data on atmospheric muon, parametrize the atmospheric flux of mesons, and apply the kinematic relations to obtain the neutrino fluxes. The best known calculation of this type was made by Volkova [193] and the most recent one was performed by Perkins [194] to check the consistency of the particle physics parameters used in calculations of atmospheric muons and neutrinos. The necessary decay kinematics is discussed in detail in the book of Gaisser [6].

Contemporary calculations usually employ the Monte Carlo technique. The knowledge of the primary cosmic ray spectrum is combined with hadronic interaction models in the atmospheric environment. The difficulty in this approach is that the observed neutrinos are produced by the whole atmosphere and one has to take into account the geomagnetic cutoffs and the atmospheric structure all over the Earth.

The two calculations that have been used for the analysis of most experimental results [195, 196] are performed in one dimension. They use a model of the average atmosphere, such as given by (6.3), and calculate the neutrino

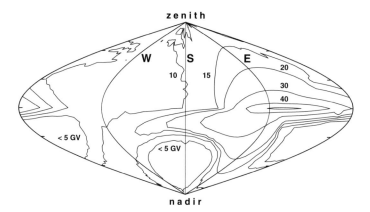

Fig. 7.11. Contour map of the geomagnetic cutoff at Kamioka. Contours connect the directions with equal rigidity cutoffs. East, south and west are indicated and north is split at the edge of the graph.

yields by protons and neutrons of fixed energy arriving at the top of the atmosphere under different zenith angles. The yield tables are then folded with the precalculated geomagnetic cutoffs for different zenith angles to obtain the spectrum of neutrinos arriving at a specific location. Figure 7.11 shows an example for the location of Kamioka, Japan.

The situation of Kamioka is typical for a location at low geomagnetic latitude. The cutoffs are high above the horizon and relatively low below the horizon. The direction of the highest cutoffs are near to the eastern horizon. One can see the direction of the magnetic poles around which the cutoffs are below 5 GeV.

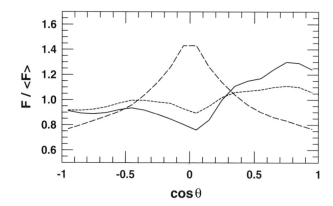

Fig. 7.12. Normalized flux of neutrinos of energy 0.5 (solid line), 1 (dotted line) and 10 GeV (dashed line) at the location of Kamioka as a function of the zenith angle.

With a given choice of cosmic ray flux and hadronic interaction model the geomagnetic field determines both the energy spectrum and the angular distribution of the atmospheric neutrinos. One would expect that at Kamioka there would be fewer low energy neutrinos than at a location with low cutoff because a smaller number of low energy cosmic rays will reach the atmosphere and interact. The lack of low energy neutrinos will be most prominent in directions with high cutoff. Figure 7.12 compares the angular distribution of neutrinos of energy 0.5, 1, and 10 GeV as a function of the zenith angle at Kamioka calculated with the described 1D approach and integrated over the azimuth angle.

The lowest energy neutrinos show the large up/down ratio, i.e. more neutrinos are produced in the hemisphere below the horizon than are produced above the horizon. There is a dip in horizontal direction where the cutoffs have the highest values. Up and down fluxes at 1 GeV are more symmetric, although they exhibit the same structure. The geomagnetic cutoffs at 10 GeV have no importance and one can see the perfectly symmetric picture with fluxes peaking at the horizon.

For the same reasons one expects more low energy neutrinos at high geomagnetic latitudes. If the fluxes of upward-going neutrinos at the two locations are about equal, at high geomagnetic latitude the flux of down-going neutrinos will be higher than at low latitudes. Figure 7.13 shows a comparison of the angle-averaged muon neutrino and antineutrino fluxes predicted for locations in Canada and for Kamioka.

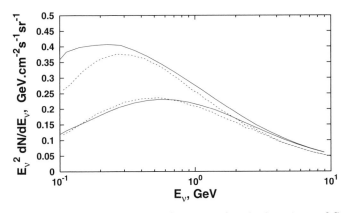

Fig. 7.13. Comparison of the fluxes of $\nu_\mu + \bar{\nu}_\mu$ for the locations of Sudbury and Kamioka. Solid lines are from Ref. [166] and the dashed lines from Ref. [197].

To give a more realistic impression of the expectation in event rate the quantity plotted is $E_\nu^2(dN\nu/dE_\nu + \frac{1}{3}dN_{\bar{\nu}}/dE_\nu)$. The $\frac{1}{3}$ factor accounts roughly for the difference between the neutrino and antineutrino cross-sections. One of the E_ν powers accounts, also very roughly, for the energy dependence of the cross-sections.

The excess of neutrinos below ~3 GeV is entirely due to the high geomagnetic latitudes at the northern site. If one were to plot separately the fluxes of down-going neutrinos the difference would double. Note also that the predictions of the two calculations for Kamioka agree quite well, but there is a significant difference for the high geomagnetic latitude site. This is due to the choice of hadronic interaction model. The models used by the two groups are closer to each other at energies above 20 GeV, where there are accelerator data. They are very different in the resonance region, close to the pion production threshold. FLUKA, the model of Ref. [197] has a very high nucleon inelasticity and correspondingly low meson production. With the same primary flux the low energy neutrino flux is not as enhanced as it is in Ref. [166]. This generates the biggest uncertainty in the neutrino flux predictions.

The situation at higher neutrino energy is somewhat better but still not perfect. Figure 7.14 compares the neutrino fluxes calculated by the three leading groups.

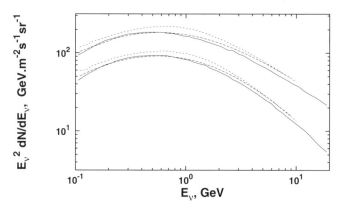

Fig. 7.14. Comparison of the angle averaged fluxes of muon neutrinos $F_\nu + F_{\bar{\nu}}/3$ for the location of Kamioka. The solid lines are from Ref. [197], the dotted lines from Ref. [198] and the dashed lines from Ref. [166] .

The comparison does not look too bad, but there are noticeable differences. The flux of Ref. [197] is lower by about 15% at an energy of 10 GeV and this difference remains constant at higher energy. These two calculations use the same cosmic ray flux model and the difference is related to the hadronic models used by the two groups. In the GeV energy range the calculation of Honda et al. [198] is significantly higher than the other two. The primary flux used by this group can explain at least a part of that difference. Figure 7.14 shows only three of many current atmospheric neutrino flux predictions.

Although some of them agree with the range presented in Fig. 7.14, there are others that give neutrino fluxes smaller by almost a factor of 2.

It is not absolutely fair to make the comparison of Fig. 7.14 because two of the calculations are made in 3D geometry and the third one [166] in 1D. One-dimensional calculations assume that all secondaries move in the same direction as the primary cosmic ray. In the decay routines only decay particles that are forward in the Lab system are retained, but this is a small correction. The correct approach is to account for the transverse momentum of the secondary particles (0.3–0.5 GeV/c on the average for different species) and for the bending of muons in the geomagnetic field. When the secondary particle is energetic ($p_\parallel \gg p_\perp$) the deflection is small and for E_ν greater than about 3 GeV the two techniques give the same result. Lipari [199] have shown analytically that for secondary particles emitted at large angles one has to account for the special geometry of the neutrino production. Neutrinos are produced in the spherical layer of the atmosphere and the detection is at a point inside that sphere. Only in the center of the sphere is the 1D approximation exactly correct. Close to the surface of the sphere the angular distribution of the secondaries causes an enhancement of the flux near the horizon. Using realistic $p_\perp$ distributions this enhancement is noticeable at $E_\nu < 3$ GeV and grows at lower neutrino energy. Reference [197] calculates the flux in 3D and this can explain the agreement with [166] at low energy in spite of the lower pion production hadronic model. Muon bending in the geomagnetic field has only second-order effect as some muons decay higher in the atmosphere and with larger energy loss than in 1D.

The enhancement also changes the angular distribution – there is no dip at horizontal direction at Kamioka as shown in Fig. 7.12, rather a peak. In practical terms this is not very important because the secondary leptons at low neutrino energy have a very wide angular distribution, which makes the dip and the peak indistinguishable.

7.2.3 Experimental data

There have been many measurements of the atmospheric neutrinos since the first detections in the 1960s. The experimental efforts were intensified in the 1980s when several experiments built to observe proton decay detected an anomaly in the flux of atmospheric neutrinos. Since both the proton decay events and the atmospheric neutrino events peak at about 1 GeV, atmospheric neutrinos are the highest background for proton decay. Initially this was the only reason for the investigation.

The data of the Kamiokande [200] and the IMB [201] detector showed that the ratio of muon to electron neutrinos is far from the canonical factor of 2, which cannot be wrong by much. New experiments were built and started operation in the 1990s. The most powerful one is the Super-Kamiokande detector, briefly described in Chap. 3. The exposure of this experiment to atmospheric neutrinos is by now over 75 Kt.yr, exceeding any other experiment by a factor of 10. We will concentrate on the results of this experiment and refer to others only to support (or contradict) the Super-K findings. Kajita & Totsuka reviewed the experimental techniques and results from studies of atmospheric neutrinos in Ref. [208]. Figure 7.15 shows the energy spectrum of fully contained neutrino events detected by Super-K [202].

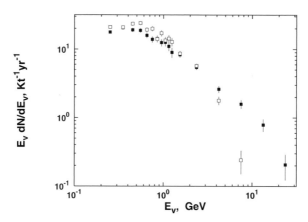

Fig. 7.15. Energy spectrum of muon-like (open symbols) and e-like (closed symbols) events detected by Super-K. E_v is the visible energy of the detected leptons.

The spectrum of Fig. 7.15 cannot be directly compared with the predictions for the neutrino fluxes. To start with, only a fraction of the neutrino energy is transferred to the leptons. This fraction varies on an event by event basis and is different for QE, 1π and DIS interactions. In addition there are detector biases that are different for muons and electrons. Muons are heavier and respectively have higher threshold for emission of Cherenkov light. Electrons lose their energy faster and are contained more easily in the detector as FC events. High energy muons leave the detector, become PC events and their energy cannot be determined. Finally there is some ambiguity in the determination the type of the lepton in a water-Cherenkov detector. Remember that this is done by the shape of the signal which is on the average different – muons produce a cleaner circle on the wall of the detector than electrons.

For all these reasons the comparison with predictions has to be done after very careful Monte Carlo simulation of the detector response. The Monte Carlo code is fed with the predicted neutrino flux, neutrino cross sections and the properties of the detector, and generates experimental-like events that are compared to the real detected events. The same Monte Carlo generator is used to study the event type and energy determination errors when events of known type and energy are scanned the same way as the experimental ones.

Figure 7.15 shows the neutrino anomaly very well. At E_ν around and below 1 GeV we expect to see twice as many muon as electron neutrino events. The observed difference is only about 25–30%. This is fully consistent with the earlier results that provoked interest in the atmospheric neutrinos together with the solar neutrino puzzle. These results were expressed in the form of the double ratio between the detected and predicted ratio of muons and electrons. The predicted ratio comes from the detector Monte Carlo.

$$R2 = \frac{R(\mu/e)_{\text{observed}}}{R(\mu/e)_{\text{predicted}}} \sim \frac{1.25}{2.00} \sim 0.6 \qquad (7.22)$$

for the GeV flux of Fig. 7.15. The ratio cannot be deduced by eye from the figure at higher energy as the experimental efficiency for muons strongly decreases. $R2$ is calculated by a comparison with the Monte Carlo predictions.

Table 7.1 compares the results from several different experiments. Note that the Kamiokande and Super-K results are separated in sub-GeV and multi-GeV samples. The division is at $E_v = 1.33$ GeV. The results for contained events in Table 7.1 are only for 'single-ring' or 'single-track' events, i.e. when only one lepton above the threshold is generated in the neutrino interactions. The analysis of 'multi-ring' events is much more difficult.

Table 7.1. Experimental data on the double ratio $R2$ in different experiments. The first error is statistical and the second one is systematic. For upward-going muons the values are for the ratio between observed and predicted rates.

Experiment	Kt.yr	double ratio
IMB	7.7	$0.54 \pm 0.05 \pm 0.11$
Frejus	2.0	$1.00 \pm 0.15 \pm 0.08$
Nusex	0.74	0.96 ± 0.30
Soudan2	5.1	$0.68 \ \pm 0.11 \pm 0.05$
sub-GeV		
Kamiokande	7.7	$0.60 \pm 0.06 \pm 0.05$
Super-K	75.5	$0.66 \pm 0.02 \pm 0.05$
multi-GeV		
Kamiokande	8.2	$0.57 \pm 0.08 \pm 0.07$
Super-K	75.5	$0.68 \pm 0.03 \pm 0.08$
Super-K (PC)	75.5	$0.71 \pm 0.05 \pm 0.08$
upward muons		
Kamiokande	14.4	$0.75 \pm 0.06 \pm 0.05$
Super-K	75.0	$0.92 \pm 0.03 \pm 0.05$
MACRO	—	$0.74 \pm 0.04 \pm 0.05$

The numbers in Table 7.1 do not include the theoretical errors due to different atmospheric flux predictions. Data for neutrino-induced upward-going muons are taken at different muon threshold energies. These are actually single ratios between the detected flux and the predictions. All GeV neutrino double ratios are consistent within 2σ, even those of the Frejus detector. The contemporary data are consistent on a much better level. Quite important is

the agreement of the latest Super-K [203] and Soudan2 [206] data because
they are taken with entirely different instruments. Soudan2 is a traditional
iron calorimeter and its double ratio, which is in agreement with Super-K,
relieves earlier fears that the neutrino cross-section in water is not well known.

There is also a general trend, although within the errors, that the double
ratio grows with the energy. The central values of Super-K for sub-GeV,
multi-GeV, PC and upward-going muon events are respectively 0.66, 0.68,
0.71 and 0.92. The comparison between the ratios for upward-going muons at
Super-K and MACRO [207] may reflect the lower muon threshold at MACRO.
The newly and better measured double ratios and the tendency for energy
dependence are consistent with the hypothesis for muon neutrino oscillations
that was raised more than 10 years ago and is now confirmed by the analysis
and interpretation of the new experimental results. Reference [208] reviews
the current experimental data on atmospheric neutrinos.

7.3 Neutrino oscillations

Neutrino oscillations were first suggested as a possibility by Pontecorvo [209].
Oscillations imply that neutrinos have nonzero mass. The known neutrino
flavor eigenstates can be represented as a linear combination of different
'basic' neutrino flavors. In the simple two-neutrino case this assumption can
be expressed as:

$$
\begin{aligned}
\nu_\alpha &= \cos\theta \times \nu_1 + \sin\theta \times \nu_2 \\
\nu_\beta &= -\sin\theta \times \nu_1 + \cos\theta \times \nu_2 \, ,
\end{aligned}
\tag{7.23}
$$

where θ is the mixing angle between ν_1 and ν_2. A calculation of the wave
functions for neutrino flavors α and β shows the development of a phase
difference $\Delta\phi = \Delta m^2 t/(2E_\nu)$, where $\Delta m^2 = |m_\alpha^2 - m_\beta^2|$ is the absolute
value of the difference in the squared masses of the two flavors. This phase
difference defines the conversion (oscillation) probability of ν_α into ν_β

$$
P_{\nu_\alpha \to \nu_\beta} = \sin^2(2\theta) \sin^2\left(\pi \frac{L}{L_{osc}}\right) \, ,
\tag{7.24}
$$

which is proportional to the strength of the mixing in (7.23). The oscillation length $L_{osc} =$ is proportional to the neutrino energy E_ν and inversely
proportional to the squared mass difference Δm^2.

Expressed in units suitable for the energy of the atmospheric neutrinos
and for the geometry of their detection the oscillation probability is

$$
P_{\nu_\alpha \to \nu_\beta} = \sin^2(2\theta) \sin^2\left(1.27 \frac{\Delta m^2(\mathrm{eV}^2) L_{\mathrm{km}}}{E_{\mathrm{GeV}}}\right) \, .
\tag{7.25}
$$

In neutrino oscillation studies the sensitivity to Δm^2 comes from the L/E ratio while the determination of the mixing strength is a matter of experimental
statistics independently of Δm^2.

Atmospheric neutrinos allow the exploration of a variety of L/E_ν values. Ignoring the depth of the underground detector and assuming that neutrinos are always generated vertically at a constant altitude h one can calculate the range of $L = [2R^2(1-\cos\theta)]^{1/2}+h(\theta)$. Downward-going vertical neutrinos are created at distances h between 10 and 20 km from the detector, while upward-going ones must propagate at distances as large as the Earth's diameter, i.e. of the order of 10^4 km. Since different types of events are generated by neutrinos of energy between 1 and 100 GeV, atmospheric neutrinos can probe L/E_ν from 1 to 10^4 and correspondingly Δm^2 values from 1 to 10^{-4} eV^2. Since the atmospheric neutrino statistics is relatively small, however, only large values of $\sin^2(2\theta)$ can be studied, close to the maximum mixing angle of $45°$.

The physical parameter L/E_ν is a combination of two observable parameters: the neutrino energy deduced from the lepton energy and propagation length, which can be experimentally estimated from the angle of the incoming lepton. The angle between the lepton and the parent neutrino is significant at low energy and decreases as $E_\nu^{-1/2}$ with energy. For the Super-K sub-GeV sample with $\langle E_\nu \rangle$ of 0.77 GeV the average $\cos(\vartheta_{\nu\mu})$ is 0.53. For the multi-GeV sample ($\langle E_\nu \rangle = 5.7$ GeV) $\cos(\vartheta_{\nu\mu}) = 0.97$. At still higher energy, for upward-going neutrino-induced muons, the angle is measured in degrees.

To demonstrate the effect I plot in Fig. 7.16 the effect of oscillations with $\sin^2(2\theta) = 1$ and $\Delta m^2 = 0.01$ eV^2 on neutrinos propagating through one Earth radius. The shaded area is the distribution of neutrino energies that initiate partially contained events. Oscillations convert a fraction of these neutrinos into different (non-observable) neutrino flavor as a function of E_ν. The effect on the total event rate is the subtraction of the part of the shaded area above the oscillation probability line. While at relatively large E_ν one can in principle observe the oscillation curve, at low energy the oscillations are very fast and one can only measure the average decrease of the rate to $1/2$ of the expected rate without oscillations. The sensitivity to Δm^2 is lost for lower energy neutrinos where all detector rates will always be reduced by a factor of two. This example is done for fixed propagation distance, while any experimental measurement integrates over energy and distance ranges. The measurable oscillation features are then reduced to different levels of suppression of the neutrino rate as a function of the energy and propagation distance.

The lower panel of Fig. 7.16 shows what the effect of oscillations is on the rate of neutrino events. A fraction of the parent neutrinos have oscillated according to (7.25) and do not reach the detector. The rate is what remains of the shaded area. It is lower and $R2$ decreases but the value of Δm^2 is not known. For the example neutrino spectrum above Δm^2 of 0.01 and 0.001 produce almost indistinguishable rate decreases. A better derivation of the oscillation parameter space requires a more detailed analysis of the experimental results, such as studies of the angular distribution of the neutrino event rate or, even better, rate vs. L/E plots.

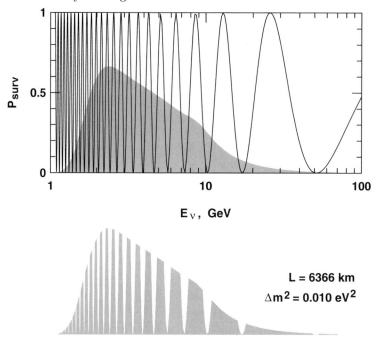

Fig. 7.16. The oscillation probability is plotted as a function of the neutrino energy for propagation on 6,366 km and for Δm^2 of 0.01 eV^2. The shaded area is the energy spectrum of neutrinos responsible for partially contained neutrino events.

7.3.1 Matter effects

The scenario of the previous section assumes that neutrinos are propagating in vacuum. In the presence of matter this scenario will be modified because different neutrino flavors have different interactions in matter [210]. Matter necessarily contains electrons and ν_es can have charge current interactions on electrons while ν_μ and ν_τ can only have weaker neutral current interactions. The extra ν_e cross-section creates an effective potential V_{ν_e} which is different from the potential V_{ν_μ, ν_τ}. For ν_e the effective potential $V_{\nu_e} = \sqrt{2}\, G_F n_e$, where n_e is the electron density of the medium and G_F is the Fermi coupling constant. This potential can be considered as a contribution to the ν_e mass and changes the neutrino mixing and transition probability.

The mixing angle term $\sin^2 2\theta$ in (7.24) is replaced by its matter value

$$\sin^2 2\theta_m = \frac{sin^2 2\theta}{[(\varphi - cos 2\theta)^2 + sin^2 2\theta]^{1/2}} \qquad (7.26)$$

and the oscillation length L_{osc} with

$$L_{osc}^m = \frac{L_{osc}}{[(\varphi - cos 2\theta)^2 + sin^2 2\theta]^{1/2}}, \qquad (7.27)$$

where $\varphi = 2V_{\alpha\beta}E_\nu/\Delta m^2$ contains the difference in the effective potential of the two neutrinos in the medium. For neutrino species with identical effective potential one recovers the expressions for vacuum oscillations.

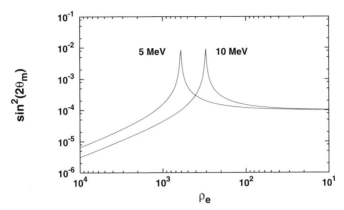

Fig. 7.17. Change of $\sin^2 2\theta_m$ with the electron density. $\Delta m^2 = 10^{-4}$ eV^2 and $\sin^2 2\theta = 10^{-4}$ for this example.

These new definitions have a very strong effect on the propagation of neutrinos in the Sun [211] – the MSW effect. The Sun contains an enormous range of electron densities and for a large range of neutrino energies the difference in the effective potential V_{ν_e,ν_μ} which is proportional to the electron density φ could become equal to $\cos 2\theta$. Even a very small mixing angle instantaneously increases and causes quick resonant oscillations. On the exit from the Sun n_e decreases and for fixed E_ν the electron neutrino flux cannot recover after the resonant transition. Figure 7.17 shows the change of $\sin^2 2\theta_m$ as a function of the electron density for a choice of oscillation parameters. The MSW effect has become a favorite explanation of the solar neutrino puzzle.

It is important to note that, while for vacuum oscillations the mass hierarchy of neutrinos does not matter, for matter oscillations it is important, because the effective potential $V_{\alpha\beta} = -V_{\beta\alpha}$.

7.3.2 Oscillation parameters

The current wealth of experimental information allows for a good determination of the oscillation parameters. For atmospheric neutrinos the strategy is to study the angular distribution of the detected neutrino events as a function of the lepton energy. The latest analysis of the Super-K group [203] divide the experimental events into 175 angular, momentum and event type bins, compare them to the predictions, and perform a χ^2 fit in the oscillation parameter space. This analysis is based on more than 7,000 FC events plus a corresponding number of PC events and upward-going muons. The fits are done for the two-neutrino oscillation scenario (as in (7.23)) or in a more complicated three [204] or four [205] neutrino oscillation scenarios. These scenarios involve two and three Δm^2 and mixing angle parameters, respectively. The idea is to give the oscillation scheme maximum freedom and to probe all available possibilities for neutrino transition. In the case of four-neutrino oscillations one of the neutrinos has to be a sterile neutrino ν_s which cannot be experimentally detected because it does not interact with matter. One of the known neutrinos could, however, oscillate to ν_s and disappear.

Figure 7.18 shows the angular distribution of the electron and muon events in two energy ranges – sub-GeV ($E_v < 1.33$ GeV) and multi-GeV ($E_v > 1.33$ GeV).

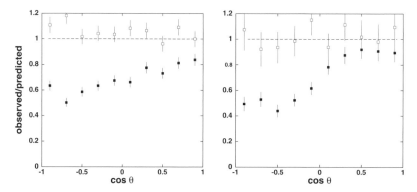

Fig. 7.18. Angular distribution of electron (open symbols) and muon (closed symbols) events in Super-Kamiokande. The points show the ratio between observed and predicted number of events. Left-hand panel is for sub-GeV events and the right-hand panel for multi-GeV events.

The sub-GeV muon events in Fig. 7.18 are much below the predictions even in an almost vertical directions. The ratio between observation and prediction decreases when the distance to the production layer increases – in directions closer to the nadir. Electron events do not show a strong angular dependence and have an angle average excess of about 7% above the prediction. The result could be interpreted in two ways: either the prediction for ν_e is too low, or a small fraction of the ν_μ convert to ν_e on top of the major conversion to tau or sterile neutrinos. Note that because of the large mass of the τ lepton, ν_τs of that energy cannot interact. At higher energy (right hand panel) the situation is somewhat better. Down-going muons are at about 90% of the predictions. Upward-going muons are at about 50% of the predictions as predicted in the oscillation scenarios. The global fit of the Super-K data gives Δm^2 of between 1.5 and 4 $\times 10^{-3}$ eV^2 and $\sin^2 2\theta > 0.9$, i.e. almost maximum mixing angle, for the disappearance of muon neutrinos. The central value of Δm^2 varies between 2.0 and 3.2 $\times 10^{-3}$ eV^2 in slightly different analyses.

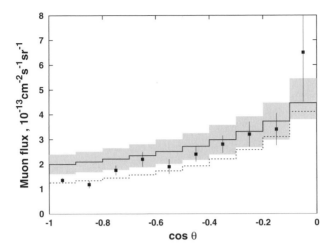

Fig. 7.19. Angular distribution of upward-going neutrino-induced muons (points) in MACRO compared to predictions without (solid histogram) and with (dashed histogram) oscillations.

This result is confirmed by other experiments. Figure 7.19 compares the rate of upward-going muons of energy >1 GeV in MACRO [207] compared to the prediction of Ref. [154]. A theoretical error of $\pm 20\%$ is indicated with a shade. Experimental points for $\cos\theta < -0.5$ lie well below the predictions. MACRO is 76.6 m long, 12 m wide, and 9.6 m tall. It is highly sensitive to upward-going muons and has almost no efficiency for horizontal events. For this reason the errors for almost horizontal muons are very large. The dashed histogram in Fig. 7.19 is the prediction with the best experimental fit – $\Delta m^2 = 2.5 \times 10^{-3}$ eV^2 and $\sin^2 2\theta = 1$. The Soudan2 detector fits slightly bigger and less certain $\Delta^2 m$ values, which is consistent with the other results [206]. The exposure of Soudan2 is much smaller (5.1 Kt.yr) than that of Super-K.

The question what the disappearing ν_μs become is more complicated. The data shown in Fig. 7.18 exclude a $\mu_\mu \to \nu_e$ scenario as a major oscillation pattern. The difference between $\nu_\mu \to \nu_\tau$ and $\nu_\mu \to \nu_s$ oscillations is in the matter effects. ν_μ and ν_τ have the same effective potential in matter, while ν_ss have $V = 0$. The potential difference $V_{\nu_\mu \nu_s}$ modifies the oscillation pattern and both Super-K [212] and MACRO [213] have reached the conclusion that the oscillations are $\nu_\mu \to \nu_\tau$.

All these results are obtained with the use of 1D predictions of the atmospheric neutrino flux and can be somewhat modified by future analyses that use 3D simulations. In addition to the enhancement of low energy neutrino fluxes around the horizon, the account for the 3D geometry changes the neutrino production height distribution [197]. The average height of production for neutrinos arriving at the detector in a horizontal direction is smaller and this changes the oscillation pattern at propagation. It is not *a priori* obvious exactly how these changes will affect the oscillation parameters – this requires massive simulations that will take years. Estimates show, however, that the changes will not be drastic and the main oscillation pattern will be confirmed.

The situation with the solar neutrino data is more complicated. Radiochemical experiments give only the total rate of neutrinos above the threshold energy of the experiment. Super-K can measure the energy spectrum only of B neutrinos. All four data sets are fitted simultaneously. The oscillation theory is more complicated and includes more free parameters. There are at least three possibilities:

- **vacuum oscillations.** Because of the large distance to the Sun and the MeV energy of the solar neutrinos Δm^2 has to be small, less than 10^{-10} eV^2. Vacuum oscillations can manifest themselves on a yearly basis because of the small eccentricity of the Earth orbit around the Sun. A seasonal change as a function of the Earth–Sun distance has been observed but there is not a clear manifestation of vacuum oscillations. Several Δm^2 values between 10^{-11} and 10^{-10} eV^2 are allowed.
- **LMA MSW.** The MSW effect branches into a large mixing angle (LMA) solution and small mixing angle (SMA) solution. The LMA solution is allowed for $\sin^2 2\theta > 0.5$ and Δm^2 between 10^{-5} and 10^{-4} eV^2.
- **SMA SMW.** The small mixing angle solution is allowed for $\sin^2 2\theta$ between 2×10^{-3} and 10^{-2} and Δm^2 between 5×10^{-6} and 10^{-5} eV^2.

Figure 7.20 shows schematically the allowed regions for atmospheric neutrino and MSW solar neutrino oscillations.

Some of the new and future experimental data will help to derive better the solar neutrino oscillation parameters. Super-K measures [214] the neutrino energy spectrum and the dependence of of the neutrino flux as a function of the zenith angle of the Sun. Solar neutrinos propagating through the Earth are subject to matter oscillations, which could help recover the ν_e flux. LMA and SMA mixing solutions have different oscillation and recovery patterns that can in principle be extracted from data. They should also cause dips in the neutrino energy spectrum, which do not exist in the case of vacuum oscillations.

Very recently the SNO detector made a very important contribution to the neutrino oscillation studies with solar neutrinos. Because of the heavy water target SNO measures three different neutrino reactions. In addition to the electron scattering (as in Super-K) and the charge current (CC) scattering, SNO can identify the NC reaction, where all neutrinos, not only the electron neutrinos contribute to the signal. In this way it can detect all solar neutrinos, with or without oscillations, although with a smaller cross-section. The signal in NC gives a total solar neutrino flux of $5.1 \pm 0.65 \times 10^6$ cm^{-2} s^{-1}, while the CC process and the electron scattering give 35% and 47% of that amount [215]. This result confirms the oscillation solution – the ν_e flux is decreased while the total flux of neutrinos is as predicted.

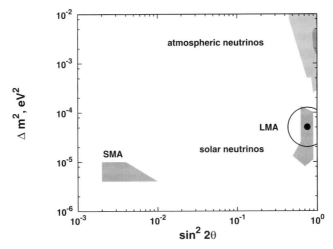

Fig. 7.20. Allowed parameter values for neutrino oscillations. Atmospheric neutrinos are all bunched in the upper right-hand part of the graph. Light shaded areas come from Soudan2 and MACRO, and the darker shading from Super-K. The SMA (lower left) and LMA solutions for MSW oscillations of solar neutrinos are also shown. The vacuum oscillations solution for solar neutrinos is not shown. The dot and the circle show the best fit of the new SNO data combined with the world data set.

In addition, SNO measures the matter effect, which is somewhat different in the three processes, by comparing neutrinos that have come directly from the Sun (detected during the day) to those that have penetrated through the Earth and are detected during the night [216]. The day/night effect results favor strongly the LMA solution of MSW. The only other oscillation model consistent with these new results is the so-called LOW solution with Δm^2 of about 10^{-7} eV^2.

All experimental facts are in favor of neutrino oscillations. Many detailed analyses follow carefully the effects of oscillations within the allowed parameter space on all possible observables and suggest new and better ways of determination of the oscillation parameters. There is only one thing missing – all current experiments are disappearance experiments. The detected rates of solar neutrinos are smaller than predicted. In the atmospheric neutrino case muon neutrinos disappear and nothing else is detected. A non-disputable result could only be an appearance experiment. If the oscillations are $\nu_\mu \rightarrow \nu_\tau$ then one should detect ν_τ interactions to definitely prove the oscillation pattern. The problem is the very short τ meson lifetime. The decay length of a 10 GeV τ meson is less than 0.5 mm. Although the ν_τ cross-section at 10 GeV is still quite low, the Super-K data set may contain several tens of ν_τ interactions that are impossible to identify. Such identification requires tracking devices with resolution better than the τ decay length and is not possible in water-Cherenkov detectors.

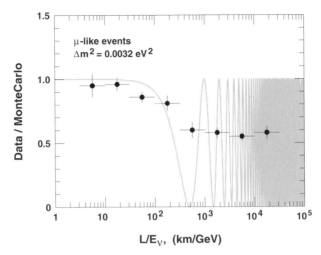

Fig. 7.21. Ratio of data to Monte Carlo predictions for muon events in Super-K. For $\Delta m^2 = 0.0032$ eV^2 the first oscillation minimum is reached at L/E_ν of 500.

Another possibility is to study the recovery of the neutrino flux after the first oscillation minimum. Figure 7.21 shows the Super-K muon neutrino flux as a function of L/E_ν. The current measurements only average over large bins of L/E_ν and thus measure the average flux decrease. If one were able to measure better the energy and angle it would be possible to use finer L/E_ν binning and detect the first oscillation minimum at L/E_ν of about 50 and the recovery after that. This would also be a proof of the oscillation scenario.

Both of these proofs of neutrino oscillations could be achieved by a long baseline (LBL) neutrino oscillation experiment. The idea is to produce a strong accelerator neutrino beam and detect the result of oscillations at a large distance from the production site. Such experiments are planned and constructed both at CERN and in Fermilab [217]. A smaller experiment of the same type already exists between the KEK 12 GeV accelerator and Super-K [218]. Future LBL experiments will control the neutrino beam with a near detector much closer to the accelerator than the oscillation length. Some of the far detectors will be sensitive enough to detect τ neutrino interactions and the known pathlength L and the better energy measurements will derive with smaller uncertainty the oscillation curve.

Part II

Contemporary Challenges

8 Cosmic ray showers

Cosmic ray showers are cascades initiated by cosmic rays interacting in the atmosphere. Cascades had already been observed in the 1920s when a single track belonging to a charged particle was observed to split into two tracks. The observations of showers led to the development of the electromagnetic cascade theory in the 1930s with the participation of many of the famous physicists of the time. Towards the end of that decade Pierre Auger and collaborators discovered the extensive air showers [219]. These are huge cascades initiated by particles of energy 10^6 GeV as estimated in that reference.

At high energy the flux of cosmic rays is so small that showers are the only way to observe them. Already Auger and collaborators discovered that the high density area of the showers is of the order of 10^4 m^2. The method of observation is to put many counters at some distance and look for coincidental hits. If only 1% of the total area is instrumented, the chance of detection increases by many orders of magnitude because of the 10^4 m^2 effective area of the high energy cosmic ray particle.

The interpretation of these events is, however, difficult. The shower array detects particles that cross almost simultaneously the observation level. There is no information on the energy and type of the primary cosmic ray – these have to be derived from the shower properties. Since even the point of the first interaction in the atmosphere is not known the observer has to first derive the degree of shower development.

The interpretation of air shower results is done on the basis of Monte Carlo calculations of the shower development. The observed air shower features are compared to shower models that assist the derivation of the primary particle energy and type. Conclusions cannot be made for individual cascades because of the fluctuations in shower development. The cosmic ray spectrum and composition are derived for a reasonably large statistical samples.

Even this is not easy, because at energies exceeding 10^6 GeV the main parameters of the hadronic interactions are not measured directly. The particle physics input in the shower models is based on extrapolations from lower energy. Different extrapolations lead to different conclusions and create a considerable systematic uncertainty in the interpretation.

We describe briefly some of the extrapolations of the hadronic interactions to high energy and show a representative set of results on the cosmic ray spectrum and composition derived from air shower observations.

8.1 Electromagnetic cascades

The idea of cascade development and the most important features of cascades are easy to understand in the toy model suggested by Heitler [220]. Heitler describes a cascade consisting of particles of the same type that interact at length λ. Two new particles are created by the interaction, each carrying one half of the primary energy. This picture is sketched in Fig. 8.1.

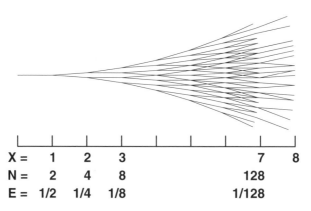

Fig. 8.1. Heitler's toy model of cascade development. In this sketch $E_c = E_0/128$.

So after one interaction length λ the cascade consists of two particles with half the primary energy, $E = E_0/2$, after 2λ the number of particles is 2^2 and the particle energy is $E_0/2^2$, etc. The number of particles doubles with every interaction length and the particle energy respectively is cut in half. At depth $X = N\lambda$ the cascade consists of 2^N particles. This growth continues until the particle energy reaches a critical energy E_c below which the interaction cross-section is zero. The number of particles in the cascade does not grow any more, and in the case of energy loss it decreases as indicated in Fig. 8.1. The maximum number of particles in the cascade is $N_{max} = E_0/E_c$. The depth of maximum is proportional to the logarithm of the ratio of the primary and the critical energies E_0/E_c.

$$X_{max} = \lambda \log_2(E_0/E_c) \tag{8.1}$$

The decrease in the number of shower particles depends on the strength of their energy loss.

Although the model is extremely simple, it describes qualitatively correctly the shower development of electromagnetic as well as of hadronic showers up to the maximum of shower development. The number of shower particles in shower maximum is always proportional to the primary energy, sometimes raised to a power α, and the depth of the maximum X_{max} is proportional to the logarithm of the primary energy.

8.1.1 Electromagnetic cascade theory

The mathematically correct treatment of cascading is the solution of transport equations, analogous to (6.4), i.e. including particle loss and particle production terms. There were many interesting solutions of the transport equations for electromagnetic cascades in the golden era of QED – the late 1930s. We shall follow the solution of Rossi and Greisen, which was carried further than other solutions and is best described and parametrized [221, 1].

The system of equations that guide the cascade development is solved by Rossi and Greisen through a Mellin transform. It has two roots, λ_1 and λ_2 which are defined by the properties of the bremsstrahlung and pair-production at full screening. There are two approximate solutions, Appr. A which is valid at high energy, where the electron ionization loss can be neglected, and Appr. B that accounts for the ionization loss.

The number of electrons p and photons g of energy E at depth X in a shower initiated by a primary particle of energy E_0 and type j are given as

$$p^j, g^j = \frac{\mathcal{H}_i^j}{\sqrt{2\pi} s^n (\lambda_1'' X + n/s^2)^{1/2}} \left(\frac{E_0}{E}\right)^s \frac{1}{E} e^{\lambda_1 X}. \tag{8.2}$$

This solution is a function of the shower age parameter s that describes the stage of shower development. It is related to the depth in radiation lengths X as

$$X = -\frac{1}{\lambda_1'(s)} \left[\beta - \frac{n}{s}\right], \tag{8.3}$$

where β is the logarithm of the ratio of primary to threshold energy – $\beta = \ln(E_0/E_{thr})$. Shower maximum is achieved at $s = 1$. The qualitative behavior of the shower is given by the $e^{\lambda_1 X}$ term in (8.2). For s less than 1 λ_1 is positive and the number of shower particles grows. After the shower maximum, λ_1 is negative and the number of shower particles decreases. The depth-dependence of the shower size $dN/dX \simeq \lambda_1(s)$.

The second root α_2 is hidden in (8.2) in the parameter $\mathcal{H}_i^j$ which has different contents for shower particles i and primary particles j. The variable n signifies the different speed of shower development that depends on the type of the primary and shower particles. Table 8.1 gives the values of n for different shower and primary particles.

Equation (8.2) is differential in shower particle energy. The total number of shower particles above energy E is given by integration in energy of (8.2), i.e. for electrons it is

Table 8.1. Values of n in (8.2) for different primary and shower particles

Shower primary	Electrons	Photons
electron	0	1/2
photon	−1/2	0

$$\mathcal{P}^j(E_0, E, X) = \int_E^{E_0} p^j(E_0, E', X)\, dE'. \tag{8.4}$$

The total number of photons above energy E is obtained in the same way. The values of n in the integral formulae grow by 1, i.e. for electrons in γ-initiated showers $n = 1/2$. In the region around the shower maximum λ_1 and its derivatives could be approximated as

$$\lambda_1(s) \approx (s - 1 - 3\ln s)/2; \quad \lambda_1'(s) \approx \frac{-\beta}{X}; \quad \lambda_1'' \approx \frac{3}{2s^2}.$$

In these approximation the relation between the shower age parameter s and the depth in radiation lengths becomes simpler

$$s = \frac{3X + 2n}{X + 2\beta}.$$

Better approximations are given by Rossi and Greisen and are presented for a set of s values in Table 8.2 together with the corresponding $\mathcal{H}$ values.

Table 8.2. Values of the parameters in the cascade theory. The superscript of $\mathcal{H}$ indicates the primary particle and the subscript indicates the shower particles.

s	λ_1	λ_1'	λ_1''	$\mathcal{H}_e^e$	$\mathcal{H}_\gamma^e$	$\mathcal{H}_\gamma^\gamma$	$\mathcal{H}_e^\gamma$
0.4	1.130	−3.654	12.500	0.536	0.508	0.464	0.489
0.5	0.813	−2.693	7.600	0.526	0.520	0.474	0.480
0.6	0.576	−2.093	4.950	0.513	0.531	0.487	0.471
0.7	0.389	−1.685	3.500	0.496	0.541	0.504	0.463
0.8	0.235	−1.389	2.550	0.477	0.551	0.523	0.453
0.9	0.108	−1.166	1.970	0.456	0.560	0.544	0.443
1.0	0.000	−0.991	1.560	0.433	0.567	0.567	0.433
1.1	−0.092	−0.850	1.280	0.408	0.573	0.592	0.422
1.2	−0.171	−0.733	1.060	0.383	0.576	0.617	0.410
1.3	−0.239	−0.636	0.893	0.357	0.578	0.643	0.397
1.4	−0.298	−0.553	0.764	0.331	0.577	0.669	0.384
1.5	−0.350	−0.482	0.655	0.306	0.574	0.694	0.370
1.6	−0.395	−0.421	0.565	0.281	0.568	0.719	0.355
1.7	−0.435	−0.369	0.487	0.257	0.561	0.743	0.340
1.8	−0.470	−0.324	0.423	0.235	0.554	0.765	0.325

In Appr. B the threshold energy E is replaced by the critical energy ε_0 where the radiation and ionization energy loss of an electron are equal and $\beta_0 = E_0/\varepsilon_0$.

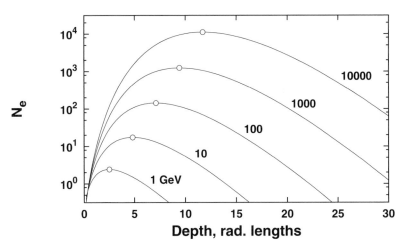

Fig. 8.2. Shower profiles for electrons in Appr. B in showers of primary photons of different energy calculated with (8.6). The primary energy is indicated by the respective profiles. The depth of shower maximum is indicated with circles.

Simple formulae describing quite well the shower longitudinal behavior were later given by Greisen [222] for the most practically important case of electrons in γ-initiated showers.

$$N_e^\gamma = \frac{0.135}{\sqrt{\beta}} \exp\left[X\left(1 - \tfrac{3}{2}\ln s\right)\right] \text{ for Appr. A} \tag{8.5}$$

$$N_e^\gamma = \frac{0.31}{\sqrt{\beta_0}} \exp\left[X\left(1 - \tfrac{3}{2}\ln s\right)\right] \text{ for Appr. B}$$

where s is related to the depth in radiation lengths X as

$$s = \frac{3X}{X + 2\beta} \tag{8.6}$$

Figure 8.2 shows the depth-dependence of the number of electrons (shower profiles) in Appr. B calculated using (8.6) for showers initiated by γ-rays of energies from 1 to 10^4 GeV. The points associated with each curve show the position of shower maximum calculated from (8.6). These formulae are widely used for quick estimates of the shower size and depth of maximum in different types of showers.

Figure 8.3 compares the more exact solutions of Rossi and Greisen with Greisen's formula (Appr. A) for primary to threshold energy ratio of 10^6. The agreement between the exact solution and Greisen's formula is excellent in the region of the shower maximum and at greater depths – for age parameter $s > 0.8$. The number of photons is higher than the number of electrons and the ratio between the two components increases with depth.

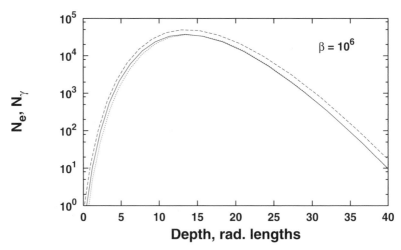

Fig. 8.3. Comparison of the solution of Appr. A with Greisen's simplified formula for $\beta = 10^6$. The solid line shows the shower profile for electrons and the dashed line for photons in Appr. A. The dotted line is the result from (8.6).

All the formulae above give the average shower behavior, which may be very different from the shower profile of an individual shower. Each shower starts at a different depth with an exponential distribution of starting points $\exp(-X_0/\lambda)$. This by itself creates significant fluctuations in the shower development. To these fluctuations one should add the intrinsic fluctuations in the shower development which depend on the interaction lengths realized in the second, third, etc. shower particle generations as well as on the kinematic parameters realized in their interactions. It is very difficult to calculate exactly the fluctuations in the shower development analytically. Estimates have been made that show the minimum of the fluctuations are in the region of the shower maximum. This is natural because at shower maximum one finds the maximum size of the shower. Good estimates are made using the Monte Carlo technique.

Another difficult problem is the lateral spread of the shower particles. In bremsstrahlung and pair-production the secondary particles are not emitted exactly in the direction of the primary one. The average transverse momentum in these processes is of the order of the electron mass m_e. In addition, electrons change their direction by Coulomb scattering. For multiple Coulomb scattering the average angle of deviation is given by Nishimura [223] as

$$\langle \delta\theta^2 \rangle = \left(\frac{E_s}{E} \right)^2 \delta X \,, \tag{8.7}$$

where $E_s = m_e c^2 \sqrt{4\pi/\alpha}$ is the effective energy (21 MeV) for multiple Coulomb scattering. This value sets the size of the Molière length r_1 that characterizes the spread of shower particles on approximately 1/4 radiation length

$$r_1 = \left(\frac{E_s}{E_c} \right) X \,. \tag{8.8}$$

Note that in this definition r_1 is in units of radiation lengths or in g/cm². Measured in meters, r_1 is different for different materials and increases with the decreasing density of the air in the atmosphere. Showers developing at higher altitudes have bigger lateral dimensions.

The theory of the lateral spread of the shower particles was developed by Greisen [222] and by Kamata & Nishimura [224]. The approximate solution for the lateral distribution of electrons is called the NKG formula, that gives the density of shower electrons as

$$\rho_e(r, X) = N_e(X) \frac{C(s)}{r r_1} \left(\frac{r}{r_1} \right)^{s-1} \left(1 + \frac{r}{r_1} \right)^{s-9/2} \,. \tag{8.9}$$

$C(s)$ is a normalization coefficient derived from the definition of the lateral spread function:

$$\frac{2\pi}{N_e(X)} \int_0^\infty r\rho(r) \, dr = 1 \,.$$

At shower maximum $C(1) \simeq \sqrt{2\pi}$.

The Molière length r_1 is defined for low energy electrons. For higher energy particles both the angle at production and the multiple Coulomb scattering angle are smaller. The lateral distribution of higher energy particles in the cascades is also narrower. One possible way to scale the characteristic spread is to weight the Molière length with the ratio of the critical energy E_c to the electron energy E. The characteristic spread then decreases with energy as $r_1 E_c/E$.

8.1.2 Monte Carlo calculations

Apart from all the difficulties with the exact implementation of the interaction properties in an analytic calculation the solutions of the cascade theory in the previous section are based on the following assumptions:

- (1) there are only two processes, bremsstrahlung and pair-production, plus ionization loss in Appr. B;
- (2) the cross-sections for these processes are energy independent;
- (3) the secondary particles are distributed as at full screening.

For these reasons the solutions of the cascade theory are very good for high energy showers, when all conditions are met. At energies below ε_0 the pair-production cross-section quickly declines and the Compton scattering becomes a more important interaction process for the shower photons. Figure 8.4 shows the energy dependence of the photon cross-section in air in the MeV–GeV range. One interaction per radiation length in air corresponds to a cross-section of 4.50×10^{-26} cm^2.

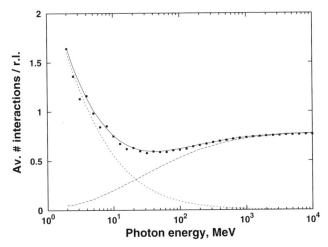

Fig. 8.4. The dashed line shows the pair-production cross-section, the dotted line the Compton cross-section and the solid line the sum of the two. The points show the cross-section realization in the Messel scheme.

With the changing cross-sections and interaction kinematics it is obvious that the Monte Carlo method, which can in principle describe exactly all processes, can give a very good description of the shower development. The first very impressive set of calculations was performed by Messel and collaborators [225]. Since the computers in the early 1960s were by today's standards very primitive, the only possibility of doing a good calculation was to go deep into the interaction physics and create an efficient sampling method. One of the serious problems solved by this group is the implementation of the bremsstrahlung cross-section, which is inversely proportional to the energy of the secondary photon and tends to infinity for very low photon energies – the infrared catastrophe. All free paths in Messel's code are sampled simultaneously with the energy of the secondary particles. The points in Fig. 8.4 show the result from Messel's sampling method compared to the process cross-sections.

Except for showers of extremely high energy, most calculations of electromagnetic cascades are now done with the Monte Carlo method. There are different professionally maintained codes, of which the best known is EGS4,

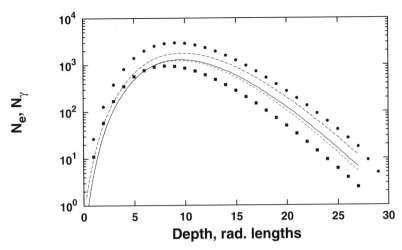

Fig. 8.5. Number of electrons (solid) and photons (dash) calculated with the cascade theory in Appr. B. The dotted line results from Greisen's formula and the points give the result of a Monte Carlo calculation with an energy threshold of 10 MeV. The primary particle is a photon of energy 1,000 GeV.

which stems from the work of Ford and Nelson [226]. We will give only a few examples for the shower longitudinal and lateral distribution that are calculated with the electromagnetic Monte Carlo code of Ref. [227].

Figure 8.5 compares the shower profiles for electrons and photons obtained in Appr. B to Monte Carlo calculations with energy threshold 10 MeV and primary photon of energy 1,000 GeV. Since Appr. B represents the total number of shower particles and the Monte Carlo calculation has a relatively high energy threshold, it is not surprising that Appr. B gives more electrons and positrons. The number of photons in the analytic calculation is obviously insufficient and less by about a factor of two in the vicinity of the shower maximum.

Another big advantage of the Monte Carlo calculations is that they give a very good description of the fluctuation in the shower development. Figure 8.6 shows the shower profiles of electrons of energy above 1 GeV in a photon-initiated shower of primary energy 10^5 GeV compared to the average shower profile, which is shown with squares. The individual profiles are very different at depths smaller than 5 and greater than 15 radiation lengths. The fluctuations are smallest in the region of the shower maximum, where nine of the ten showers have about the same size. The tenth shower starts developing very late – at about 5 radiation lengths.

Since every cascade particle is individually tracked in a Monte Carlo calculation, they naturally give a good description of the shower lateral spread. Figure 8.7 gives the lateral distribution of electrons and photons with energy above 1.5, 10, and 100 MeV in a γ-initiated air shower of primary energy

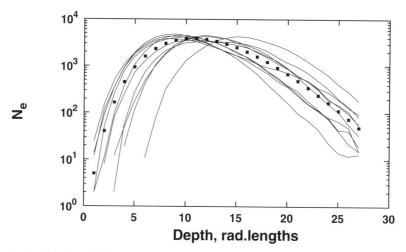

Fig. 8.6. Number of electrons with energy above 1 GeV in a shower from primary photon of energy 10^5 GeV. The lines show the shower profiles for 10 individual showers and the squares show the average profile from 100 showers.

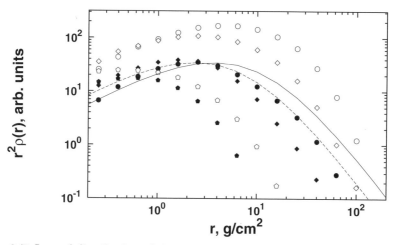

Fig. 8.7. Lateral distribution of electrons and photons of energy above 1.5 (circles), 10 (diamonds), and 100 (pentagons) MeV in the maximum of photon-initiated shower of energy 1,000 GeV. The electron points are filled. The solid line shows the NKG formula with r_1 of 9.3 g/cm^2 and the dashed line is for the fitted characteristic distance of 5.94 g/cm^2.

10^3 GeV. The lateral distributions are calculated at depths that correspond to shower maximum for all three threshold energies, i.e. 497, 427, and 340 g/cm^2. The electron density is multiplied by the squared distance from the shower axis r^2 to represent the total number of electrons at a distance and to emphasize the shape of the lateral spread. The result of (8.9) is shown with a solid line. There are two interesting points that Fig. 8.7 raises, apart from the fact that the analytic solution is only approximately correct.

The lateral spread is smaller than the NKG formula even for the low threshold energy of 1.5 MeV. A fit of the Monte Carlo calculated lateral distribution gives $r_1 = 5.9$ r.l. instead of the Molière radius of 9.3 g/cm^2. This was long ago noticed by A.M. Hillas [228], and he recommends the use of the NKG formula with characteristic distance of $r_1/2$ for practical estimates of the electron lateral spread. The fitted values for electrons of threshold energies of 10 and 100 MeV are respectively 4.10 and 1.41 radiation lengths.

The lateral spread of the shower photons is significantly flatter than that of the electrons, especially at low energy. One can understand that effect in terms of energy loss. Electrons lose ε_0 on ionization in one radiation length. The particles at great distance from the shower axis have to move under larger zenith angles. Electrons lose energy and stop, positrons in addition annihilate, while photons remain intact and dominate at large distance from the axis. The fitted characteristic distances for photons above the three threshold energies are respectively 12.1, 6.48, and 1.52 radiation lengths.

8.2 Hadronic showers

Hadronic showers are cascades that are initiated by a hadronic interaction of a nucleon or a heavier nucleus. In the first interaction a large fraction of the nucleon energy (about one-half) is transferred to secondary mesons, both charged and neutral. Apart from second-order collective effects the same happens in nucleus–nucleus interactions where more than one nucleon from the primary nucleus can participate in the interaction. In this introduction we assume that all secondaries are either charged or neutral pions. The ratio of the number of charged to neutral pions is approximately 2. The rest of the energy is retained by a secondary nucleon which, after traversing on the average one more interaction length, interacts again and generates a second generation of mesons. Meanwhile some of the secondary mesons interact themselves to generate the next generation of the hadronic cascade. This process continues until there is target material or the hadron energy is below the interaction threshold that is almost equal to the hadron mass.

Secondary neutral pions decay immediately into two γ-rays unless their energy is extremely high – the neutral pion decay length $l_d = \gamma_{\pi^0} \times 2.51 \times 10^{-6}$ cm. The daughter γ-rays start electromagnetic cascades practically at the interaction point. Charged pions have a much longer decay length ($l_d = \gamma_{\pi^\pm} \times 780$ cm) – they can either decay or re-interact. This interaction/decay competition of all charged mesons determines the details of the development of hadronic showers. High energy charged pions, with their large decay length because of time dilation, almost exclusively interact. Low energy pions decay into muons and muon neutrinos.

The decay/interaction competition is most complicated in atmospheric cascades that develop in a medium with constantly changing density. At high altitudes pions are more likely to decay than at low altitudes, where the atmosphere is considerably denser.

Interacting charged pions generate again two-thirds charged and one-third neutral pions that also belong to the second shower generation. The process continues until the charged pion energy decreases so much that they almost always decay. The hadronic cascade thus consists of two interrelated processes each one of which is not dissimilar to the electromagnetic cascade – the primary hadronic cascade that generates overlapping electromagnetic cascades which are also the main observable feature of the hadronic showers. The development of the hadronic skeleton of the shower can be observed by its muons component. There is of course a truly hadronic component consisting of nucleons and mesons, but the number of hadrons is significantly smaller than the muons, electrons and photons.

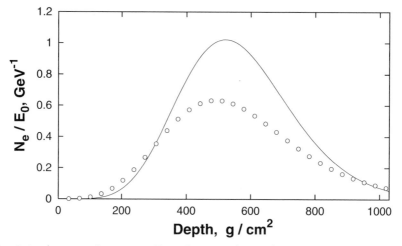

Fig. 8.8. Average shower profiles of proton (points) and photon (line) initiated showers. Both primary particles are of energy 10^5 GeV. The quantity shown is the total number of electrons divided by the primary energy in GeV.

One important fact is that the interaction lengths of the participating hadrons are longer than the the radiation length. They are also energy-dependent, but at energy below 10^5 GeV good approximate numbers for the nucleon and pion interaction lengths in air are 85 and 120 g/cm^2 respectively. For this reason the shower hadronic component carries a large fraction of the primary particle energy deeper than electromagnetic cascade would.

On the other hand the secondary particles multiplicity in hadronic interactions is higher than the effective multiplicity of two secondaries in electromagnetic interactions and this faster energy dissipation compensates for the longer interaction lengths.

Figure 8.8 compares the average shower profiles of proton and photon initiated showers of energy 10^5 GeV. There are several important differences. The proton initiated shower peaks somewhat earlier than the photon shower. The depths of maximum are respectively 506 and 520 g/cm^2. This is due to the higher secondary multiplicity in the hadronic interactions, which is also responsible for the higher number of electrons at depths of about 200 g/cm^2. One the other hand the proton shower is not absorbed as easily as the photon one – this shows the feeding of the electromagnetic component by the hadronic component, that has a longer mean free path. The biggest difference is the number of particles at shower maximum. For proton showers in this energy range N_e^{max} is about 0.6 E_0/GeV, i.e. the average energy per electron in shower maximum is about 1.6 GeV. The rest of the primary energy goes to the hadronic component – true hadrons, muons and neutrinos.

Although hadronic cascades are much more complicated, one can still use Heitler's toy model [220] to illustrate the development of their electromagnetic and hadronic components. The zeroth order estimate can be made under the assumption that only the first generation neutral pions contribute to the shower electromagnetic component. Imagine that a nucleon of energy E_0 GeV interacts in the atmosphere at depth λ_N. In the interaction it loses $(1 - K_{el})$ fraction of its energy and generates $\langle m \rangle$ secondary pions, one-third of which are neutral. The neutral pions immediately decay into two γ-rays of equal energy as in Heitler's model. One can then write the depth of maximum as the sum of the depths of maximum of the electromagnetic showers and the interaction length of the primary nucleon

$$X_{max} = X_0 \ln \left[\frac{2(1 - K_{el})E_0}{(\langle m \rangle / 3)\varepsilon_0} \right] + \lambda_N(E_0) , \qquad (8.10)$$

and the number of electrons at shower maximum as

$$N_e^{max} = \frac{1}{2} \frac{\langle m \rangle}{3} \frac{(1 - K_{el})E_0}{\varepsilon_0} . \qquad (8.11)$$

The factors of 3 (or 1/3) in these estimates account for the the fraction of the multiplicity in neutral pions and the factors of 2 (1/2) are for the splitting of the neutral pion energy. With a reasonable choice of values for $K_{el} = 0.5$, $\langle m \rangle = 12$ and $\lambda_N = 80$ g/cm^2, the depth of maximum of a 10^5 GeV proton initiated shower is estimated at about 500 g/cm^2 and the size at maximum at about 8×10^4 electrons. Since all parameters in Eqs. (8.10) and (8.11) are energy dependent and the equations account only for the first-generation pions these estimates can not be very exact, but they still contain the main features of the shower development – the size at maximum is proportional to the primary energy and X_{max} is proportional to its logarithm.

Gaisser [6] has parametrized the longitudinal development of hadronic showers as a function of the point of first interaction X_1, the depth X_{max}

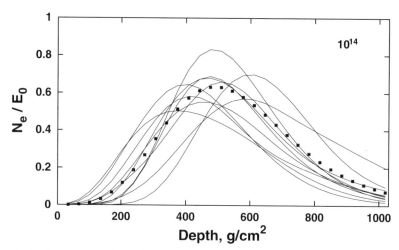

Fig. 8.9. Shower profiles of ten individual showers (lines) are compared to the average shower profile of 10^5 GeV vertical proton shower.

and size N_e^{max} at maximum and the mean free path λ.

$$N_e(X) = N_e^{max} \left(\frac{X - X_1}{X_{max} - \lambda} \right)^{\frac{X_{max} - \lambda}{\lambda}} \exp - \left(\frac{X - X_1}{\lambda} \right). \qquad (8.12)$$

The size at maximum N_e^{max} is a function of the primary energy E_0 and in the case of Feynman scaling is related to the critical energy as $0.045E_0/\epsilon$. In the contemporary hadronic interaction models the size at maximum is energy-dependent. Equation (8.12) is used as a standard fit for the shower longitudinal development and is usually called the Gaisser-Hillas formula.

The depth of the first interaction X_1 is distributed as $\exp -(X_1/\lambda)$ and is one of the main sources of shower fluctuations. Figure 8.9 compares the shower profiles of ten individual proton showers with the average shower profile (squares) of 10^5 GeV vertical proton showers. The importance of the first interaction point X_1 is easy to see in the beginning of the shower development.

This is not the only source of shower fluctuations. One can study the shower fluctuations for simulated showers if the shower profiles are plotted as a function of the atmospheric depth after the first interaction points X_1. Alternatively, one can study the shower profiles after sliding the depth with respect to X_{max}, as is shown in Fig. 8.10 for the same ten showers. This technique can also be used with experimentally detected showers in experiments that can determine X_{max}. The figure illustrates the intrinsic shower fluctuations that are due to the individual realizations of the hadronic interaction properties – interaction length, inelasticity, secondary particles multiplicity and energy spectra of the secondaries.

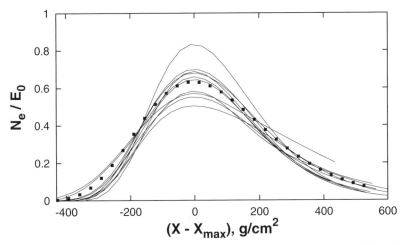

Fig. 8.10. The shower profiles of the same ten individual showers (lines) as in Fig. 8.9 are arranged so that their depths of maximum coincide. The points again show the average shower behavior.

Although the visual impression of Fig. 8.10 is different, the fluctuations are smallest in the vicinity of the shower maximum. In this random example the size at maximum varies by less than a factor of 2, while the variations are much bigger away from X_{max}. The intrinsic shower fluctuations make the analysis of the shower properties much more difficult. They also require the observation of several shower components, not only the shower electrons, that are shown in Figs. 8.9 and 8.10. The reason for the big fluctuations shown is that shower electrons only depend on the fraction of the primary energy that has been transferred to the electromagnetic component during the shower development. Because of energy conservation, showers with lower electromagnetic content must have higher hadronic content. Measuring at least two shower components accounts much better for the total shower energy, i.e. for the energy of the primary particle E_0, and is an important tool in the interpretation of air shower data.

Matthews [229] has adjusted Heitler's approach to hadronic showers by the introduction of physically motivated effective parameters. The number of muons in a shower is related to the total number of decaying pions, i.e.

$$\ln N_\mu = \ln N_\pi = n_c 2\langle m\rangle/3 = \beta \ln(E_0/\varepsilon_\pi) , \qquad (8.13)$$

where ε_π is an effective energy. The number of muons in a shower is then

$$N_\mu = (E_0/\varepsilon_\pi)^\beta , \qquad (8.14)$$

where β reflects the fraction of the multiplicity in charged pions. To describe realistically the total number of muons, Matthews uses ε_π of 20 GeV and

$$\beta = \ln(\frac{2}{3}\langle m \rangle)/\ln\langle m \rangle = 0.85 \qquad (8.15)$$

The number of muons in the shower thus increases more slowly than S_{max} with the primary energy.

The simple formulae above can be used to represent the development of showers initiated by heavier nuclei in the *superposition approximation.* It represents the shower initiated by a nucleus of mass A and total energy E_0 as a superposition of A showers initiated by nucleons of energy E_0/A. The average depth of maximum for showers from a nucleus of mass A then becomes

$$X_{max}^A = X_0 \ln \left[\frac{2(1 - K_{el})E_0}{(\langle m \rangle/3)\varepsilon_0 A} \right] + \lambda_N(E_0) = X_{max}^p - X_0 \ln A . \qquad (8.16)$$

Air showers initiated by He, O, and Fe nuclei of the same total energy will reach maximum 50, 100, and 150 g/cm^2 earlier than proton showers. These differences in the position of X_{max} are exaggerated by our oversimplification but they demonstrate the sensitivity of the shower development to the cosmic ray composition. Note that the shower size at maximum does not change as we multiply by A both the numerator and the denominator of (8.11).

With the same substitutions

$$N_\mu^A = A\left[(E_0/A)/\varepsilon_\pi\right]^\beta = A^{1-\beta} N_\mu^p , \qquad (8.17)$$

i.e. showers initiated by heavy nuclei generate more muons than proton showers. The muon excess over proton showers is 23%, 52%, and 83% respectively for He, O, and Fe showers and $\beta = 0.85$.

Equations (8.16) and (8.17) demonstrate the sensitivity of air shower measurements to the mass of the primary particle, i.e. to the cosmic ray composition. Showers initiated by heavy primary nuclei (large A) will reach their maximum development significantly higher in the atmosphere than proton showers. Although they will have the same N_e^{max}, such showers will be strongly absorbed by reaching the observation level. Their shower size at observation level will be smaller than that of proton induced showers. On the other hand showers generated by heavy primary nuclei generate more muons. By measuring the ratio of N_e/N_μ at the observation level one can hope to distinguish between showers of the same energy generated by different primary nuclei.

The fluctuations in the shower development have to be much smaller for heavier primary nuclei. In the case of superposition all fluctuations decrease by $\sqrt{A}$, as the total primary energy is split in A independent cascades. In more realistic estimates [230] the fluctuations are higher, but still significantly lower than for proton showers.

Apart from the oversimplified treatment of the shower development, Eqs. (8.14) and (8.17) do not account for the energy loss and decay of the muons after their production, which change the effective value of β in air showers as a function of the muon energy. Higher energy showers develop deeper in the atmosphere, i.e. closer to the observation level. When X_{max} is close to the observation level, low energy muons are less likely to decay and β increases by an amount $\delta\beta$ that depends on the muon energy E_μ and the altitude of the observation level. From more realistic estimates we know that the modified β value is still less than 1.

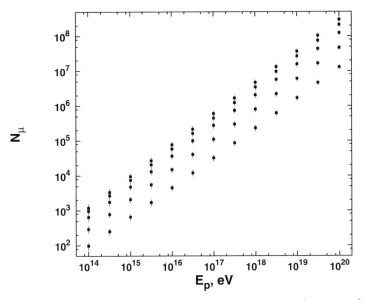

Fig. 8.11. Average number of muons at sea level in vertical proton showers as a function of the primary energy in GeV. From top to bottom the muon energy thresholds are 0.3, 1, 3, 10, and 30 GeV.

Figure 8.11 shows the N_μ dependence on the the primary energy in vertical proton showers calculated with a Monte Carlo code [231]. There is a slight variation of the power law index β with the muon threshold energy. The values of β monotonically decrease from 0.90 ($E_\mu > 0.3$ GeV) to 0.86 for $E_\mu > 30$ GeV. These variations are due to the hadronic interaction model – different models predict varying normalizations and power law dependences. The qualitative behavior is very similar to that shown in Fig. 8.11.

Another important feature of the air shower muon content is its zenith angle dependence, which is also a function of the muon energy in the same way as discussed in Sect. 6.2.1. Inclined showers develop in more tenuous atmosphere where pions and other mesons are more likely to decay than interact, i.e. they always generate more muons independently of the muon energy. Low energy muons, however, have much higher decay probability on the increased tracklength, and thus inclined air showers have less GeV muons than vertical showers of the same primary mass and energy. At muon energy about 10 GeV the production and the decay compensate each other and the number of muons does not depend strongly on the shower zenith angle. At still higher primary energy the influence of muon decay is insignificant and inclined showers are richer in muons.

8.2.1 Air shower detection

As a rule the air shower detectors are on the surface of the Earth. Depending on the pursued energy range they could be located at high altitude or at sea level. For showers in the $10^{14} - 10^{15}$ eV range a high-altitude location is preferable because then the observation level is close to shower maximum: shower size is bigger and the shower fluctuations are small. For the highest energy showers smaller altitudes are better, because there is no danger of observing showers before their maximum development.

Figure 8.12 shows a cartoon of shower development and of the two main detector techniques. The cartoon shows several generations of hadronic interactions and several muon decays. The curve on the right-hand side is the developing electron component of the shower that reaches maximum some place above the observation level and then starts to decrease. On the observation level there are several types of detectors – the squares represent counters of the electromagnetic component. The little house indicates a hadronic calorimeter. The two underground detectors use the rock as shielding and contain muon counters. The shallow one is for GeV muons – the shielding should be of the order of 500 g/cm^2 or 1.5 to 2 meters of rock. The TeV muon detector has to be located much deeper – according to (7.4) the shielding should be more than 2.7 km.w.e. – at a kilometer depth in rock.

At the left-hand side of the cartoon there is an optical detector observing the shower development around the maximum development. This small symbol represents the second observational technique – detection of the light emitted during the shower development. This could be either Cherenkov light emitted by the shower electrons or fluorescent light emitted by the atmospheric nitrogen atoms excited by the passage of the shower.

The two types of light are useful in different energy ranges. Because of the small refraction index of the atmosphere the Cherenkov cone is narrow and most of the light is emitted along the shower axis and is quite intense. Even 1 TeV showers, that are fully absorbed before they reach the observation level, produce observable Cherenkov signals.

The nitrogen fluorescence is on the other hand isotropic. Every electron produces about four photons on one meter pathlength. Only very high energy showers generate enough light to generate a signal above the night sky background. Both detection methods require clean and dry atmospheric conditions, no cloud coverage and moonless nights. Even at the best locations observations can be made during less than 10% of the time.

As an example for an air shower array we will describe the Karlsruhe Shower Core and Array DEtector array KASCADE [232], which is built slightly above (100 m) sea level. The shower detector of KASCADE consists

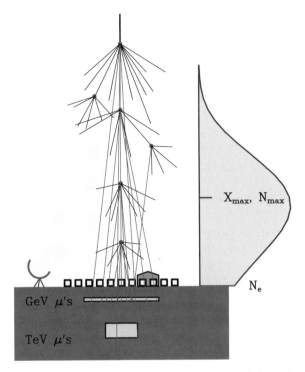

Fig. 8.12. Cartoon of shower development and detection.

of 252 scintillator counters that register and distinguish between electrons and muons. The total area of the array is 4×10^4 m^2 of which 500 m^2 is sensitive to electrons and 600 m^2 is sensitive to muons of energy above 250 MeV. The sensitive fraction of the total area of the array is thus slightly more than 1%, a number that is better than most other shower arrays. This array is coupled with a large, 320 m^2 hadronic calorimeter in its center. This device measures the energy and number of hadrons associated with the shower and is also sensitive to muons of energy above 2 GeV. In a tunnel close to the center of the array there is another muon detector of area about 250 m^2 and energy threshold of 0.8 GeV.

KASCADE measures simultaneously the electron and muon densities over a large area. The large number of shower detectors in the array provide a good statistical accuracy of the measurement and successful reconstruction of the total number of electrons and muons in the shower. The additional muon detector and the calorimeter measure the muon density at a different energy threshold and the hadronic accompaniment for a fraction of the experimental statistics. These measurements can be used as additional parameters in the reconstruction of the energy and the type of the primary cosmic ray nucleus.

As an example of an optical air shower detector we shall briefly describe the high resolution Fly's Eye detector HiRes [233]. HiRes detects the nitrogen fluorescence of very energetic showers – above 10^{17} eV. It consists of two detectors located on hilltops above the desert in Utah, USA, and separated by 12.6 km. The optical systems of the detectors consist of mirrors of effective area 3.8 m^2 that reflect the light on clusters of 256 photomultipliers. Each phototube views about 1° cone in the sky. The first detector, HiRes-1, consists of 21 mirrors that view almost the full 360° range from 3° to 17° above the horizon. HiRes-2 consists of 42 mirrors that observe the 3–31° angle above the horizon. Coincidences between the two detectors are possible because of the good optical quality of the desert atmosphere. The detector can work only during clear moonless nights and it can observe during less than 10% of the time.

Cherenkov shower detectors are very different. BLANCA [234], for example, consists of 144 separate detectors at an average distance of 40 m from each other. Each detector is a cone that concentrates the Cherenkov light emitted by the shower on a phototube. BLANCA detectors were inside the CASA air shower array and were triggered by it.

8.2.2 Reconstruction of the shower parameters from the observations

The reconstruction of the shower parameters relies on simulations of the shower development and the detection process. These simulations give the relation between the primary particle energy and type and the observable shower parameters at the altitude and for the specific types of shower detectors. Before the analysis is done in terms of the primary cosmic ray particle, however, one has to analyze the observed particle densities and derive the shower parameters themselves. The reconstruction procedure is very different for the different types of shower detectors.

Shower arrays

Shower arrays usually trigger when several detectors are hit within a short interval of time Δt. The exact value of Δt depends on the size of the shower array and on the way its data acquisition system works. If an experiment needs to be sensitive to showers of zenith angle up to θ, $\Delta t = d\cos\theta/c$ where d is the diameter of the array and c is the speed of light. Once the experiment triggers, the particle densities and their arrival time in all detectors are recorded. The density map of a shower may look like the one presented in Fig. 8.13. This particular map is not of any experimental event. It is based on Greisen's [235] parametrization of the lateral distribution of charged particles in air showers. The parametrization is similar to the NKG parametrization 8.9 for electromagnetic showers with a correction factor. The density distribution depends on the shower are parameter s and the Molière length r_1 at the observation level as

$$\rho(r) = \frac{C_1(s)N_e}{2\pi r_1^2}\left(\frac{r}{r_1}\right)^{s-2}\left(1+\left(\frac{r}{r_1}\right)\right)^{s-9/2}\left(1+C_2(\frac{r}{r_1})^{\delta}\right). \qquad (8.18)$$

Greisen recommends the use of $s = 1.25$, $\delta = 1$ and $C_2 = 0.088$ for showers of $N_e = 10^6$ particles at sea level.

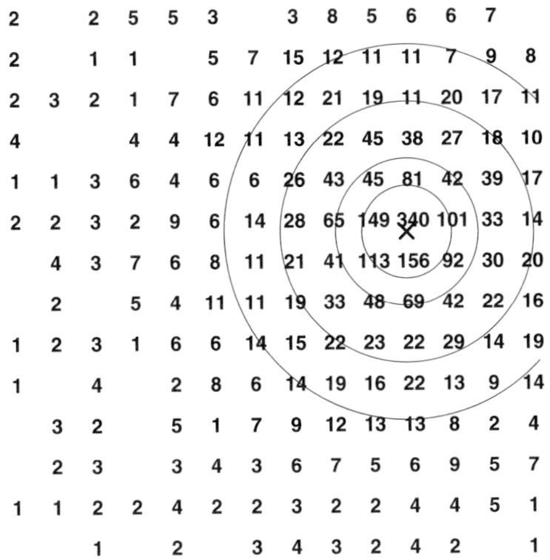

```
2         2  5  5  3      3  8  5  6  6  7
2         1  1         5  7  15 12 11 11 7  9  8
2  3  2  1  7  6  11 12 21 19 11 20 17 11
4            4  4  12 11 13 22 45 38 27 18 10
1  1  3  6  4  6  6  26 43 45 81 42 39 17
2  2  3  2  9  6  14 28 65 149 340 101 33 14
   4  3  7  6  8  11 21 41 113 156 92 30 20
   2      5  4  11 11 19 33 48 69 42 22 16
1  2  3  1  6  6  14 15 22 23 22 29 14 19
1     4      2  8  6  14 19 16 22 13 9  14
   3  2      5  1  7  9  12 13 13 8  2  4
   2  3      3  4  3  6  7  5  6  9  5  7
1  1  2  2  4  2  2  3  2  2  4  4  5  1
      1      2      3  4  3  2  4  2      1
```

Fig. 8.13. Map of an air shower in a 196 detector array on a 15 m grid. The densities from (8.18) fluctuate with a Gaussian distribution with $\sigma = \sqrt{\rho}$. Because of that some detectors do not trigger.

The concentric circles in Fig. 8.13 are distances where the average particle density exceeds 10, 20, 50, and 100 m^{-2} with the parameters above. The first step of the reconstruction is to find the position of the shower core, which in Fig. 8.13 is shown with an ×. In real life one attempts to fit the particle densities with formulae of the type of (8.18) using the gradient in density towards the core and assuming azimuthal symmetry.

To achieve this, however, one has to first determine the shower arrival angle because the distance r in (8.18) is in the plane normal to the shower axis. This is done by fitting the arrival time of the shower particles. The shower is approximately a relatively thin disk of charged particles that propagates with the speed of light. The time delay between the hits in different detectors marks the shower direction. The first detector hit by the particle disk points at its direction and for a perfectly thin disk one can fit the direction with only three counters. The complications stem from the thickness of the shower front, which increases with the distance to the shower axis and its shape that is different from a flat disk. Different groups parametrize the shape of the shower front either by an arc or by a cone, as shown in Fig. 8.14. The figure is a projection of the shower front on the plane defined by its azimuthal angle. The time delay between the two detectors at distance Δl

would be $\Delta t = \Delta l \times \tan\theta/c$, where c is the velocity of light. The shower front curvature has to be accounted for before the direction is fitted for a flat shower front. The resolution becomes worse far from the shower axis where the thickness of the disk grows while the particle density decreases.

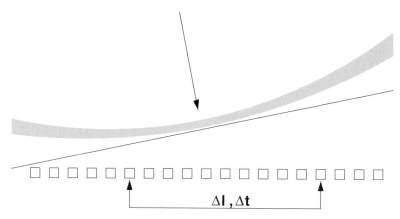

Fig. 8.14. Simplified one-dimensional sketch of the derivation of the shower direction. The shower front is projected on the plane defined by the shower azimuthal angle.

The fitting of the position of shower core and its direction is difficult to do in one step. The fluctuations in the recorded densities are large, plus the inevitable coincidental hits of unrelated particles and some usual problems with the performance of a few detectors. One has to iterate a number of times and to eliminate the data from some of the detectors which generate high χ^2 values until the solution does not change any more. Only at that time can one study the lateral distribution of the shower particles and determine their total number N_e.

Figure 8.15 plots the density distribution for the shower mapped in Fig. 8.13. The density fluctuations are significant but the large number of points allows for a good fit of the lateral distribution and the total number of charged particles. It is not a priori obvious which is the best way to fit the lateral distribution. Some experiments prefer the theoretically better way of fitting the shower age parameter, s, together with the shower axis position and its direction. Others find that the fitting of the shower age introduces too much scattering in the fit performance and use only the average s parameter appropriate for the altitude and design of the array.

I do not give any quantitative description of the particle density distribution and the shower front parameters other than Greisen's classical formula because all of them depend strongly on the design of the experiment. Thin scintillators, for example, would measure only the shower electrons and muons. The shower photons, however, convert to electron–positron pairs in

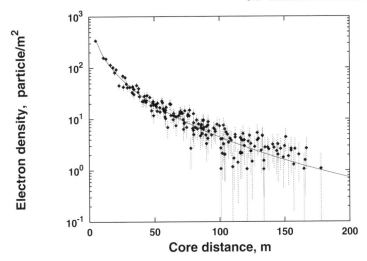

Fig. 8.15. The density distribution for the shower mapped in Fig. 8.13. The line shows the average charged particle lateral distribution.

thicker detectors, increase the total observed number of particles and also change the particle lateral distribution and the shape of the shower front. The lateral distribution of the shower photons is steeper around the axis because photons do not scatter in the propagation in the atmosphere. At large distances from the shower axis the photon lateral distribution is flatter than that of electrons, because the low energy photons do not lose energy and the inclined electrons have large energy loss on ionization. In practice each experiment measures the lateral distribution that it detects before the analysis procedure is finalized.

Not all experiments present their results in the terms of number of electrons N_e, number of muons N_μ, etc. and prefer the direct relations of particle densities with primary energy. Long ago A.M. Hillas noticed that at certain distances from the shower core the particle density could be directly related to the primary energy [236]. The charged particle density at 600 meters from the shower axis ρ_{600} is widely used in the analysis of giant air showers. The charged particle density at this distance from the shower axis does not depend much on the mass of the primary cosmic ray nucleus. The small variations with A are compensated by the simplicity of extraction ρ_{600}, that avoids the introduction of errors associated with the fitting of the particle lateral distributions. Other distances are more appropriate for lower shower energy and detectors at higher elevations. Parameters such as ρ_{600} are now the result of Monte Carlo shower calculations.

Cherenkov detectors

Cherenkov air shower detectors measure the lateral distribution of the Cherenkov light emitted by the shower. Since the opening angle of the Cherenkov light in air is small, all photons are emitted in the direction of the shower axis. The Cherenkov threshold for electrons at sea level is 21 MeV and increases at higher altitudes. Tens of MeV electrons do not scatter too much during the cascade development.

The detector thus detects showers which are pointing at it, pretty much like a traditional shower array. Each shower electron above the Cherenkov threshold creates some number of Cherenkov photons and a fraction of them reaches the observation level. The more electrons there are in the shower (the higher its energy is), the stronger the Cherenkov flash.

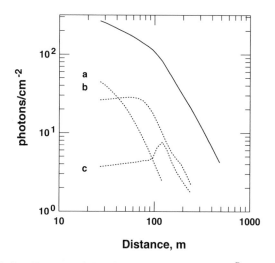

Fig. 8.16. Lateral distribution of the Cherenkov light in a 10^7 GeV proton shower at sea level as calculated in Ref. [237] (solid line). The dotted lines show the contributions of three electromagnetic sub-showers that develop at different atmospheric depths.

The principles of the analysis of a shower's Cherenkov light are presented in the paper of Patterson and Hillas [237] and summarized in Fig. 1 of that paper, which I reproduce here in Fig. 8.16. It shows with a solid line the lateral distribution of the Cherenkov light at sea level from a 10^7 GeV proton shower. The dotted lines show the lateral distribution of the light from three of the 17 electromagnetic sub-showers that are summed to generate the shower's total Cherenkov light signature. Sub-shower a is the closest to the observation level and has the steepest lateral distribution. Sub-shower b has a maximum coinciding with that of the proton shower X_{max}. It has a very flat lateral spread up to about 100 meters that steepens to the same as of sub-shower a at larger distances. Sub-shower c has peaked well above X_{max}. It has a prominent ring near a radius of 150 meters and exhibits the same steep lateral spread as the other two sub-showers at larger distances.

The proton shower Cherenkov signal, which is the sum of all sub-showers follows the lateral spread of all sub-showers at distances above 150–200 meters. This is the print of the Cherenkov cone in air. Expressed in terms of a power law with distance ($\rho(r) = Ar^{-\beta}$) the spread corresponds to $\beta > 2$. In this particular shower $\beta = 2.16$. Inside 125 meters the shape of the lateral distribution reflects the distance between X_{max} and the observation level. Early-developing showers have a flat lateral distribution. Late-developing showers are steep.

The shape of the lateral distribution inside the 150-meter circle reflects the position of the maximum of the shower development. The intensity of light at this circle is proportional to the total electron pathlength, i.e. nearly proportional to the shower energy.

Exact calculations of these quantities have to account for the observation depth of the concrete experiments, for the change of the threshold energy for Cherenkov radiation with altitude, and for the absorption of the Cherenkov light in the atmosphere. This is done on the basis of Monte Carlo calculations performed specifically for the conditions and the design of the experiment. The lateral spread inside the 150 meters is not easy to fit with a single function. The HEGRA AIROBICC array [238], for example, fits the lateral distribution between 20 and 100 meters with an exponential. The depth of maximum is then obtained as $X_{max} = X_{obs}/\cos\theta - (680 - 20880\text{m} \times \beta)$ g/cm^2. The shower energy is determined from the Cherenkov photons intensity C_{90} at 90 meters from the shower axis with mass-dependent formulae that are derived from the Monte Carlo calculations. These have, of course, also some model-dependence.

The VULCAN detector at the South Pole [239] use the ratio of the Cherenkov light intensity at 40 (C_{40}) and 100 (C_{100}) meters from the shower core to describe the lateral distribution. The parameter used is κ where $\exp\kappa = C_{40}/C_{100}$. The depth of shower maximum is evaluated as $X_{max} = X_{obs}/\cos\theta - (463 - 76\kappa - 97\kappa^2)$ g/cm^2. The primary energy for proton showers is obtained from C_{100} as $E_0 = 423(C_{100}/\text{m}^2)^{0.91}$ GeV. Slightly different expressions can be used for other primary nuclei.

The biggest advantage of the detection of Cherenkov light is its high photon density, compared to the electron density at the observation level. The lateral distribution of the Cherenkov light is measured as a smooth function, unlike the ragged data points in Fig. 8.15. The depth of shower maximum is determined with an error of 20 to 40 g/cm^2, excluding the systematic error from the Monte Carlo hadronic model. The error in determining the primary energy is of the same order. All Cherenkov air shower detectors are positioned within traditional air shower arrays and are usually triggered by them. The shower axis is usually determined by the measured electron density and the parameters measured by the two arrangements are correlated for a better understanding of the shower properties.

Fluorescent detectors

While the Cherenkov light is concentrated in the direction of the shower, the nitrogen fluorescent light is isotropic. In terms of detection this means that the optical detector looks for showers that do not hit the ground close to it. Corrections have to be applied for showers that land within the Cherenkov light radius from the fluorescent detector.

Fluorescent detectors trigger when many photomultiplier tubes (each of field of view of approximately $1°$) detect a light signal within a short time window. To a fluorescent detector the shower looks like a point moving through the sky with the speed of light. The first question to ask is: how far away is the shower? What is its impact parameter $R_\perp$, i.e. the closest distance of the shower trajectory to the detector? The impact parameter tells the analysis what fraction of the isotropically emitted fluorescent light has reached the detector. A precise reconstruction of the shower trajectory is the first step in the reconstruction of the shower longitudinal profile and the primary cosmic ray energy. Figure 8.17 gives an idea what fluorescent detectors see.

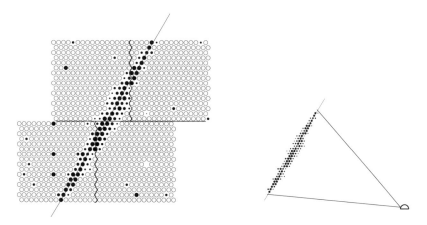

Fig. 8.17. The left-hand panel shows an image of a shower detected by a fluorescent detector. PMTs fire in brief intervals to outline the shower trajectory. Some PMTs are fired because of chance coincidences. The right-hand panel shows how the observed shower trajectory and the detector itself (the little hemisphere at lower right) define the shower-detector plane.

The PMTs that are triggered by the shower determine an arc in celestial coordinates. Together with the position of the detector the arc determines a plane. The reconstruction of the shower trajectory becomes in principle very easy for stereoscopic observations. If two fluorescent detectors observe the shower development, each of them determines a shower-detector plane and the shower trajectory is the intersection of the two planes. Since the detector positions are well known the geometry in this case is easy to solve.

Much more difficult is the problem when only one detector has detected the air shower. In principle the distance can be determined by the time differences in triggering consequent PMTs. Since the angular opening of each phototube is known, showers far away from the detector move more slowly from one PMT to the next, i.e the fixed opening angle corresponds to bigger linear distance along the shower trajectory. For example at a distance of 20 km a trajectory of length 2° is about 700 m and the shower will take 2.3 μs to traverse it. At a distance of 10 km the time will be cut in half. Even in this case one solves for the shower trajectory, however, the practical problems are numerous, and the solutions are usually less certain.

Contemporary experiments prefer to have either stereoscopic detection or some additional information, such as the position of the shower axis from a conventional air shower array. Fixing one point on the shower trajectory makes its parameters more exact.

The next task is to convert the detected fluorescent light into a shower profile. Each charged particle of the shower excites the nitrogen in the air and creates fluorescent light. The rule of thumb is that an electron creates four photons per meter pathlength. This number depends on the atmospheric density and temperature, and on the electron energy, although the variation is not very strong. For simplicity we will use it in the estimates below. Another important parameter is the absorption length of the fluorescent light in the atmosphere. Absorption is a part of the analysis and depends on the impact parameter and on the atmospheric conditions.

Optical air shower detectors are built at locations where the atmospheric quality is good and the absorption low. Monitoring the atmospheric conditions is a difficult and complicated everyday task, which we will leave to the experimentalists. The average live time of fluorescent detectors is less than 10% of the real time.

Because of the isotropic light emission, fluorescent detection becomes possible only at very high energy. Imagine that one billion electrons pass through 700 m of atmosphere at a distance 20 km from the detector. The total number of emitted photons is 2.8×10^{12} but the area of the 20 km sphere is 5×10^{13} cm^2 and the photon intensity at the detector is 0.056 cm^{-2}. One needs mirrors of large collection area to detect a significant number of photons. The exact energy threshold depends of course on the design of the detector and is generally in the vicinity of 10^{17}–10^{18} eV. The effective aperture of the detector (solid angle times area) grows with the shower energy. For this reason fluorescent detectors are much more efficient at high energy.

The amount of light observed by each phototube is converted to number of electrons as a function of the atmospheric depth and this is a direct observation of the shower profile. X_{max} can be determined within the limits of the statistical errors of each point. For the new HiRes detector X_{max} is determined with a statistical error of about ± 20 g/cm^2. The next step is the determination of the shower energy. Since the shower profile is known, one can determine the total electromagnetic energy E_{em} that goes into electrons and positrons. $E_{em} = \alpha \int_0^\infty N_e(X)dX$, where α is a constant that expresses the average ionization energy loss rate for the shower and is of the order of 2.2 MeV/(g/cm^2). This number should be equal to the ratio of the critical energy ε to the radiation length X_0, but modern Monte Carlo calculations, that follow the electrons down to very low kinetic energies, usually give a slightly higher number.

To obtain the total energy of the primary particle the 'missing energy' has to be added to E_{em}. This is the energy that is transferred to hadrons, muons and neutrinos during the shower development. Individual hadrons and muons carry much more energy than the electrons, and the neutrinos are not detected at all. The fraction of missing energy is energy-dependent. It also varies with the atomic mass of the progenitor nucleus and depends on the hadronic interaction model. From the example of Fig. 8.8 the fraction of missing energy is about 30–40% at 10^{14} eV. With the increase of the shower energy that fraction decreases. The reason is that the energy of the secondary mesons in the hadronic interactions increases and they are most likely to interact rather than decay. When they interact a fraction of their energy is transferred to E_{em}. Figure 8.18 shows the fraction of missing energy as a function of the primary energy for different nuclei.

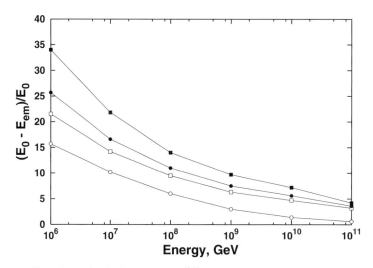

Fig. 8.18. Fraction of missing energy (%) in extensive air showers as a function of the primary energy. Circles indicate proton showers, squares are for primary He, dots for CNO, and filled squares for Fe.

Heavier primary nuclei also show a higher fraction of missing energy than proton showers. Because of the lower energy per nucleon, the secondary mesons in these showers tend to decay and increase the fraction of missing energy. The amount of missing energy depends also on the hadronic interaction model. Models that produce many soft, low energy secondaries, increase the fraction of missing energy.

At very high energy, where the fluorescence detectors are effective, the electromagnetic energy is a very good approximation for the total shower energy and this simplifies the reconstruction of the shower properties. Showers can be binned in E_{em} with a small average correction for the missing energy and the distribution of X_{max} can be then studied. Showers from heavy nuclei develop earlier, so the width and the shape of the X_{max} distribution can be used as a statistical estimate of the chemical composition of the cosmic rays at very high energy.

8.3 Extension of the hadronic interaction models

Our brief description of the properties of hadronic interactions in Sect. 2.4 dealt with interactions at relatively low energy, up to several hundred GeV in the Lab system. In the 1960s, when such interactions were mostly studied, the main achievement in the description of hadronic interactions was the definition and the experimental studies of Feynman scaling. Soon afterwards, however, after the construction of the ISR at CERN, physicists realized that Feynman scaling does not work at higher energies. ISR was the first of the new type of accelerators, called colliders. The classical, fixed target experiments accelerate particles and let them interact on stationary targets. Colliders accelerate two particle beams and let them interact head to head. It is obvious that the design and construction of a collider experiment is much more difficult: one has to steer the beams of interacting particles to a precision exceeding 1/10 mm. The advantage is that the CMS energy of the interaction is twice the energy of each of the beams. So two proton beams, each of energy 25 GeV, colliding head to head will have $\sqrt{s}$ of 50 GeV, which translates into equivalent Lab energy of of 1,200 GeV. The rapid increase in energy at which hadronic interactions are studied was only possible because of the colliders.

If Feynman scaling were holding up at high energy, ISR experiments would have measured the same interaction properties as at lower energy. It is easier to explain these properties in terms of the rapidity, y. Figure 8.19 shows the expected rapidity distributions in pp interactions with $\sqrt{s}$ of 10 and 50 GeV. Feynman scaling predicts a flat rapidity distribution with density of about two particles per rapidity unit in the central region. The maximum rapidity for pions is $\ln(\sqrt{s}/m_\pi)$, about 4.3 for $\sqrt{s}$ of 10 GeV and about 5.9 for 50 GeV. The secondary multiplicity increases only as the allowed rapidity phase space increases, i.e. it is proportional to $\ln\sqrt{s}$.

The experiments instead found several violations of Feynman scaling. The central rapidity density increased with energy and the shape of the rapidity distribution became energy-dependent. Instead of being flat the rapidity density peaks at $y_{CMS} = 0$. The total interaction cross-section and the charged

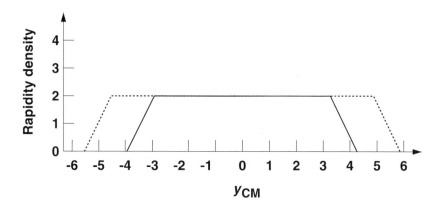

Fig. 8.19. Expected rapidity distribution if Feynman scaling were valid for interactions at CMS energies above $\sqrt{s}$ of 10 GeV. The solid line is the charged particles distribution at that energy and the dotted line is for $\sqrt{s}$ of 50 GeV.

particle multiplicity increase faster than $\sqrt{s}$. The average transverse momentum $p_\perp$ also increases with energy. None of the main interactions features scales exactly with energy.

8.3.1 The minijet model

One of the first suggestions for the physics of the change of the interaction properties was made by Cline, Halzen & Luthe [240]. They noticed that cosmic ray data, taken at much higher energy than the more exact accelerator measurements, suggest an even bigger increase of the average transverse momentum. Moreover, the $p_\perp$ distributions have different shape at low and high $p_\perp$. This feature is consistent with the structure of the hadrons that consist of quarks. The suggestion is that at low $p_\perp$ interactions are periferal and the constituent quarks participate collectively in them. When the interactions go through the individual constituent quarks, high $p_\perp$ secondaries are produced in hadron jets. These are the deep inelastic interactions of the partons. Using a specific model for the energy distribution of the components of a hadron (partons) these authors fit the accelerator experimental results and predict the behavior of the transverse momentum at much higher energy. They also suggest that the total interaction cross-section increases because of the increasing importance of that second interaction process, where individual partons from the colliding hadrons interact with each other.

This is the beginning of the minijet model of the hadronic interactions that went through different stages of development for almost twenty years [241, 242, 243]. The model separates the interaction cross-section into two parts: the low energy cross-section which is carried through low transverse momentum exchange and cannot be properly calculated using perturbative quantum chromodynamics (QCD) and the QCD minijet cross-section.

This second part can be calculated by [244] integration of differential QCD cross-section over the energy individual partons carry:

$$\sigma_{QCD} = \int dx_1 \int dx_2 \int_{Q^2_{min}}^{Q^2_{max}} d\hat{t} \times \frac{d\sigma_{QCD}}{dx_1\, dx_2\, dQ^2}(\sqrt{s}) , \qquad (8.19)$$

where $\hat{s} = x_1 x_2 s < 2Q^2_{min}$ reflects the minimum momentum transfer at which the theory can be applied. The interacting partons take x_1 and x_2 of the interaction energy, s, and

$$\frac{d\sigma_{QCD}}{dx_1\, dx_2\, dQ^2}(\sqrt{s})$$

reflects the distribution of the fraction of energy carried by partons of different type. Such distributions are called structure functions.

Calculated in this way σ_{QCD} grows very rapidly (depending on the choice of Q^2_{min} and the structure functions) and soon starts exceeding the measured total inelastic interaction cross-section. The physical meaning of $\sigma_{QCD} > \sigma_{inel}$ is that [243] σ_{QCD} is the total inclusive cross-section for the production of minijets and the high cross-section suggests that more than one minijet pair is created in the interaction. The number of such pairs n can be calculated as $n(b) = A(b)\sigma_{QCD}$, where $A(b)$ is the probability for the collision between the partons at impact parameter b.

The total inelastic cross-section σ_{inel} is obtained through an integration of a real eikonal function χ in the impact parameter space:

$$\sigma_{inel} = \int d^2b \left[1 - e^{-2\chi(b,s)} \right] . \qquad (8.20)$$

The eikonal function is split in two parts corresponding to the two interaction processes:

$$\chi(b, s) = \chi_{soft}(b, s) + \chi_{QCD}(b, s) , \qquad (8.21)$$

where $\chi_{QCD}(b, s) = n(b, s)/2$ and χ_{soft} is a constant chosen so that it reproduces the 'low energy' inelastic hadronic cross-section of 32 mb, where Feynman scaling is valid and no hard (high $p_\perp$) interactions are present.

The soft (low $p_\perp$) interactions that are a vital part of the whole interaction model are modeled according to the dual parton model (DPM) of QCD string production and fragmentation [245, 246]. The idea is that in a nucleon–nucleon interaction the three valence quarks in each nucleon are split into a quark and a diquark. A proton (uud), for example, could be split into uu or ud diquarks, the first one being twice as probable. The target nucleon is split in the same way. The diquark from the projectile combines with the quark from the target and vice versa. Each diquark–quark combination creates a string that carries the energy which its constituents take from their parent nucleons.

The fractional energy x_q that a quark takes from the parent nucleon is usually calculated as

$$f(x_q) = \frac{(1 - x_q)^\alpha}{\left[x_q^2 + \mu^2/s\right]^{1/4}} , \qquad (8.22)$$

where μ is the effective mass of the quark. The effective mass and the parameter α are chosen so that they fit the measured secondary particle distributions: μ is of the order of 300 MeV/c^2 and α is around 3.0. The diquark from the fragmentation takes the rest of the energy, i.e. $x_{dq} = 1 - x_q$. The energy of each string is then $\sqrt{s}(x_{dq}^p + x_{dq}^t)/2$, where x_{dq}^p is the fraction of energy of the diquark from the projectile nucleon and x_{dq}^t that of the target. The momenta of the two strings are $\sqrt{s}(x_{dq}^p - x_{dq}^t)/2$. In the center of mass system the two strings move in the opposite directions depending on the energy of the diquarks in that system. In meson interactions the two mesons are split between the constituent quark–antiquark pairs and all procedures are very similar.

8.3.2 Monte Carlo realization of QCD models

The theoretical model introduced in the previous subsection does not have that many physical free parameters, but the desire to reproduce well the experimental results introduces many 'switches' that are very important in its practical implementation. The first consistent Monte Carlo implementation of a QCD interaction model is the Pythia [247] code. All later implementations borrow at least some of the solutions developed in Pythia.

The general approach is as follows:

– All cross-sections are usually calculated and tabulated for the use in the code. These calculations are CPU-intensive and lead to a long initialization time of the code.
– The code is initialized with the current values of all 'switches'.
– The number of collisions is chosen from the precalculated distribution as a function of $\sqrt{s}$ – in the case of the minijet model this is the number of minijet pairs n.
– The values of n pairs of x_1 and x_2 are chosen from the structure functions used in the model.
– The total interaction energy left for the 'soft' interaction is decreased to $\sqrt{s}(1 - \sum x_i)$.
– $n + 1$ string pairs are created, n of them for the minijet pairs and one for the soft DPM component. The DPM string pair uses the fraction of the interaction energy left from the n other strings.
– The strings are fragmented and hadronized. The type and momentum of each secondary are calculated

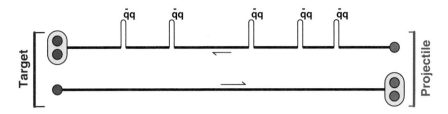

Fig. 8.20. Pictorial representation of the string creation and fragmentation process.

Figure 8.20 demonstrates how the strings are fragmented and hadronized. Each string is fragmented in its center of mass. At this time the string is fully described by its energy, momentum, and quark types on its ends. To start the fragmentation, one of the string ends is chosen at random. One quark–antiquark pair is created that combines with a fraction of the string to create a hadron. All conservation laws are applied to determine the type of the new hadron. The new pair could also be diquark–diantiquark, but with much smaller probability of order 4–5%. In such cases the created new particles are a baryon–antibaryon pair.

The fraction of the string energy that goes into the new particle is chosen by a fragmentation function. Pythia uses the Lund fragmentation function

$$F(z) = \frac{(1-z)^a}{z} \exp\left(\frac{-bm_T^2}{z}\right) , \qquad (8.23)$$

where z is the fractional energy of the produced hadron relative to its parent quark or diquark and the transverse mass $m_T = \sqrt{m^2 + p_\perp^2}$. The transverse momentum is sampled independently for every quark, antiquark or diquark. The transverse momentum of the generated hadron is the vector sum of the $p_\perp$ of its constituents.

Quarks of all flavors are sampled, but with different probability. Up and down quark pairs, for example, are about four times more abundant than $s\bar{s}$ pairs.

After the creation of the new hadron is completed one is left with a shorter (less energetic) string, that is also fully defined. The previous parent quark is now replaced with the member of the sampled quark pair that does not participate in the newly created hadron. The same algorithm is applied to this shorter string. The process continues until the string is not long enough to generate any particles, i.e. its total energy is less than the masses of two particles. This final step is very important and its solution is usually quite complicated. In general terms the solution is to use a threshold mass of the string to stop the fragmentation process and create two final particles, that obey all conservation laws.

All strings are fragmented in the same way, and the secondary particles energy and momenta are boosted from the string rest frame to CMS or Lab systems.

Much more attention to detail than the brief description above is required for a good description of the hadronic interactions. Of special importance is the calculation of the diffractive cross-sections and the implementation of diffractive interactions. These are interactions where one or both of the interacting nucleons becomes an excited state that decays into a small number of particles. Diffractive interactions are a non-negligible part of all inelastic interactions of about 20%. They are treated in a nonstandard way in different theoretical models and are often the reason for disagreements between the models.

Different type of problems are created and solved for interactions on nuclear target. It is well known that the cross-section for inelastic interactions of hadrons on target of mass A is proportional approximately to the 'area' $A^{2/3}$ of the target nucleus. This high cross-section indicates that the projectile nucleon interacts with more than one of the target nucleons. These are called 'wounded' nucleons. The average number of wounded nucleons $\langle \nu \rangle$ can easily be calculated by the ratio of the cross-sections, i.e.

$$\langle \nu \rangle = \frac{A\sigma_{inel}^{hp}}{\sigma_{inel}^{hA}} , \tag{8.24}$$

where we define σ_{inel}^{hA} in the same way as the proton target inelastic cross-section, i.e. when secondary particles are generated. As an example we can calculate $\langle \nu \rangle$ for proton interactions on nitrogen. The inelastic cross-section for protons on nitrogen σ_{inel}^{pN} is 265 mb at about 100 GeV, while σ_{inel}^{pp} is 32 mb. The average number of wounded nucleons is $\langle \nu \rangle = 1.7$, which means that most of the time the interactions are peripheral with only one wounded nucleon, while occasionally they could be as many as four or five wounded nucleons.

The application of the string model to hadron nucleus collisions was discussed in Ref. [248] and is illustrated in Fig. 8.21. The first step is to calculate the number of wounded nucleons, n_W, in the target. Once this is done, the question is how to create the strings, i.e. how to connect the single projectile nucleon to several wounded nucleons in the target. The suggestion is that one of the string pairs is, as in Fig. 8.20, drawn between the projectile and one of the target nucleons. All other $n_W - 1$ string pairs are between the target nucleons and $q\bar{q}$ pairs of the sea associated with the projectile. In Fig. 8.21 the top and bottom strings connect the projectile with one of the wounded nucleons of the target. The two strings in the middle connect another target nucleon to a sea $q\bar{q}$ pair.

The main difference between the first string pair and all others comes from the structure functions for valence (constituent) quarks and sea quarks. The valence quarks take most of the projectile hadron energy, sea quarks take on the average much less – they have much steeper structure functions. If the structure function for valence quarks is for example $f_v = [x^2 + \mu^2/s]^{-1/4}$ as it is in the original SIBYLL model [249], the one for sea quarks is almost

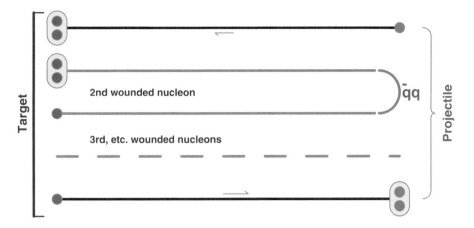

Fig. 8.21. Pictorial representation of the string creation and fragmentation process in hadron–nucleus interactions.

twice as soft – $f_s = [x^2 + \mu^2/s]^{-1/2}$. Because of this difference, most of the secondaries generated in the strings connected to the projectile sea are emitted in the backwards and central regions of the interaction, i.e. at zero or negative CM rapidities as is observed in experiments. At higher energy some minijet production would be associated with the individual soft strings.

The same principles can be used for modeling nucleus–nucleus interactions. The picture becomes very complicated as many nucleons can participate in more than one collision.

8.3.3 Contemporary models used for the analysis of air shower data

The minijet model is only one of the models that are used in simulations of the hadronic interactions at high energy. Some features of its implementation discussed in Sect. 8.3.2 were shown several years ago not to be very exact. All the faults of the model are related to the use a constant term for the soft processes in (8.21). The QCD term in the same equation depends very strongly on the parton structure functions. The HERA collider measured the structure functions to small x values down to almost $x = 10^{-6}$, almost two orders of magnitude better than in previous measurements [250, 251]. HERA measures the structure function by studies of the differential proton–photon cross-sections. Since at low x values most of the interaction cross-section comes from the partons with the softest structure function, the measured structure function is the gluon one. It was demonstrated that the structure function rises more rapidly than had been assumed in previous estimates. Applied in (8.19) the new structure functions give to σ_{QCD}, and thus to the total inelastic cross-section, values that are higher than those measured.

That gives some advantage to alternative models that have two main differences from the one described above: (1) the soft cross-section is not energy-independent; and (2) there is a factor that compensates for the very rapid increase of σ_{QCD}.

Theoretically most of these models are based on the Gribov–Regge [252] theory and the exchange of pomerons [253]. A pomeron is a hypothetical exchange particle, which in its mathematical definition is the pole of a partial wave in the scattering process. Theory gives the tools for the calculation of the amplitudes of the pomeron exchanges.

There are soft and hard pomerons, that serve the same purpose as χ_{soft} and χ_{QCD} above. The cross-section for exchange of hard pomerons is related to the gluon structure functions and its theoretical description. Pomeron exchanges can be used to predict the high energy behavior of the diffractive cross-section.

Apart from the mathematical structure of the theory, its practical results are that both the soft and the hard interactions are energy-dependent. This leads to a better fit of the measured cross-sections in the accelerator energy range. The other advantage is that the minimum momentum exchange Q^2_{min} is also energy-dependent. From practical point of view the decrease of the available phase space in integrations like (8.19) lead to a smaller QCD cross-section that fits the experimental data even with the very steep structure functions.

In physical terms it may serve another purpose – it may represent the shadowing of the partons that are associated with a high energy hadron. When the parton density (at low x values and high energy) reaches a very high value, the individual partons cannot see each other and thus interact; they are obscured by intervening particles. This is obvious in the simple geometrical definition of a cross-section, but certainly also happens in the real world.

There are several codes that have been used in the last ten years for analysis of experimental data from air showers. The real testing of these models against each other, accelerator data, and cosmic ray data started with the CORSIKA project of the KASCADE collaboration. CORSIKA [254] is a general-purpose Monte Carlo code, created and maintained by the KASCADE group. The use of a single code with different hadronic interaction models allowed for the first time a real comparison between different hadronic interaction models and between different sets of experimental results. Another similar code, AIRES [255], which is a descendant of Hillas' code MOCCA was more recently developed.

The interaction models used vary from those based on the pomeron exchange: DPMJET [256] and QGSJET [257], the original and updated versions of SIBYLL [258], to a simpler parametrization of experimental results – HDPM [259]. All models are used with the same independent low energy code at energy below 80 GeV. A recent joint publication [260] compares the

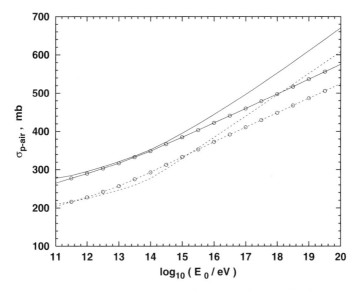

Fig. 8.22. Inelastic cross-section for proton (solid lines) and pion (dash lines) inter-actions in air in the SIBYLL and QGSJET (points and lines) hadronic interaction codes.

two shower simulation systems and the currently implemented interaction models in detail. The general conclusion is that with the development of the testing technique and the constant improvement of the models the differences between them have very much decreased.

They have not disappeared, though. As an example, I show a comparison of some of the most important model features in QGSJET and the current version of SIBYLL 2.1. Figure 8.22 compares the inelastic cross-sections for protons and pions in air and Fig. 8.23 compares their charged particle multi-plicities as a function of the energy. Both of these models have been developed specifically for calculations of cosmic ray showers. They have sacrificed some small cross-section processes and secondary particles to achieve execution time that allows calculation of the highest energy cosmic ray showers.

One conclusion is that in the energy range where models have to reproduce experimental data they are more or less in agreement. At energies above 10^6 GeV, however, where there are no data, the models begin to extrapolate the interaction features very differently. SIBYLL predicts a much higher inter-action cross-section, by almost 100 mb at 10^{11} GeV. QGSJET, on the other hand, generates more than twice as many secondaries at the same energy.

In spite of these huge differences these two models generate showers that are not very different up to energies of 10^{18} eV. The reason is that both the large multiplicity and the large cross-section speed up the shower develop-ment. What SIBYLL does with its large cross-section, QGSJET does with its high multiplicity. For example, the depth of shower maximum at 10^{18} eV

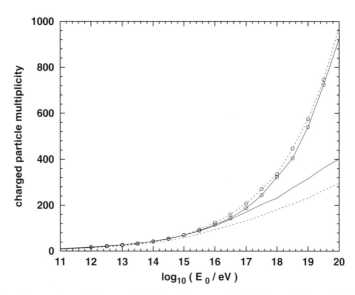

Fig. 8.23. Charged particles multiplicity in proton (solid) and pion (dash) interactions in air predicted by SIBYLL and QGSJET (points and lines).

is 740 g/cm^2 for SIBYLL and 730 g/cm^2 for QGSJET. This difference is practically the same as at all energies up to 10^{18} eV and grows to 24 g/cm^2 at 10^{20} eV.

The compensation does not affect all shower components. The much bigger charged multiplicity of QGSJET increases the number of GeV muons in the shower, which is consistently higher than in SIBYLL. The difference reaches 30% at 10^{20} eV.

8.4 Energy spectrum and composition at the knee

The energy range where the cosmic ray spectrum changes its slope is called 'the knee'. Its existence was first suggested by the Moscow State University group on the basis of their air shower data. Many groups have studied the knee region and the change of the cosmic ray spectrum is well established. Up to an energy of 10^6 GeV the spectrum of all cosmic ray nuclei is a power law with differential spectral index α of 2.70–2.75. The spectral index increases by $\Delta\alpha$ of about 0.3. It is possible that just before the change of the spectral index there is a flattening of the spectrum. A collection of experimental data on the cosmic ray spectrum in this region is shown in Fig. 8.24.

The full triangles in Fig. 8.24 are results from direct observations. They come from the satellite experiment Proton-4 [261] and the inverted ones are from JACEE [133]. The data of these two experiments to the all-particle flux are more reliable than the data on its individual components because of their

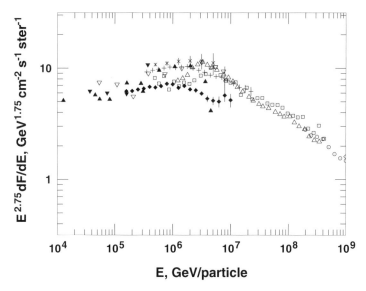

Fig. 8.24. Energy spectrum of all cosmic ray nuclei in the region of the knee. The differential spectrum is multiplied by $E^{2.75}$ to make the details in its shape more obvious. See text for references for the experimental work.

low charge resolution. The inverted open triangles are from the Tien-Shan experiment [262] and the open triangles show a more recent result of the Moscow University group [263]. The Tien-Shan experiment is at an altitude above 3,000 meters, close to the shower maximum, and can thus measure the spectrum at low energy. Such an overlap with direct measurements is extremely valuable, because it can be used for calibration of the air shower energy and composition estimates. The crosses are results of the Tibet air shower array [264] and the 'x's from the HEGRA experiment [238]. The Tibet array is at an extremely high altitude of 4,300 meters above sea level, which is almost at shower maximum. HEGRA is located at a more moderate altitude in the Canary islands. The full diamonds are the results of the CASA/MIA air shower array in Utah [265] and the open squares from the the Akeno 1 km^2 array [266]. The open circles at high energy come from the stereo measurement of the Fly's Eye fluorescence detector [267]. We do not include the spectrum given in Fig. 5 of Ref. [268] only because it is labeled as a preliminary result. If we put it in Fig. 8.24 the value at 10^6 GeV would be at the level of the HEGRA flux.

Different data sets in Fig. 8.24 disagree as much as factor of two. Part of the difference comes from relatively small differences in the energy estimates. Since the differential spectrum is multiplied by $E^{2.75}$ the vertical normalization is very sensitive to the energy assignment. If all other experimental efficiencies are well understood, one can estimate the difference in the energy assignment assuming that experiment 1 assigns energy which is k the energy of experiment 2 ($E_1 = kE_2$). In this case the ratio of the two energy assignments is $(\mathcal{F}_2/\mathcal{F}_1)^{1/1.75}$. For the maximum difference of factor of 2 this gives energy assignments different by almost 50%.

In addition to the difference in the normalization different experiments show different positions and shapes of the knee. CASA-MIA and Tibet measure the change of the spectral index at an energy slightly above 10^6 GeV, while many other experiments see at energy exceeding 3×10^6 GeV. Akeno measures a break of the spectrum, while Tibet sees a very broad transition region.

The inconsistency of the experimental data makes the interpretation of the knee difficult. There is no lack of theoretical ideas. As early as 1959 Peters [269] suggested that the knee is a rigidity-dependent effect. It could be related to a maximum rigidity that can be achieved in acceleration processes in the most numerous type of source, rigidity-dependent escape or a change of the propagation pattern in the Galaxy. It is difficult to assume that the change of the spectral index is indeed related to escape. The gyroradius of a 10^6 GeV proton in 1 μG field is only one parsec. The typical galactic fields are stronger and the gyroradii of heavier nuclei are significantly smaller as well. One parsec is the typical diameter of supernova shells 1,000 years after the explosion, which should be able to contain 10^6 GeV protons.

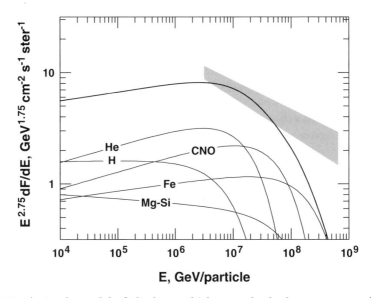

Fig. 8.25. A simple model of the knee which extends the low energy model presented in Chapter 5 to high energy with an exponential cutoff at 10^7 GeV. The shaded area shows the range populated by data in Fig. 8.24.

On the other hand, the rigidity-dependent effect is an attractive idea. It could indeed be related to the maximum acceleration energy in the cosmic ray sources. Figure 8.25 shows a very simple flux model with rigidity-dependence. It uses the spectrum and composition shown in Chap. 5 and extends it to high energy with exponential cutoff in rigidity at 10^7 GV. The thin lines show the contribution of different nuclear groups to the all-particle spectrum. The proton spectrum turns over at 10^7 GV and those of heavier nuclei turn over at energies of $Z \times 10^7$ GeV. At energies above 10^8 GeV there are only heavy, high Z, nuclei in the cosmic ray flux. The end of the modeled spectrum is where the Fe component is also exponentially cut off.

The shaded area outlines the experimental data at $\log E/\text{GeV} > 6.5$. Some of the features of the measured spectrum are well reproduced. There is, for example, a flattening of the spectrum at about 10^6 GeV. The mild turnover of the spectrum at 4×10^6 GeV is consistent with measurements that show a smooth transition range. Above 3×10^7 GeV, however, the model is not consistent with anything. The experimental data show that the spectrum continues to higher energy and to fit both the knee and the spectrum above 10^8 GeV one needs to introduce a new component. The main difficulty with a new component is that it should have a steeper energy spectrum. It is easy to fit together a new component when it is flatter than the lower energy one – it is not very important at low energy and can be easily 'hidden' under the dominating steeper spectrum. In this case, however, one has to 'fine tune' the component so that it does not dominate at low energy.

One such component could be the acceleration on supernova shocks in the WR star winds suggested by Völk and Biermann [61]. The higher energy is achieved because of the higher magnetic field values in the WR star environment. The number of SNR from stars with very high surface magnetic fields and dense wind environment is very small. The spectrum could be steeper because of the decreasing number of SNR that can reach higher energy. No finetuning is necessary because the type of cosmic ray sources is the same. It is possible that the application of the contemporary acceleration models to a distribution of SNR with precursors of high surface magnetic field could fit the observed spectrum. I believe that such a model has not been developed because the required astronomical information does not yet exist.

The rigidity-dependent models naturally predict the cosmic ray chemical composition. At energy above 10^7 GeV the spectrum of the simple model of Fig. 8.25 is totally free of protons (H nuclei). At 2×10^7 the He nuclei start disappearing. The He spectrum seems much higher than the proton one because of its flatter spectr al index. The composition will be constantly changing in the knee region and these changes are predicted by the rigidity dependent models.

A classical way to present results on the cosmic ray composition estimated from air shower data is to give the average value of the logarithm of the primary particle mass $\langle \ln A \rangle$. Different composition estimates are not in better agreement than the results on the spectrum. As an illustration Fig. 8.26

presents the results from the analyses of data from the KASCADE [268] and EASTOP [270] experiments.

The EASTOP result comes from a measurement of the muon density at 180–210 meters from the shower core ρ_μ^{200} as a function of the total number of electrons in the shower, N_e. The authors prefer to deal with muon density rather than total number of muons, to avoid the fitting of the muon lateral distribution and the related errors. The measured density of muons (ρ_μ^{200}) of energy above 1 GeV (for vertical showers) is compared to Monte Carlo simulations at energies between 2×10^6 and 8×10^6 GeV, covering the knee range. The comparison shows that the composition becomes significantly heavier. It is consistent with the Monte Carlo predictions for He showers at the beginning of the range, while towards its end it is better consistent with showers of CNO origin. In addition to the comparisons of the average values of ρ_μ^{200}, the distributions of those parameters are presented and fitted with calculated ones for single components. The fits agree well with the average estimates. In a fraction of a decade in energy $\langle \ln A \rangle$ increases by 0.9 ± 0.1. This result is shown in Fig. 8.26 with a box.

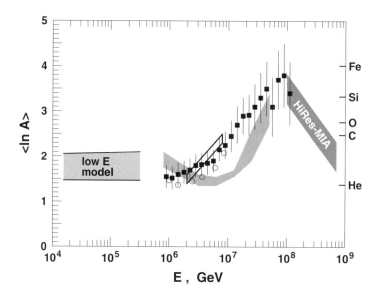

Fig. 8.26. Results from studies of the cosmic ray composition in the region of the knee. See the text for explanation of the symbols and for references.

The three other results are obtained by the KASCADE experiment. We prefer showing different results from the same experiment to emphasize the difficulty in the derivation of the cosmic ray composition from air shower data. The basis for all derivations of the composition are Monte Carlo calculations. The U-shaped shaded band comes from a neural network analysis. A neural network is trained to recognize showers initiated by different primary nuclei at different zenith angles from their electron N_e and muon N_μ sizes. When given real detected events the neural network returns the primary mass. The shaded band in Fig. 8.26 shows the results for three bins in zenith angle from $0°$ to $32°$.

The squares come from a different type of analysis of N_e and N_μ called 'unfolding' by the collaboration. I will define this procedure better because it is actually the basis of all attempts to determine the composition, independently of what it is called by the experimental groups.

$$\frac{d\mathcal{F}}{d\ln N_e d\ln N_\mu} = \sum_A \int \frac{d\mathcal{F}_A(\ln E)}{d\ln E} p_A(\ln N_e, \ln N_\mu, \ln E) \, d\ln E \ . \quad (8.25)$$

The sum is in principle over all nuclei that are present in the cosmic ray flux, although experiments use not more than five different chemical groups. The probability that a shower of primary mass A and energy E will create N_e and N_μ at the observation level is the crucial quantity $p_A(\ln N_e, \ln N_\mu, \ln E)$ that is derived from Monte Carlo calculations. The derivatives and the integration in (8.25) are logarithmic, as they are in the KASCADE analysis, to coincide with the resolution of air shower arrays. In practical terms, what was done by KASCADE is that N_e and N_μ distributions were calculated for showers of primary mass A and energy E and a range of zenith angles. These distributions are parametrized and the primary energy, primary mass, angle and size variations of the parameters are used to analyze the experimental data with four mass groups. The results are not very different from those of the neural networks, except maybe at energy about 10^7 GeV, where they show slightly heavier composition. The results from the unfolding are in very good agreement with an earlier [271] analysis of the same experimental parameters that is shown with circles. All four analyses agree that the cosmic ray composition is becoming heavier in the region of the knee. This is consistent also with the direct measurements of JACEE and RUNJOB that approach (with much lower statistics) this energy range.

The band above 10^8 GeV that shows a different trend is obtained by the HiRes-MIA hybrid experiment [272]. Hybrid means that the fluorescent experiment uses the GeV muons detected by MIA to improve the mono reconstruction of the showers. The shower reconstruction gives the two main parameters used in the composition study: the primary energy from the integral observed pathlength and depth of maximum $X_{\max}$ from the shower profile. The energy-dependence of $X_{\max}$ is compared to calculations for protons and Fe nuclei done with different hadronic interaction models. Because of the model differences the value of $\langle \ln A \rangle$ cannot be determined but a clear trend exists for a composition that is becoming lighter above 10^8 GeV.

The trend is obvious from the value of the observed elongation rate (ER). ER is the change of $X_{\max}$ per decade of energy, i.e. $dX_{\max}/d\log E$. The elongation rates calculated with the QGSJET model are 58.5 g/cm^2

for proton showers and 61 g/cm² for iron showers. There is not a strong model-dependence in the elongation rate in this energy region, even though the predicted X_{max} values are different. The measured elongation rate is $93.0 \pm 8.5 \pm 10.5\,g/cm^2$, where the second error is systematic. Since this number is larger than the calculated ones, it means a transition from early developing shower to deep showers. This can only be caused by the emergence of a light cosmic ray component. If a smaller ER than predicted were measured, it would indicate a shift to heavier composition. The HiRes-MIA group estimates $\Delta\langle \ln A \rangle = -1.5$. I have normalized by eye $\langle \ln A \rangle$ at 10^8 GeV to the KASCADE value to illustrate the dramatic change in the composition suggested by the HiRes-MIA experiment.

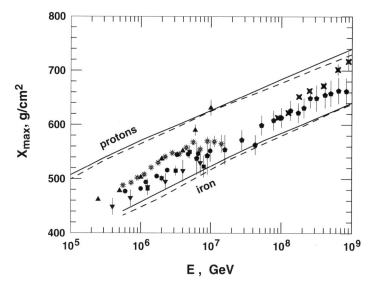

Fig. 8.27. Average depth of shower maximum as a function of the shower energy measured by several experiments. See text for symbols and references.

The study of the elongation rate is a widely used method for the derivation of the cosmic ray composition. Figure 8.27 shows a set of results on X_{max} as a function of the shower energy. The 'x's show the result of HiRes-MIA [272] and the pentagons are from the Yakutsk air shower array [273]. All other results are strictly from observations of the shower's Cherenkov light emission. The asterisks are from CASA-BLANCA [274], the triangles from DICE [275], and the inverted triangles from HEGRA [238]. This last data set includes the systematic errors in the data analysis. The dots come from the SPASE-Vulcan group [239] and the squares from CACTI [276]. The DICE experiment uses the time profile of the Cherenkov signal to extract the depth of maximum, while all other groups use the lateral distribution of the Cherenkov light.

The solid lines are the average depths of proton and Fe-initiated showers calculated with SIBYLL 2.1. The dashed lines are calculated with QGSJET. The two models agree quite well on the value of ER and the differences in the calculated X_{max} are minimal.

All data sets except for DICE show the same trend – elongation rate consistent with more or less constant (or becoming lighter) composition and a strong decrease of the ER around or above 10^7 GeV, indicating a heavier composition. There is a difference in the X_{max} values that are of the order of 20–30 g/cm^2, which are barely consistent with the systematic errors derived by the HEGRA group. It is difficult to discuss the result of DICE, where the experimental technique is still under development. Only two of its points are notably different from the rest of the sample.

The majority of the experiments measure a cosmic ray composition that becomes heavier between $5{\times}10^6$ and 10^7 GeV. The differences in the absolute X_{max} values are large, but still significantly smaller than the predicted values for proton and Fe showers. The few points above 10^7 GeV agree with the measurement of the Yakutsk array. The measurement of HiRes/MIA also agrees with Yakutsk around 10^8 GeV. Soon afterwards, however, the HiRes-MIA data show a definite trend to a lighter composition which the Yakutsk set does not confirm.

It is difficult to draw a definite conclusion about the exact changes of the cosmic ray spectrum and composition at the knee. All experiments agree that the cosmic ray spectrum steepens above 10^6 GeV. The exact position of the spectral change and the width of the transition region are not yet well determined. The composition studies, both with surface air shower arrays and with optical detectors, indicate a change in the average mass of the cosmic ray nuclei after the steepening of the spectrum, once again with large uncertainty in the energy range and shape. All these numerous data sets are consistent with a rigidity effect, either in the cosmic ray acceleration or in their propagation.

A very interesting result – the emergence of a light cosmic ray component – is published by the HiRes-MIA group. It is not confirmed by other measurements in the same energy range but is fully consistent with the data at higher energies that are discussed in the next chapter.

9 The end of the cosmic ray spectrum

The cosmic ray spectrum is so steep that even with the relatively big and stable air shower arrays that were built in 1950's it was not obvious whether it ends or it continues forever to really huge energies. It seemed quite possible that there are showers of energy 10^{20} eV and higher that cannot be seen only because of their very low flux. With a spectral index of 2.7 each higher decade of energy decreases the flux by a factor of 50. After the knee, which has already been suggested, and the steepening of the spectrum each decade would decrease the flux by more than a factor of 100.

The scientists involved in cosmic ray research became very optimistic when in 1963 John Linsley reported on a shower of energy more than 10^{20} eV detected by the Volcano Ranch array [277]. During the next couple of years many projects for building huge air shower arrays to detect similar showers were proposed and innovative approaches were discussed. Although today's sophisticated particle acceleration theory had not yet been developed, the general assumption was that these ultra high energy cosmic rays (UHECR) are of extragalactic origin. The gyroradius of a 10^{20} eV proton is of the order of the dimension of the whole Galaxy and it cannot even contain such events. So they have to be accelerated in much bigger and more powerful astrophysical systems [278].

It took an important discovery of an entirely different character to change the tone of the discussion. In 1965 Penzias & Wilson [279] announced the discovery of the cosmic microwave background (CMB) radiation. It is the remnant of the big bang, of the time when the Universe was very hot, which now has cooled down by its expansion.

Soon after this discovery two papers on cosmic rays came out almost simultaneously in the Soviet Union and in the USA [280, 281]. Both of them predicted that the cosmic ray spectrum would end at an energy just below 10^{20} eV because the high energy nuclei would interact with the CMB and lose energy. This process would cut off the cosmic ray spectrum and, even if cosmic ray particles were accelerated to higher energies, they would not be able to survive the propagation from their sources to us. This prediction has since been called the Greisen–Zatsepin–Kuzmin (GZK) cutoff.

In his article 'End to the cosmic ray spectrum?' Greisen [281] uses the 3 K temperature of the CMB and calculates the energy loss distance of 10^{20} eV protons to be about 13 Mpc because of photoproduction interactions. The CMB photons would excite the giant dipole resonance in heavy nuclei and disintegrate them at similar lengths. Greisen also points out that in addition to the photoproduction interactions lower energy protons will produce e^+e^- pairs in the CMB and this process would peak at about 3×10^{19} eV.

These estimates are still valid today. Recent, much more accurate, data on the CMB and the improved knowledge of photoproduction interactions allow for more exact predictions, but there is still experimental evidence that cosmic rays of energy exceeding the GZK cutoff exist.

9.1 Cosmic microwave background

When Penzias and Wilson discovered that the noise in their antenna was not due to a technical problem they proved the existence of the CMB and started a new era of cosmology. The existence of CMB is the strongest evidence of the existence of the big bang and the related processes of nucleosynthesis. It was suggested on such grounds by George Gamow.

It is now proven that the cosmic microwave background has a perfect blackbody spectrum with temperature of 2.73 K. The best estimate of the temperature is 2.725 ± 0.002 K [282]. Figure 9.1 shows the energy density of the microwave background in units of eV/cm^{-3}. In the same units the energy density $\rho_E(\epsilon)$ of the blackbody (Planck) spectrum is given as

$$\rho_E(\epsilon) = \frac{1.32 \times 10^{13} \epsilon^4}{[\exp(kT/\epsilon) - 1]} , \qquad (9.1)$$

where k is Boltzmann's constant. For a temperature of 2.725 K the number density of the microwave background is 411 cm^{-3}, its total energy density is 0.26 eV cm^{-3} and the average energy of the microwave photons is 6.34×10^{-4} eV.

The microwave background is universal and isotropic. The anisotropy of its temperature is on the 10^{-5} level after the effects of the Earth's, the solar system's and the galactic motion are subtracted.

It is important to note that, because of the expansion of the Universe, the temperature of the microwave background increases with the redshift, z, as $(1+z)$. For the same reason its number density increases as $(1+z)^3$. The total energy density increases correspondingly as $(1+z)^4$ and the energy spectrum of CMB at any redshift can obtained by shifting the curve in Fig. 9.1 to higher energy with $(1+z)$ and scaling it up by $(1+z)^3$.

9.2 UHECR interactions on the microwave background

The energy threshold for secondary particle production is the center of mass energy that equals the masses of all produced particles. A proton interacting

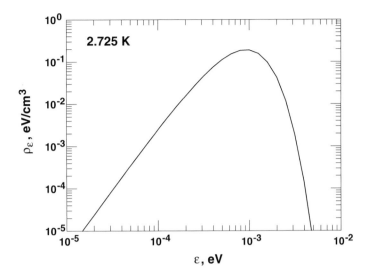

Fig. 9.1. Energy density of the microwave background.

with a photon could produce a π^0 if the CM energy of the two interacting particles equals the sum of their masses – $\sqrt{s} = m_p + m_{\pi^0}$. Such reactions are possible if every proton has some hadronic content. Figuratively speaking the photon behaves like a pion 1/100 of the time.

One can calculate the proton energy at which such reactions are possible with the photons of the cosmic microwave background. In the Lab system the square of the CM energy is

$$s = m_p^2 + 2E_p\epsilon(1 - \beta_p \cos\theta) \,, \tag{9.2}$$

where ϵ is the energy of the photon and $\cos\theta$ is the angle between the two particles. In face to face collisions $\cos\theta = -1$ and for the average energy of the CMB (6.34×10^{-4} eV) the threshold energy becomes

$$E_p = \frac{m_{\pi^0}}{4\epsilon}(2m_p + m_{\pi^0}) \simeq 10^{20} \text{ eV} \,.$$

Interactions are possible at energies smaller by a factor of 3 or 4 as the microwave spectrum extends to higher energy, as shown in Fig. 9.1. Since the proton energies are very high, in the future we will replace the proton velocity β_p with unity.

These interactions are called photoproduction and are very well studied in accelerator physics. In laboratory experiments one uses a proton target and high energy photons and the reaction is described in the nucleon rest frame (NRF). In this frame the square of the CM energy is $s = m_p(m_p+2\epsilon')$, where ϵ' is the photon energy in the nucleon rest frame. The threshold photon energy in NRF could be similarly calculated: $\epsilon' = m_{\pi^0}(1 + \frac{m_{\pi^0}}{2m_p})$ and is 145 MeV. A

collection of experimental data on the photoproduction cross-section is shown
in Fig. 9.2 together with a fit of its energy-dependence [283].

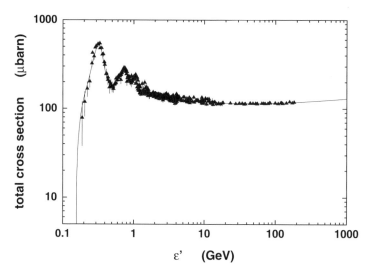

Fig. 9.2. Energy dependence of the total photoproduction cross-section as a function of the photon energy in the nucleon rest frame.

There are several processes that contribute to the total cross-section. The
most important is the $\Delta^+(1,232)$ resonance production that peaks at about ϵ'
$= 340$ MeV. At this energy the proton photoproduction cross-section exceeds
0.5 mb. At slightly higher ϵ' the cross-section decreases and is dominated by
higher mass resonances. For this reason it has a very complicated shape. At
still higher energy, above 5 GeV, the cross-section becomes very smooth –
this is the multiparticle production region that can be described with the
same QCD models that are used for pp interactions. At very high energy the
photoproduction cross-section has a slight logarithmic (or $(\epsilon')^\alpha$) increase.

The most important region for astrophysical applications is the threshold
and the Δ^+ region, where the cross-section is the highest. The steeply falling
cosmic ray proton spectrum makes the high energy behavior irrelevant for the
interaction rate in the cosmic microwave background. It becomes important in
interactions with much more energetic photons – optical, UV and X-ray fields.
Some details of the energy spectra of the final state particles are, however,
influenced by the correct description of the interaction process, which is in
general different from Δ^+ production.

The cross-section for photon interactions on neutrons $\sigma_{\gamma n}$ is quite similar
to $\sigma_{\gamma p}$ except for ϵ' around 1 GeV, where it is lower.

Another interaction parameter that is as important as the total interaction cross-section is the proton inelasticity – the fraction of the energy that the proton loses in the interaction. K_{inel} can not be measured as directly as the total cross-section and has to be derived from the composition of the final state particles in interactions of different energy.

To give the reader an idea of the importance of this quantity we show in Fig. 9.3 the inelasticity distribution for interactions of protons of different energy on the CMB. The inelasticity distributions are integrated over the CMB energy spectrum and all possible proton–photon angles. The interactions are simulated with the photoproduction event generator SOPHIA [283].

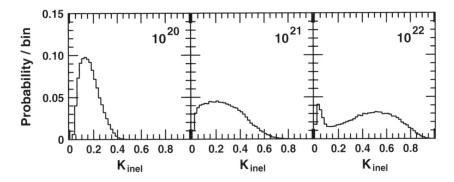

Fig. 9.3. Distribution of the inelasticity coefficient K_{inel} for interactions of protons of three different energies in the cosmic microwave background.

All interactions of 10^{20} eV protons are in the threshold region of the cross-section. Their energy loss (mostly in the Δ^+ energy range) is limited by the ratio of the proton and pion masses. The average inelasticity is 0.17. With the increase of the proton energy the contribution of heavier resonances and the multiparticle production process increases and protons start losing more and more energy – the average K_{inel} becomes $0.27\, E_p$ at 10^{21} eV. At still higher energy the multiparticle production process dominates with its almost symmetric K_{inel} distribution peaked at 0.5. The other notable feature is a diffractive channel in which only one fast pion is produced. The average inelasticity is 0.42 and does not increase more with the proton energy.

Another energy loss process is the production of electron–positron pairs. This is electromagnetic process in which the photon generates a pair in the proton nuclear field, similarly to Sect. 2.2.6. The proton threshold energy is much lower, since only two electron masses have to be added to the proton mass in the center of mass system. For face to face collisions the threshold is about 4×10^{17} eV. Berezinsky [5] gives a simple approximate formula for the cross-section at the threshold, for $E_p \ll m_e m_p / kT \simeq 2 \times 10^{18}$ eV.

$$\sigma_{pair} = \frac{\pi}{12}\alpha r_e^2 \left(\frac{\epsilon'}{m_e} - 2\right), \tag{9.3}$$

where ϵ' is the photon energy in NRF. The proton energy loss per interaction is small, of the order of $2m_e/m_p$ or about $0.001E_p$ and for this reason the pair-production energy loss is usually treated as a continuous energy loss.

Nuclei heavier than protons have also another source of energy loss – photodisintegration. The dominant process is the giant dipole resonance induced in the nuclei by the microwave background or any other photon field. The giant dipole resonance cross-section peaks in the ϵ' energy range 10–30 MeV. The nucleus absorbs the photon and forms an excited state, which decays, releasing one or two nucleons. The photoabsorption cross-section roughly obeys the Thomas–Reiche–Kuhn sum rule. It is usually defined as

$$\sigma_{\text{phabs}} \equiv \int_0^\infty \sigma(\epsilon')\,d\epsilon' \simeq 60\frac{NZ}{A}. \tag{9.4}$$

The photoabsorption cross-section σ_{phabs} is measured in mb.MeV. In (9.4) A is the mass number, Z is the charge and N is the number of neutrons.

This is only a rough approximation of the real cross-section that depends on the stability of the nucleus. At energies, ϵ', lower than about 30 MeV the disintegration is dominated by the emission of one or two nucleons. At higher energy the emission of more than two nucleons is possible. Table 9.2, compiled from data given in Ref. [284], gives the threshold energies ϵ'_{thr} for the emission of different combinations of nucleons by four common and stable nuclei (Fe, Si, O and He) and, for comparison, Be.

Table 9.1. Energy thresholds (in MeV) for emission of different combinations of nucleons by several different nuclei [284].

	Z	A	n	p	2n	np	2p	α
Fe	26	56	11.2	10.2	20.5	20.4	18.3	7.6
Si	14	28	17.2	11.6	30.5	24.6	19.9	10.0
O	8	16	15.7	12.1	28.9	23.0	22.3	7.2
He	2	4	20.6	19.8	28.3	26.1	–	–
Be	4	9	1.7	16.9	20.6	18.9	29.3	2.5

Generally, because of its charge, it appears easier to emit a proton than a neutron. Stable nuclei are more difficult to disintegrate, although Table 9.2 shows no absolute rules. The five nuclei shown are part of the ^{56}Fe disintegration chain together with 55 more isotopes. All of their disintegration cross-sections and emission thresholds have to be accounted for in a ^{56}Fe photodisintegration calculation. We encourage the reader to go to the original papers, such as Refs. [285, 284] to learn more about the photodisintegration process.

9.2.1 Propagation of UHE protons in the Universe

The first step in a propagation calculation is to find the energy-dependence of the proton mean free path, $\lambda_{p\gamma}$, in the microwave background as a function of the proton energy, E_p. The mean free path is defined [286] as

$$\lambda_{p\gamma}^{-1}(E_p) \;=\; \frac{1}{8E_p^2}\int_{\epsilon_{thr}}^{\infty} d\epsilon\,\frac{n(\epsilon)}{\epsilon^2}\int_{s_{min}}^{s_{max}} ds(s-m_p^2)\sigma_{p\gamma}(s)\,. \qquad (9.5)$$

Here ϵ is the photon energy in eV and $n(\epsilon)$ is the photon number density in units of cm^{-3} eV^{-1}, s_{min} is the square of the minimum center of mass energy $(m_p + m_{\pi^0})^2$, and $s_{max} = m_p^2 + 4E_p\epsilon$ assuming that $\beta_p = 1$. The threshold photon energy

$$\epsilon_{thr} \;=\; \frac{s_{min} - m_p^2}{4E_p}$$

with the same assumption. The units of $\lambda_{p\gamma}$ are cm. Figure 9.4 shows the energy-dependence of $\lambda_{p\gamma}$ using the photoproduction cross-section from Fig. 9.2 with a dashed line. The mean free path reaches a minimum at about 5×10^{20} eV and slightly increases at higher energy as the peak of the photoproduction cross-section no longer coincides with the maximum photon density of the thermal distribution.

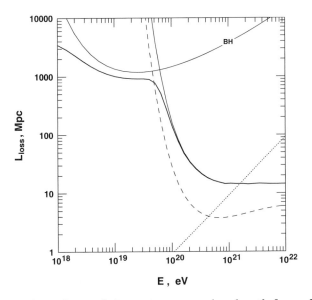

Fig. 9.4. Energy-dependence of the proton energy loss length from photoproduction, pair production (marked BH), and the total energy loss length L_{loss} in Mpc (thick line). The dashed line shows the proton interaction length in the microwave background, $\lambda_{p\gamma}$, and the dotted line shows the neutron decay length.

The mean energy loss length, L_{loss}, is the measure of the distance at which a particle loses its energy. It is defined as

$$L_{loss} = \frac{E_p}{dE_p/dx} = \frac{\lambda_{p\gamma}(E_p)}{K_{inel}(E_p)} . \tag{9.6}$$

Figure 9.4 shows with thin solid lines the energy loss length due to photo-production and for production of electron positron pairs (BH). Note that the difference between $\lambda_{p\gamma}$ and L_{loss} decreases from a factor of more than 5 at 10^{20} eV to about a factor of 2 at 10^{22} eV as the inelasticity coefficient increases as shown in Fig. 9.3. For energy above 10^{20} eV the photoproduction energy loss dominates and above 8×10^{20} the energy loss length is almost constant at about 15 Mpc. Below about 5×10^{19} eV the pair-production energy loss dominates, reaching a minimum L_{loss} at about 2×10^{19}. The energy loss on pair production in this figure is calculated following the recipe given in Ref. [288]. At the low energy end of the graph the energy loss length tends to become constant and equal to the adiabatic energy loss, which is due to the expansion of the Universe. For the Einstein–de Sitter model of a flat, matter dominated Universe the current energy loss length is

$$L_{loss}^{ad}(z = 0) = \frac{c}{H_0} \simeq 4,000 \text{ Mpc} \tag{9.7}$$

for $H_0 = 75$ km s^{-1} Mpc^{-1} which we use in this section. The best current estimate for the Hubble constant is $H_0 = 72 \pm 6$ km s^{-1} Mpc^{-1}. The redshift dependence of the adiabatic energy loss is

$$L_{loss}^{ad}(z) = L_{loss}^{ad}(z = 0)(1 + z)^{-3/2} \tag{9.8}$$

in the same model. The energy loss lengths for the particle production processes scale differently because of the changes of the energy and number density of the photon field. Their redshift dependence is

$$L_{loss}(E_p, z) = (1 + z)^{-3} L_{loss}[(1 + z)E_p, z = 0] . \tag{9.9}$$

The decay length of neutrons is $9 \times 10^{-12} \gamma_n$ Mpc, where γ_n is the neutron Lorentz factor. Neutrons of energy less than about 4×10^{20} eV on the average decay before they interact. Above that energy neutrons interact with about the same mean free path as the protons.

Figure 9.5 shows the spectrum of protons injected with energy between 10^{21} and $10^{21.1}$ eV after propagation at different distances from 10 to 4,000 Mpc.

The first propagation distance, 10 Mpc is about $2\lambda_{p\gamma}$ at that energy. So about 90% of the protons interact, some of them more than once, and produce a wide spectrum covering more than one order of magnitude. Slightly more than 10% of the injected protons have not yet lost any energy and populate the injection bin. At 40 Mpc the spectrum is even wider with a minor fraction

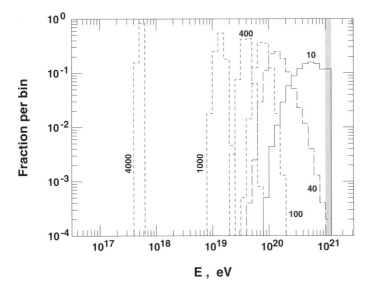

Fig. 9.5. Evolution of the proton energy after propagation in the microwave background on distances of 10, 40, 100, 400, 1000, and 4000 Mpc. The injection energy of the protons is shown with a shaded area - a E_p^{-2} spectrum between 10^{21} and $10^{21.1}$ eV.

of the protons still in the injection bin while some others have lost more than 90% of their energy. The process continues until most of the protons arrive at 100 Mpc with energy lower than 10^{20} eV. At that point the photoproduction interaction length has a steep energy-dependence – higher energy protons still interact, while the lower energy protons suffer mostly smaller pair production losses. As a result the higher energy protons lose more energy than the lower energy ones and the proton energy distribution becomes much narrower. At 1 Gpc the energy distribution peaks at about 10^{19} eV, just below the peak of the pair-production losses. The tightening of the energy spectrum continues as the higher energy protons suffer pair-production losses as the lower energy ones have only adiabatic loss. At that point the redshift is already significant ($z = 0.25$ for $H_0 = 75$ km s^{-1} Mpc^{-1}) and the adiabatic energy loss length decreases as in (9.8). At a distance of 4 Gpc all protons have energy lower than 10^{18} eV and in further propagation would only experience adiabatic losses.

These features of the proton energy evolution explain the formation of the expected proton spectrum at large distances from its source. Figure 9.6 shows the spectra of protons at different distances compared to their injection spectrum $E_p^{-2} \exp(-E_p/10^{21.5})$, which is shown with a smooth curve. As in the previous figure the propagation to 10 Mpc does not change drastically the proton spectrum. The highest energy particles do not survive, but otherwise the spectrum is similar to the injected one. The changes become obvious at

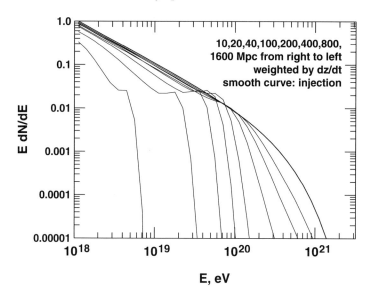

Fig. 9.6. Proton spectra at different distances from their source. The smooth curve shows the proton acceleration spectrum. Other curves, from right to left, show the same protons after propagation to 10, 20, 40, 100, 200, 400, 800, and 1,600 Mpc. All spectra are weighted with the redshift-dependent injection time corresponding to distance.

the distance of 40 Mpc. The maximum proton energy decreases by almost a full order of magnitude. A bump in the spectrum starts appearing at about 10^{20} eV and a slight dip starts to develop at energy about $10^{19.5}$ eV. The bump is caused by the higher energy protons that have lost energy and pile up where the energy loss length becomes significantly bigger. At 100 Mpc these features are fully developed and at higher distances the whole spectrum appears to move to lower energy, maintaining the pile-up and the dip at correspondingly lower energies. The pile-up and the dip were first identified as important features of the extragalactic proton spectrum by Berezinsky & Grigorieva [287].

Part of the lower normalization of the fluxes at 10^{18} eV for the large propagation distances are due to the leftwards shift of the whole spectrum due to adiabatic losses. Another part, however, is caused by the expansion of the Universe. A unit injection time scales with redshift and maintaining constant injection spectrum requires weighting of the spectrum with the metric element $\eta(z) = dt/dz$. The general definition is

$$\frac{dt}{dz} = \frac{1}{H_0(1+z)} \left[\Omega_M(1+z)^3 + \Omega_\Lambda + (1 - \Omega_M - \Omega_\Lambda)(1+z)^2 \right]^{-1/2} .$$

(9.10)

In the case of the Einstein–de Sitter model ($\Omega_M = 1$) the expression simplifies to $dt/dz = H_0^{-1}(1+z)^{-2.5}$, which is used for the normalization of the spectra shown in Fig. 9.6. The cosmological model most consistent with the current observational results is that with $\Omega_\Lambda = 0.7$ and $\Omega_M = 0.3$.

The curves shown in Fig. 9.6 are the input which is necessary for the calculation of the extragalactic proton spectrum. All that is needed is to integrate these yields in redshift using a particular cosmological model. The difference in cosmology is not very important because high redshifts do not contribute much to the UHECR spectrum detected at Earth. Even a propagation on 3,200 Mpc ($z = 0.8$ for $H_0 = 75$ km s^{-1} Mpc^{-1}) does not produce protons above 10^{18} eV. This is already the energy range where we believe that cosmic rays of galactic origin dominate.

The problem is the unknown source distribution. Since the possible luminous astrophysical systems are far apart, and the spectrum changes rapidly on propagation, it is important to know if the closest UHECR source is 10 Mpc or 100 Mpc away. Then there is the unknown cosmic ray acceleration spectrum. Many authors use a flat ($\gamma = 2$) power law spectrum with cutoff at some high energy similar to the one shown in Fig. 9.6. The acceleration experts, however, believe that the relativistic shocks observed in the powerful astrophysical systems would accelerate particles to a power law with spectral index of 2.2 to 2.3 [289].

9.2.2 Propagation of UHE nuclei

To calculate the energy loss distance for heavy nuclei due to photodisintegration one has to sum over the photodisintegration cross-sections of the nuclei of mass lower than the injection nucleus. In all papers available in the literature such calculations include the photodisintegration on photons with ϵ' between the threshold and 150 MeV. The assumption is that at higher energy the main energy loss process is photoproduction. In order to have a better impression of the propagation nuclei heavier than H we show in Fig. 9.7 the energy loss distance from Fig. 1 of Ref. [290] combined with the energy loss on photoproduction and the adiabatic energy loss from Fig. 9.4. Reference [290] calculates the energy loss distance for photodisintegration and pair production. The lines shown in Fig. 9.7 give the total energy loss distance as

$$\frac{1}{L_{loss}^{tot}} = \frac{1}{L_{loss}^{phdis,Pair}} + \frac{1}{L_{loss}^{p\gamma}} + \frac{1}{L_{loss}^{ad}} . \tag{9.11}$$

The wiggles of the curves at high energy are caused by inaccuracy in reading Fig. 1 of Ref. [290].

If one considers only the photodisintegration and pair production energy losses, little happens at total energy of the nucleus below 10^{19} eV. The energy per nucleon is below the pair production energy even for He nuclei, and only half or less of the nucleons are charged. He nuclei are affected first by pair production and soon after that by disintegration losses. The minimum energy loss distance of about 8 Mpc is reached at energies between 1 and 2×10^{20}

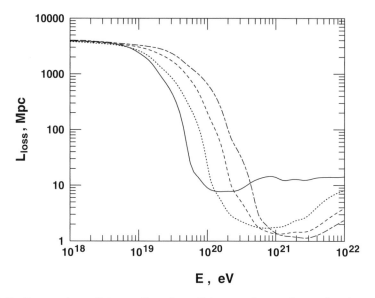

Fig. 9.7. Energy loss distance for photodisintegration, pair production, photo-production and adiabatic expansion of He (solid), O (dots), Si (dashes) and Fe (dash-dash) nuclei is plotted as a function of the total energy of the nuclei.

eV. At that point the photoproduction losses take over and the energy loss distance at high energy approaches 15 Mpc, the same as for protons.

With the increasing mass of the nuclei the thresholds for all these pro-cesses are increased. At the same time the minimum distance decreases with the nuclear mass. The minimum energy loss distance for O nuclei is reached at about 7×10^{20} eV but is less than 2 Mpc. Iron nuclei reach their minimum L_{loss} at energy above 10^{21} eV but it is slightly bigger than 1 Mpc. If the calculation were carried to much higher energy, all nuclei would have had the same energy loss distance as protons.

The photodisintegration is also influenced by the existence of photon fields different from the microwave background, namely the existence of universal optical/infrared background. The influence is small and negligible in compar-ison of the uncertainties in the disintegration cross-sections.

Reference [284] shows the spectra of Fe nuclei after propagation at dif-ferent distances. The picture is qualitatively similar to the one shown in Fig. 9.6 – a pile-up appears after propagation at only 3 Mpc. The location of the pile-up is slightly above 10^{21} eV, where the photodisintegration energy loss distance starts to increase. The total shape of the spectrum does not change as much as it does for protons. The position of the pile-up moves to lower energy with the propagation distance and is about 10^{20} eV at 100 Mpc.

This picture only concerns the injected primary nuclei. The nucleons emit-ted in the disintegration process and the secondary low mass nuclei create a second population of particles. The average mass number of this population decreases with distance and tends to unity at cosmological distances.

The general conclusion from the propagation of heavy nuclei in the CMB is that they lose energy even more rapidly than protons. The energy loss happens at somewhat higher energy, especially in the case of iron.

9.3 Experimental results and implications

9.3.1 Giant air shower detectors

There are seven different detectors that have published results on the highest energy cosmic rays. Five of them are giant air shower arrays and two are fluorescent detectors. we will very briefly describe these detectors individually.

Volcano Ranch
This is the pioneering detector that was designed and built by Linsley, Scarsi & Rossi in New Mexico, USA. This array consists of nineteen plastic scintillators of area 3.3 m^2, arranged in the shape of a hexagon. The largest distance between the individual counters is 880 m, which gives the array an enclosed area of 8 km^2. The first shower of energy exceeding 10^{20} eV was detected by this array [277] prior to the discovery of the microwave background. In spite of the relatively small size of the array, Volcano Ranch contributed a lot of knowledge to the field in terms of a careful description of the lateral distribution of the shower electrons.

SUGAR
The Sydney University Giant Air Shower array was one of the ambitious projects of the 1960s, which may have been constructed well ahead of the technological progress necessary for it. It consisted of 54 autonomous stations, each with its own power source and data recorder. The enclosed area is more than 60 km^2. Individual stations were triggered independently and the trigger time was recorded relatively to a timing signal that was sent to them. The trigger involved coincidences between two 6 m^2 scintillators that were buried 1.7 m below ground, and were thus mostly sensitive to GeV muons. The tapes from the individual stations were collected on a weekly basis and sent to Sydney for analysis. Showers were identified by time coincidence. SUGAR is the only giant air shower detector operated in the southern hemisphere. It did not produce significant results but some of its design concepts are revived in modern arrays.

Haverah Park
The Haverah Park array was designed and built by the University of Leeds in England. Since the usage of land in England is much higher than in New Mexico or Australia, it was not possible to arrange the detectors in a definite geometric structure. The detectors were of a novel design, which is also

currently revived – tanks full of water with a photomultiplier that detects the Cherenkov light of the shower particles. These tanks provide more than three radiation lengths, so most of the shower electrons are fully absorbed, and most of the shower photons convert into electron–positron pairs. Thus the detector measures not only the electron (or muon) density but the total energy flow in the shower disk. Such a measurement makes the estimate of the primary cosmic ray energy somewhat less model dependent. The total enclosed area of the Haverah Park array is 12 km^2.

Yakutsk

This is a complex air shower array run by the Institute for Cosmophysical Research and Aeronomy in Yakutsk, Siberia. It consists of three nested scintillator arrays, that used to enclose an area of 18 km^2. Today's size is about 10 km^2. In addition the arrays have muon detectors that measure muons of energy above 0.5 GeV, and a set of photomultipliers that observe the sky and measure the air-Cherenkov signal of the showers. The timing information from the detectors is transmitted via a microwave link, while power and all other information is carried by cables. The ability to detect different shower components gives this array the advantage of simultaneous measurement of the shower at two development stages – shower maximum can be derived from the air-Cherenkov measurement and the size at the surface is directly measured.

AGASA

The Akeno Giant Air Shower Array is operated in Akeno, Japan. This is the biggest existing air shower array with an enclosed array of 100 km^2. It consists of 111 scintillation detectors of area 2.2 m^2 spaced about 1 km apart. 27 of these stations are also equipped with muon detectors of various size. AGASA has operated in this way since 1995, after several years of operation as four different branches. The oldest (Akeno) branch contains 1 km^2 area with much more densely positioned detectors. Timing signals and data are transmitted to groups of individual detectors and back to a central station via optical cables. AGASA has an excellent angular resolution – the typical error at 10^{19} eV is 3° and decreases to 1.5° at 10^{20} eV. Currently AGASA dominates the shower statistics of all ground arrays.

Fly's Eye

Fly's Eye detector is a fluorescent detector operated by the University of Utah group. This is the first realization of a fluorescent detector. It started in 1981 with a single (FE I) detector consisting of 67 mirrors of diameter 1.6 m, that observes the whole sky with 880 photomultipliers. The second detector, FE II, became operational in 1986. It is 3.4 km away from FE I and views half of the sky in the direction of the first detector. Events detected with only one of the detectors were analyzed as 'monocular' events (see Sect. 8.2.1. The stereo view with both detectors gave the Fly's Eye much lower uncertainty in the analysis. The highest energy cosmic ray was detected by FE I in monocular mode. The best estimate of its energy is 3×10^{20} eV [291]. Figure 9.8 shows the shower profile of this event.

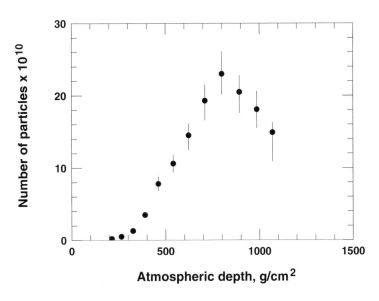

Fig. 9.8. Shower profile of the highest energy cosmic ray event detected by the Fly's Eye. Note that the number of electrons at maximum exceeds 2×10^{11}.

HiRes
HiRes is the second-generation Fly's Eye detector that is built close to the location of the original detector and is described briefly in section 8.2.1. HiRes has run in monocular mode since 1999 and has already generated results on the cosmic ray spectrum. Because of its better angular resolution, HiRes is able to look for showers much further away than the original Fly's Eye and has accordingly much higher aperture.

Table 9.2. Giant air shower detectors.

Experiment	Operation	Latitude	Longitude	Atm. depth g/cm^2
Volcano Ranch	1959 – 1963	35°09′ N	106°47′ W	834
SUGAR	1968 – 1979	30°32′ S	149°43′ E	1015
Haverah Park	1968 – 1987	53°58′ N	1°38′ W	1016
Yakutsk	1974 –	61°36′ N	129°24′ E	1020
AGASA	1990 –	35°47′ N	138°30′ E	920
Fly's Eye	1981 – 1992	40°00′ N	113°00′ W	870
HiRes	1999 –	″ ″	″ ″	870

Table 9.3.1 gives the location, atmospheric depth and the period of operation for all detectors listed above. The exposure for the ground arrays is relatively easy to estimate. Ground arrays are usually operated 90% of the time and until recently have only analyzed showers with zenith angle less than 45°. This would correspond to exposure of 0.9×12 yr $\times \Omega \times$ area $= 1,900$ km^2 ster, which is pretty close to the real number. For fluorescent detectors such estimates are impossible as the lifetime is only a small fraction of the real time and the aperture is energy-dependent. A better description of these arrays and an estimate of their accuracy is given in Ref. [292]

9.3.2 Spectrum and composition of UHECR

The measurement of the cosmic ray spectrum at the highest energy end is even more difficult than in the region of the knee. The detectors in the giant air shower arrays are widely spread, often at distances of 1 km or more. The determination of the air shower core position is less exact than at lower energy. The hadronic interaction models are further away from the energies at which they have been well tested. Different experimental data sets are thus not expected to be in a very good agreement. Figure 9.9 shows several sets of data on the cosmic ray spectrum, coming from AGASA [330], the Fly's Eye [267], HiRes [294], and Yakutsk [295]. The measured differential energy spectrum is multiplied by E^3 to show better different features. Note that a small difference in the energy assignment creates a large difference in the normalization of the spectra. The spectrum measured by the Haverah Park array is not shown because the original result [296] is now being re-analyzed with modern hadronic interaction models. We show the part of the spectrum that has already been re-analyzed reanalized [297] assuming that the showers are initiated by protons or by Fe (inverted triangles) nuclei.

The two air shower arrays determine the primary energy from the density measured at 600 meters from the shower core. This is a method suggested and developed by Hillas [298]. They demonstrate on the basis of shower simulations that the particle density at that distance depends weakly on the hadronic interaction model and on the mass of the primary nucleus. The original calculation is for Haverah Park and the density in the tanks includes the signals from electrons, muons and photons. Contemporary simulations confirm the idea also for scintillation detectors. This procedure has to also account for the type of individual detectors. AGASA uses air shower simulations and arrives at the formula

$$E_0(\text{eV}) = 2 \times 10^{17} \times S_{600} , \qquad (9.12)$$

where S_{600} is the density measured by the AGASA detectors at 600 meters. Yakutsk normalizes a similar relation to the total amount of Cherenkov light carried by the shower. Converted to the altitude of AGASA the formula of Yakutsk gives an energy estimate higher by about 15%. The AGASA estimate

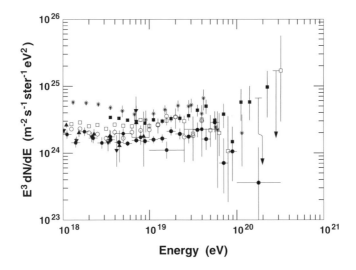

Fig. 9.9. The cosmic ray energy spectrum, multiplied by E^3 as measured by the Yakutsk experiments [295] (asterisks), AGASA [330] (filled circles), Fly's Eye in monocular (empty squares) and stereo (empty circles) mode [267], HiRes in monocular mode (filled circles and hexagons for HiRes 2 and HiRes 1) [294] and the re-analysis of the Haverah Park data [297] (filled triangles).

by itself is 10% higher than the energy estimates made with its 1 km^2 array. If brought down by that amount the AGASA spectrum is in a very good agreement with the published Haverah Park spectrum [296].

The spectrum derived from the monocular observations with the Fly's Eye (squares) is in a good agreement with the current spectrum of AGASA. The newest monocular HiRes [294] spectrum, however, is very different in overall normalization, and in its shape at the highest energies.

The stereo data of the Fly's Eye suggest that the shape of the spectrum changes at energy of about $10^{18.5}$ eV [299] The data is consistent with a power law slope of 3.25 below that energy and 2.75 above it, as indicated in Fig. 9.10. This feature of the cosmic ray spectrum is known as the 'ankle' of the spectrum, following the knee definition at three orders of magnitude lower energy. The 2002 monocular data of HiRes is consistent with similar interpretation although the monocular detection is less reliable than the stereo one. The AGASA data do not require a change of the spectral shape. If there is an ankle, it is most likely at somewhat higher energy as shown with the dashed line in Fig. 9.10.

There are two distinctively different features in the spectra shown in Fig. 9.10. One of them is the difference in the overall normalization. The other one is the position of the possible change in the spectral shape. In principle both of them require a higher energy assignment by the AGASA

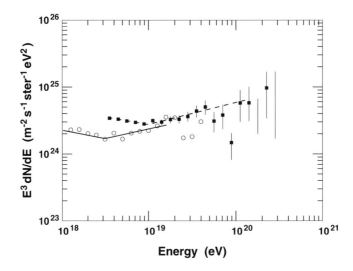

Fig. 9.10. The UHECR spectrum as measured by the Fly's Eye in stereo mode (empty circles) and AGASA (filled squares) as shown in Fig. 9.9. The solid line shows the break in the spectrum as reported by the Fly's Eye group. The dashed line shows a possible break in the AGASA spectrum.

experiment. The difference of the energy assignment derived from the spectral feature, however, is not consistent with the one determined by the flux normalization – a factor of 3 derived from the spectral feature would have resulted in one order of magnitude in the normalization. In fact the difference in the normalization of the two spectra is consistent with a constant difference of about 40% in the energy assigment of the two experiments,

The third difference between all results of the AGASA group and those of the Fly's Eye (and HiRes) is the number of events above 10^{20} eV. Table 9.3 lists the events above 10^{20} eV which are now claimed by all experiments. The events from the original Haverah Park analysis [296] are not included in the table, because these are now being re-analyzed.

Eleven of a total of seventeen events come from the AGASA experiment [300]. Using the total exposure of 1,460 km^2 yr sr this comes to about one event per 130 km^2 yr sr, a number consistent with the rate from the new analysis [301] of inclined showers recorded by Haverah Park – two events in 280 km^2 yr sr. On the other hand HiRes 1 has exposure estimated at 2,200 km^2 yr sr for events above 10^{20} eV and only sees one event above 10^{20} eV. Yakutsk has exposure of 450 km^2 yr sr and also sees only one event. So also does Fly's Eye in monocular mode with exposure of 870 km^2 yr sr. The total exposure of Volcano Ranch is only 67 km^2 yr sr.

One has to realize that at very high energy the aperture of a ground array is fixed by its total area and the maximum zenith angle that the experimen-

Table 9.3. List of events with energy above 10^{20} eV claimed by different experiments. Note that only two events from Haverah Park which come from a recent analysis of inclined showers are included. 1 EeV = 10^{18} eV. The arrival direction of the HiRes event is not yet published.

Experiment	E, EeV	l	b	RA	Dec
Volcano Ranch					
	139	84.3	4.8	306.7	46.8
Fly's Eye					
	320	163.5	9.7	85.2	48.0
Yakutsk					
	147	162.2	2.6	75.2	45.5
AGASA					
	101	207.1	26.8	124.3	16.8
	213	131.1	−41.1	18.8	21.1
	134	77.7	20.7	281.3	48.3
	144	38.8	45.5	241.5	23.0
	105	57.1	−5.0	298.5	18.7
	150	33.4	−13.5	294.5	-5.8
	120	90.4	−44.3	349.0	12.3
	104	99.1	−23.8	345.8	33.9
	122	176.1	73.6	176.0	36.3
	246	106.7	−38.6	358.5	22.3
	121	179.2	−1.1	84.0	29.0
HiRes 1					
	180				
Haverah Park (inclined)					
	123	178.1	2.2	86.7	31.7
	115	54.4	−29.8	318.3	3.0

tal group has decided to analyze. For AGASA this is 45°. For fluorescent detectors, however, the exposure is not fixed. One of the greatest advantages of a fluorescent detector is that its aperture grows with energy until the absorption of the fluorescent light in the atmosphere cuts it to a constant value. Combined with the limited fraction of time during which the fluorescent detector is active the total exposure is more difficult to estimate without a proper knowledge of the atmospheric conditions.

The measurement of the energy spectrum is related to the composition of the UHE cosmic rays. Using its data in stereo mode the Fly's Eye group concluded that at energy about 3×10^{18} eV, where the slope of the spectrum changes, there is also a change of the cosmic ray composition. The composition was studied by an analysis of the X_{max} distribution for showers in different energy bins. Figure 9.11 shows the X_{max} distribution for showers of energy 3–5×10^{17} eV (a) and above 10^{18} eV (b) as presented in Ref. [302].

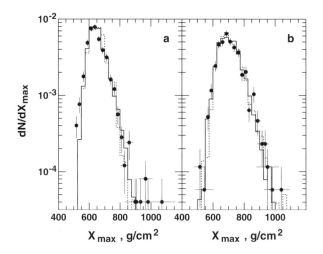

Fig. 9.11. Distributions of X_{max} measured by the Fly's Eye in showers of energy 3–5×10^{17} eV (a) and above 10^{18} eV (b). The histograms are fits performed with two different interaction models. The fits predict 12–21% proton showers for the distribution in (a) and 39–41% protons for (b).

The solid and the dashed histograms show the fits of the distributions performed with two interaction models. The model presented with the solid line is closer to the contemporary models described in Chap. 8. For the lower energy bin the fit predicts that between 12 and 21% of all showers are generated by light nuclei (H, He). In the higher energy sample the fraction of protons is 39–41%. If the same trend continues [299] the change of the spectrum shape occurs when protons start dominating the UHE cosmic ray composition.

The same trend can be derived from the average X_{max} as a function of energy, as shown in Fig. 9.12. The measured X_{max} values increase much more rapidly than the predictions of the contemporary hadronic interaction models. This can only be explained with a change of the composition. The situation is similar to that in the knee region, although the change of the composition is in the opposite direction – from heavy to light. The derivation of the exact fraction of heavy and light nuclei has model-dependence. In Fig. 9.12 for example the composition that can be derived from the QGSjet98 model is lighter than the one predicted by SIBYLL 2.1 – the two points at 2–3×10^{19} eV are fully consistent with pure proton composition. The point at 3×10^{20} eV shows the depth of maximum for the highest energy event measured by the Fly's Eye detector. The data of the two detectors are in a good agreement in this respect.

These indications for the change of the composition in the region of the ankle of the cosmic ray spectrum are not confirmed by the observations of AGASA. AGASA studies the composition by correlating the muon content of the showers to the total energy. A change to a lighter composition would

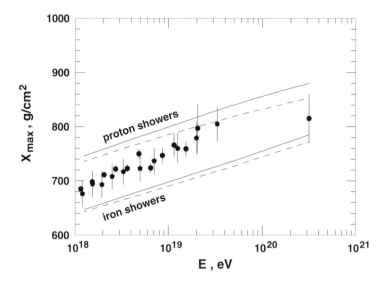

Fig. 9.12. Correlation of $\langle X_{max} \rangle$ with energy as measured by Fly's Eye [299] and Yakutsk [295]. The solid line shows the predictions of SIBYLL 2.1 and QGSjet98 for proton and iron initiated showers.

lead to a decrease of the muon/electron size ratio with energy. The data of AGASA do not contradict to the suggested composition change, but are also consistent with a constant cosmic ray composition.

9.3.3 Possible astrophysical sources of UHECR

A natural way to imagine the sources of the highest energy cosmic rays is to extend the models that explain the acceleration of the galactic cosmic rays and look for larger and more luminous astrophysical objects. An additional requirement is that these objects cannot be very far away. If they were, the requirements of the luminosity of the source would grow tremendously because of the proton energy loss on propagation.

The minimum requirement for an acceleration site is the containment of the accelerated cosmic ray – let's think of protons – at the site. This leads to the relation $E_{max} = \gamma e Z B R$ that relates the absolute maximum acceleration energy to the strength B of the magnetic field at the acceleration site and its linear dimension R. The factor γ accounts for the bulk Lorentz factor of the medium where the acceleration proceeds. Hillas [303] develops this requirement including the effect of the average velocity of the scattering centers $\beta_s c$ and obtains the condition

$$\left(\frac{B}{\mathrm{G}}\right)\left(\frac{R}{\mathrm{pc}}\right) > \frac{0.2}{\beta_{sc}Z}\left(\frac{E}{10^{20}\,\mathrm{eV}}\right)\,, \qquad (9.13)$$

where Z is the charge of the accelerated particle. Figure 9.13 sketches the graph that illustrates this requirement which is now usually referred to as a Hillas plot.

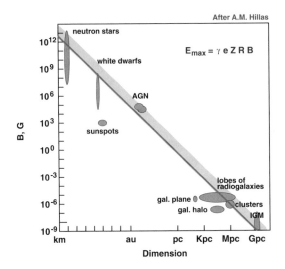

Fig. 9.13. Hillas plot [303] showing the size and magnetic field strength of sites that can accelerate protons (top of shaded strip) and iron nuclei to an energy of 10^{20} eV.

Sites that can in principle accelerate protons to an energy above 10^{20} eV are on the upper right-hand part of the graph above the shaded band. The lower edge of the shaded area defines sources that can accelerate iron nuclei above 10^{20} eV. There are only four types of systems that might be able to accelerate protons above 10^{20} eV: high magnetic field neutron stars with surface magnetic field exceeding 10^{13} G and linear dimension of 10 km, active galactic nuclei, lobes of giant radio galaxies and gigaparsec shocks in the extragalactic medium. Let us go through a list of possible sources identified in the literature.

Shocks from structure formation Very large-scale shocks of dimension exceeding 10 Mpc could exist from structure formation due to gravitational attraction [304]. Such shocks form in the accretion flows during structure formation and can in principle generate magnetic fields. The maximum energy that is achievable at such shocks depends on the shock dimension. For 50 Mpc and larger shocks the maximum energy can exceed 10^{20} eV if the average strength of the extragalactic magnetic fields is 10^{-9} G [305] and the shocks themselves generate μG fields. One of the problems with the acceleration at large-scale shocks may be that the process is slow and the energy loss on pair production and photoproduction during acceleration would restrict the maximum achievable energy.

Clusters of galaxies Average magnetic fields of 5 μG and extension up to 500 kpc have been observed [306] in clusters of galaxies. According to (9.13) this strength of the magnetic field over a huge volume allows the acceleration to energies well above 10^{20} eV. This possibility has been examined in some detail [307] with the conclusion that the large dimensions of the system and the related energy loss limits the maximum energy at acceleration to about 10^{19} eV.

Radio galaxies One of the most attractive suggestions for the acceleration site of UHECR is that of radio galaxies. Rachen & Biermann [308] suggested the 'hot spots' of the FR (Fanaroff–Riley) type II galaxies as the source of the highest energy cosmic rays. FR II type galaxies are giant radio galaxies that exhibit two jets going in opposite directions. The 'hot spot' is the termination shock of the jet in its propagation in the extragalactic medium. The extension of the jets and their hot spots is well known, and may reach up to 100 kpc. The value of the magnetic field at the spot cannot be directly measured, but estimates give fields exceeding 10 μG. Using efficient acceleration the FR II galaxies can accelerate protons to $\sim 10^{21}$ eV. The energy loss is not expected to be a significant problem. Since the 'hot spots' are already a part of the intergalactic structure there would be no adiabatic loss in the injection of the accelerated particles into it.

Active galactic nuclei All AGN could in principle be sources of the highest energy cosmic rays. The central engine of the AGN, where the magnetic fields are estimated to reach 5 G in volumes of linear dimension 0.02 pc [309], can easily contain particles of up to 10^{20} eV. The main problem is again the large energy losses in the very dense radiation fields in the central region of AGN. Reference [304] argues that no protons can leave the central region of AGN without severe energy degradation. Heavier nuclei will be affected even more. One possibility would be that neutrons created in photoproduction interactions of the parent protons leak out of the central AGN region since they are not magnetically contained [310]. These neutrons decay to protons once they are far enough from the AGN. For a neutron to escape, however, its energy does not have to be much larger than 10^{18} eV. The AGN jets are a different possible site of UHECR acceleration. The advantage of the jet site is that the accelerated protons would be injected with the Lorentz factor of the jet, which may easily be of the order of 10 or higher. This decreases the maximum acceleration energy in the jet frame, but may cause adiabatic loss in the transition to the intergalactic space.

Gamma-ray bursts The extreme case of acceleration in a jet is the acceleration in GRB. Gamma-ray bursts are brief outbursts of mostly MeV gamma-rays. Their spectra extend to higher energy – the highest energy photon detected from a GRB is 20 GeV. The burst duration varies from a fraction of a second to hundreds of seconds. It is now proven that the GRB are of cosmological origin (at average redshift $z = 1$) and that they originate in jets. Their total luminosity is 10^{53-54} ergs in the case of isotropic emission and for a jet scenario this number has to be scaled down with the jet opening angle. All contemporary GRB models use Lorentz factors of the GRB jets of order 100 to 1,000. The first suggestions [311, 312, 313] that GRB are the sources of the highest energy cosmic rays were based on the directional coincidence between the two highest energy cosmic rays and the two most powerful GRB detected. In addition, the total GRB luminosity coincides within one order of magnitude with that of the cosmic rays of energy above 10^{19} eV. The doubts about the validity of the GRB origin are related to the cosmological distances to these objects. The argument is that very distant GRB cannot contribute to the observed cosmic rays above 10^{19} eV and close by gamma-ray bursts do not happen often enough to match the luminosity requirements for the sources of UHECR.

Colliding galaxies The movement of galaxies through clusters, as well as galaxy–galaxy collisions, produce large-scale shocks that are suitable for particle acceleration [314]. These shocks are easily visible at radio frequencies. A shock of dimension 30 kpc for the colliding galaxies and a shock field of 20 μG could provide the conditions for acceleration to above 10^{20} eV. Colliding galaxies are thus also a suitable candidate for the acceleration site of the highest energy cosmic rays.

Quiet black holes The suggestion is that UHE protons can be accelerated at the event horizon of spinning massive black holes associated with currently non-active galaxies. This suggestion disconnects the acceleration site of UHECR from the existence of powerful astrophysical systems in our cosmological neighborhood. The model requires $10^9 M_\odot$ black holes within 50 Mpc of our Galaxy.

Pulsars Pulsars are the smallest objects in the Hillas plot that could accelerate protons to energies above 10^{20} eV. In this case models do not usually use shock acceleration, rather direct acceleration in the strong electrostatic potential drop induced at the surface of the neutron star. Another model, that is very interesting with its specific predictions is that of Ref. [315]. The suggestion is that iron ions from the surface of a young neutron star are accelerated by MHD winds. The model requires specific strength of the pulsar magnetic field, but predict that the acceleration spectrum is very flat (E^{-1}) and that the UHECR are heavy (iron) nuclei. Since the UHECR would in this case be of galactic origin, this suggestion eliminates all problems related to propagation through extragalactic distances.

All the sources listed above can in principle contain protons of energy around and above 10^{20} eV. The estimates are made on the basis of the most favorable possible parameters and assuming very efficient acceleration process. The creation of the more detailed models is not possible because of the big uncertainty in the required astrophysical input.

9.3.4 Exotic models

Although most of the astrophysical systems listed in Sect. 9.3.3 can contain the detected UHE cosmic rays, the acceleration to energies above 10^{20} eV is obviously a very unlikely process that requires extremely favorable parameter space and very efficient acceleration scenarios. The realization of the difficulties in extending the galactic cosmic rays acceleration mechanism by many orders of magnitude led to the development of exotic particle physics models of 'top-down' production of the highest energy cosmic rays. The basic idea is that the observed cosmic rays are decay products of very massive X particles with M_X as high as 10^{25} eV.

All top-down models UHECR models have two distinctive features – flat injection spectrum and a particle composition different from the 'bottom-up' acceleration scenarios. In acceleration scenarios the particles accelerated at a source and injected in the intergalactic medium are protons, higher mass charged nuclei, or neutrons created in interactions in the source. In the top-down models the massive X particles decay into a chain of all known elementary particles with nucleons and mesons as final products. The more numerous mesons decay into neutrinos and electrons or into γ-rays depending on their charge. As a result the injection fluxes of γ-rays and neutrinos exceed the nucleon fluxes by a factor of about 30, except at energies close to M_X.

The first calculation of the injection spectrum, which has been improved on but is still valid, was performed in 1983 by Hill [316] for the case of monopole annihilation. Monopoles are point-like topological defects that may have been produced in the earliest phases of the evolution of the Universe. A monopole and an antimonopole can form a bound state and then annihilate. In the process a spectrum of particles is generated, as is shown in Fig. 9.14, which represents Hill's spectra of γ-rays and nucleons.

The spectrum shown in Fig. 9.14 can be approximated as $E^{-3/2}$ with a cutoff at the approach of M_X. It was emphasized almost immediately by Schramm & Hill [317] that top-down models always have a flatter injection spectrum because of the properties of the QCD fragmentation functions. The flat E^{-2} acceleration spectrum is the dividing line between all shock acceleration models that have such or steeper power law spectra, and top-down models, that have flatter ones. More recent studies of the fragmentation process, that include much more physics input and utilize contemporary computer codes modify the injection spectrum [318]. The spectra cannot be fitted with a single power law, but are still much flatter than the ones in shock acceleration scenarios.

The neutrino injection spectra are not changed during propagation, nucleon spectra evolve as shown in Sect. 9.2.1 and the γ-ray spectra are also very strongly affected by interactions on the microwave background. The main process for photons is the pair production $\gamma\gamma \longrightarrow e^+e^-$ on the extragalactic radiation fields. The threshold energy for the process is $m_e^2/\epsilon/(1 - \cos\theta)$ where ϵ is the energy of the background photon and $\cos\theta$ is the angle between

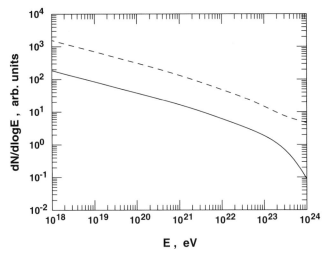

Fig. 9.14. Energy spectrum of photons (dashed line) and nucleons (solid line) as calculated in Ref. [316].

the two photons. At higher energy other processes such as double pair production $\gamma\gamma \longrightarrow e^+e^-e^+e^-$ start dominating. A pair-production interaction starts the development of an electromagnetic cascade – the electron and the positron suffer inverse Compton interactions on the radiation field and boost these photons to very high energy. Figure 9.15 shows the photon and electron interaction lengths for pair production and for IC scattering.

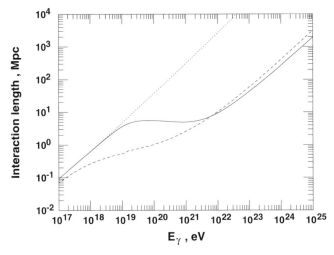

Fig. 9.15. Interaction lengths for pair production by high energy gamma-rays (dotted line is if only the microwave background were present and the solid line includes a model of the radio background) and for inverse Compton scattering of the electrons (dashed line).

The cross-sections in Fig. 9.15 are plotted up to γ-ray energy of 10^{25} eV because such energies are typical for top-down scenarios. For photon energies that high the microwave background is no longer the only important background. The extragalactic radio background becomes equally important. Figure 9.15 shows the interaction length of gamma-rays only on the microwave background (dotted line) and with the lower model of of the radio background from Ref. [319]. At γ-ray energies above 10^{23} eV the double-pair production cross-section with asymptotic interaction length of 120 Mpc restricts further the gamma-ray free path.

The dashed line in Fig. 9.15 shows the IC interaction length for electrons in the presence of the same radio background. In the absence of magnetic field the electromagnetic cascade would carry the high energy electromagnetic particles a long way without a significant energy loss. In the presence of even modest magnetic fields the ultra high energy electrons would quickly lose energy on synchrotron radiation and the cascading will be interrupted. The energy loss distance of 10^{22} eV electrons in 10^{-10} G field is only 20 kpc. In the presence of magnetic fields γ-rays can reach only distances shorter than their interaction length.

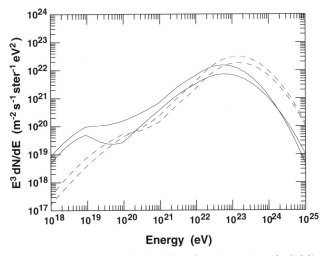

Fig. 9.16. Differential gamma-ray (dashed lines) and nucleon (solid lines) spectra multiplied by E^3 from a particular top-down model and different magnetic field values.

The injection spectra are thus very much modified. Even though at injection the γ-rays dominate the nucleons, at modest distances from injection the high energy nucleons could be much more abundant as shown in Fig. 9.16, which is extracted from Figs. 28 and 29 of Ref. [320]. At injection the photon spectrum is higher than the nucleon spectrum by about a factor of 10. At energies close to those observed at Earth the nucleon spectra dominate in this particular model.

Let us list different types of top-down models that can be found in the literature. There are generally two classes – one related to topological defects and another related to cold dark matter. In addition to monopoles the topological defects scenarios include many types of cosmic strings. Cosmic strings are one-dimensional topological defects. The mass of the strings can be as high as $3 \times 10^{10} M_\odot$ per parsec length. There are many models of potential cosmic string UHECR sources. Superconducting strings [321] generate UHE energy packages when their electric current reaches a critical value. Ordinary strings can emit X particles at their cusps, or during their intersection and final stages of evolution. There are also models of hybrid topological defects where monopoles are connected with strings [322]

The second type of top-down scenarios uses the decay of quasi-stable massive X particles produced in the early Universe [323]. These particles must have lifetime comparable with Hubble time. They may, or may not, be a substantial part of the dark matter. The maximum energy achievable in this case is also the mass of the particle. Reference [320] gives a detailed review of the top-down models.

There are also general hybrid models that involve processes that have never been observed because of their very high energy threshold. One of them is the Z-burst model [324, 325], which is based on the newly discovered neutrino oscillations (see Chapter 7) that require massive neutrinos. The massive relic neutrinos, a remnant of the big bang similar to the microwave background, will to certain extend be gravitationally attracted to concentrations of matter, say to the local supercluster of galaxies, within about 50 Mpc from us. The increase of the neutrino density makes possible the annihilation of ultra high energy cosmic neutrinos ($E_\nu > 4 \times 10^{21}$ eV) on the relic ones. The annihilation generates Z_0 bosons that immediately decay into about three nucleons and 30 photons with energies close to those of the UHECR. Since the production is not very far away these Z-bursts may be sources of the highest energy cosmic rays. Apart from the very low neutrino masses suggested by the observed oscillations, which would limit the neutrino density, the problem is that the possible source of the UHE cosmic neutrinos is not defined.

9.4 UHECR astronomy

The conclusion that the highest energy cosmic rays are produced outside the Galaxy comes from the knowledge that their gyroradii are much too large and the particles cannot be contained in it. This also means that if UHECR are generated in our cosmological neighborhood they will not be bent too much in the low extragalactic magnetic fields. A convenient way to remember the gyroradii of the highest energy particles is

$$r_g = 100 \left(\frac{E_{20}}{ZB_{-9}} \right) \text{ Mpc,}$$

where Z is the particle charge, E_{20} is its energy in units of 10^{20} eV, and B_{-9} is the magnetic field strength in unite of 10^{-9} G (nG). If the magnetic field were uniform, the cosmic rays would bend on propagation on D Mpc at angle $\theta \approx 0.53° ZDB_{-9}/E_{20}$. For a source at 20 Mpc and a nG field this would mean about 10°.

The first approximation of the extragalactic fields is that they are random fields with a length scale equal to the average distance between galaxies – 1 Mpc. In such fields the cosmic rays would experience random walk and the r.m.s. deflection angle would be [81]

$$\theta_{rms} = 0.35° \frac{ZB_{-9}}{E_{20}} (Dl_0)^{1/2} , \qquad (9.14)$$

where l_0 is the coherence length of the random field measured in Mpc as well as the distance D. Equation (9.14) is valid if each of the individual random walk scatterings are small-angle scatterings. For a distance of 20 Mpc and l_0 of 1 Mpc now the average angle for 10^{20} eV protons is less than 2°. An average angle of 10° corresponds to a distance of almost a Gpc, which should contain the sources of all extragalactic protons that could reach us.

Although there are observations of several microgauss fields in clusters of galaxies [306], the average fields are very difficult to deduce from Faraday rotation measurements. The estimates of the upper limit for random extragalactic fields range from 3×10^{-9} G to 10^{-8} G. The lower limit is derived from the isotropy of the microwave background and the higher one from the median value of the Faraday rotation measures of distant quasars assuming a coherence $l = 1$ Mpc [326].

Even with the higher upper limit of 10 nG the highest energy cosmic rays would point within several degrees of their sources within a 100 Mpc radius. No astrophysical sources or even top-down sources could contribute to the events above 10^{20} eV outside that radius, unless the UHECR are neutrinos.

9.4.1 Arrival directions of UHECR

There are two energy ranges where the arrival directions of the highest energy cosmic rays have been recently observed: in the region around 10^{18} eV, where the Fly's Eye results suggest the galactic cosmic rays start becoming challenged by a lighter extragalactic component, and at energies above 4×10^{19} eV, just before the expected GZK cutoff.

The world statistics of 10^{18} eV events is large enough for an anisotropy analysis. The AGASA experiment used 114,000 showers of energy above 10^{17} eV and found anisotropy of about 4% – there were significantly more 10^{18} eV showers coming from the direction of the galactic center and along the local galactic arm than there were showers coming from the direction of the galactic anticenter [327]. The chance coincidence for such a result, including the number of trials is about 0.2%. The current higher AGASA statistics increase the statistical significance of the effect, which is an excess of 4σ from the direction of the galactic center and a dearth of -4σ from the Galactic anticenter. The excess from the direction of Cygnus (along the galactic arm) is somewhat smaller. The effect peaks for showers of energy 1-2 × 10^{18}.

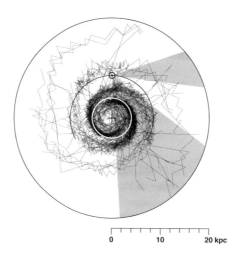

Fig. 9.17. Simulation of the movement of 10^{18} protons injected in the galactic plane isotropically on the 4 kpc circle (which is indicated with a white circle) in the BSS model of the galactic magnetic field. The shaded areas show the directions in which the AGASA experiment detects an excess of cosmic rays.

Although this observation was made for the first time by the AGASA collaboration, the result seems natural if most of the high energy galactic cosmic rays are accelerated inward of the solar circle. Figure 9.17 shows the result of a simulation where 10^{18} eV protons are injected isotropically at the 4 Mpc circle of the galactic plane in the BSS model described in Chap. 4. In a 5×10^{-6} G field the gyroradius of the 10^{18} eV protons is about 200 pc, of the order of the thickness of the galactic plane. Protons are caught in the field and start gyrating along the magnetic field lines. Since there is also a random field component, the protons are occasionally kicked out of the galactic arms and one can see them propagating in almost straight lines.

This simple simulation suggests that if a large fraction of the cosmic rays are accelerated inside the solar circle one can expect a flow of cosmic rays outwards and along the spiral arms. The only problem is why these cosmic rays point in the direction of the flow. In the simulations, where charged particles gyrate along the field line, they very seldom point in its direction. A large event sample will, however, exhibit the general pattern. For this reason the statistical analysis of a large event sample is very valuable. Similar studies are made with smaller statistics by other experimental groups and the results are not always compatible with each other.

The situation at the highest energy end is very different. Currently the largest published event sample comes also from the AGASA experiment which has collected 59 events above 4×10^{19} eV as of 2002. All other ground arrays currently have smaller statistics. The Fly's Eye events are not always easy to analyze with those of the ground arrays because they have asymmetric errors even in stereo mode. This is typical for all optical shower detectors that have to smooth their event arrival direction with the appropriate errors. There are two significant features in the arrival direction distributions of the events above 4×10^{19} eV:

- large-scale isotropy, and
- small-scale clustering.

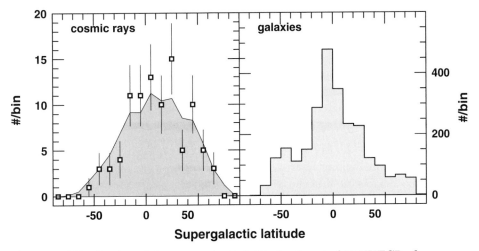

Fig. 9.18. Distribution of the supergalactic latitude of a set of 92 UHECR of energy above 4×10^{19} eV (left-hand panel) and the positions of nearby galaxies ($z < 0.02$) in the field of view of the air shower arrays.

Figures 9.18 and 9.19 demonstrate these two effects. The plane of weight of the cosmologically nearby galaxies is defined as the supergalactic plane [328] (SGP). Figure 9.18 shows the distribution of the UHECR and galaxies within $z = 0.02$ (distance of 80 Mpc for $H_0 = 75$ km/s/Mpc) in supergalactic latitude in 10-degree bins. The galaxies shown are in the field of view of the air shower arrays. The cosmic ray events are of energy higher than 4×10^{19} eV. The statistics includes AGASA (47), Haverah Park (27), Yakutsk (12), and Volcano Ranch (6). The shaded area in the left-hand panel shows the joint exposure of all experiments. The right-hand panel shows a significant excess of the galaxies within 20 degrees of the supergalactic plane, while the UHECR distribution more or less agrees with expectations from the field of view of the detectors. Figure 9.19 plots the same data on a 1-degree scale. The result is the opposite. While the nearby galaxies do not peak within degrees of the supergalactic plane, UHECR show a large peak within 1 degree of SGP.

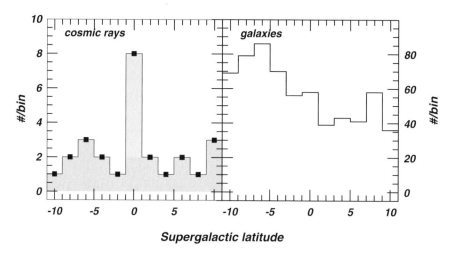

Fig. 9.19. Same as in Fig. 9.18 in 1 degree bins.

Figure 9.20 shows the arrival directions of individual events. Showers of energy above 10^{20} eV are plotted with larger circles. The plot is for a sample of 114 events that include 22 from Fly's Eye. The average angular resolution of the ground detectors is of the order of 3°. One can easily see the concentration of events almost exactly on the supergalactic plane. Not all of these events come from the AGASA experiment. Several years ago, when the experimental statistics was dominated by Haverah Park, the average proximity to the SGP was closer [329] and led to the conclusion of an association of UHECR with SGP. On the basis of the worldwide sample of UHECR available in 1995 the analysis concluded that at energy about 4×10^{19} eV cosmic rays come from directions close to the SGP and do not cluster around the galactic plane. With the increase of the experimental statistics this association became more remote.

A careful examination of Fig. 9.20 shows another feature – there are several directions in the sky that contain two or three cosmic rays. This fact was first reported by the AGASA group [330], which found one triplet and three doublets among the 47 events above 4×10^{19} eV. The separation angle between events in these groups is less than 2.5°, about the error in the direction determination by AGASA. Some of these clusters also contains a number of slightly lower energy events. The probability of that happening from an isotropic arrival direction distribution was estimated to be less than 1%. There is no obvious association of the clusters with a potential cosmic ray source, although there is an AGN and a couple of colliding galaxies within sev-

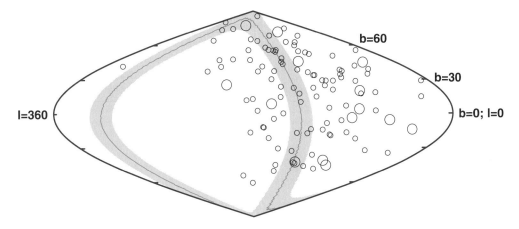

Fig. 9.20. Arrival directions of a sample of 114 cosmic ray events of energy above 4×10^{19} eV. Events of energy above 10^{20} eV are plotted with bigger circles. The shaded are shows the supergalactic plane. The plot is centered on the Galactic anticenter - the Galactic center is split on the two sides at $b=0°,l=0,360°$.

eral degrees of a couple of clusters. A joint analysis of the AGASA, Haverah Park, Yakutsk and Volcano Ranch events was performed soon afterwards, revealing six doublets and two triplets in the sample of 92 events [331]. Since the angular resolution for four different experiments is difficult to assign, and the probability of clustering depends very strongly on the separation angle, the significance of the clustering is of the same order as for the AGASA data by themselves.

The authors of Ref. [331] stress that the two triplets and two of the doublets are within 10 degrees of the supergalactic plane. A Monte Carlo study using the experimental zenith angle distributions shows a probability of 0.1–0.2% for the observation. One has to take these probabilities only as order of magnitude estimates because of the difficulties in assessing the possible experimental errors and biases and the low number statistics. The association with SGP does not necessarily favor the astrophysical acceleration or the top-down models of the UHECR origin. It is well known that this is a direction of concentration of matter, that will not only contain many active galaxies, but also topological defects. Defects could indeed be the seeds for structure formation in the Universe.

Since the observed clusters do not point at any type of known or suspected cosmic ray sources, this result is also presented in terms of self-correlation, as shown in Fig. 9.21. The figure shows the deviation from isotropic arrival direction distribution in units of standard deviations σ. This demonstrates in a very general way the character of the anisotropy observed by AGASA. At $\Delta\alpha$ (distance between the arrival direction of two events) the distribution is flat with statistical fluctuations above and below 0σ. At angles of about $2°$ the excess in close to 5σ, i.e. the detected anisotropy is small-scale. The

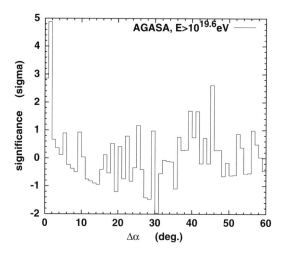

Fig. 9.21. Significance of the self-correlation in the AGASA data above 4×10^{19} eV in standard deviations σ.

same effect has been recently reported in two-dimensional self-correlation studies, where the events above 10^{19} eV form significant 10 degree spots in the difference in galactic coordinates $(\Delta l, \Delta b)$. Since these spots do not point in any particular direction the only possible explanation is that there are hundreds of sources of UHECR, some of which contribute two or three cosmic rays.

9.4.2 UHECR and the cosmic magnetic fields

The influence of the astrophysical magnetic fields on the proton propagation is a topic which is very relevant to the possibility of UHECR astronomy. We already know what the influence of the galactic magnetic fields should be using the formulae given in the previous chapter. If we assume that the average regular galactic field is about 1 μG and it extends to 20 kpc from the galactic center (which is an overestimate) then the scattering angle of a 10^{20} eV proton would be about 10°. The field is not constant and most probably reverses its direction in the different galactic arms, so the actual scattering angle is much smaller. Figure 9.22 gives an idea of the magnitude of the deflection in a particular field model.

The BSS model with a 0.3 μG z component is chosen as an illustration because it generates the largest scattering angles. The total length of the arrow corresponds to proton energy of 10^{19} eV and the first kick in the trajectory (when visible) is for 10^{20} eV protons. Protons of intermediate energies scatter between these two extremes. Note that this figure is a result of backtracking antiprotons from the Earth, and the direction of scattering is the opposite of what particles coming to the Earth from extragalactic space would experience. The calculation is not shown for directions close to the

BSS_A + B_z

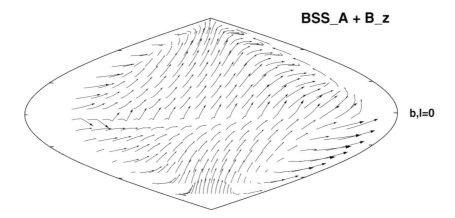

b,l=0

Fig. 9.22. Scattering of antiprotons emitted at Earth in the galactic magnetic field in model BSS (see Chap. 5) with the addition of a 0.3 μG component normal to the galactic plane.

galactic center because of the uncertainty of the field strength and shape in that region.

The random component of the galactic fields cannot contribute much to the change of the UHECR direction, as (9.14) with l_0 of less than 100 pc indicates. Its main contribution is to smear the regular picture shown in Fig. 9.22. If the UHECR were protons emitted at a few sources outside the Galaxy, they would preserve the general direction of the sources, although it might be slightly skewed by the galactic magnetic field. The detection of even a small statistic would show the different degrees of scattering as a function of the UHECR rigidity, and help to reveal the source.

Some authors relate the scattering in the GMF to the self-correlations seen by the AGASA detector. The idea is that protons coming from different directions would be bent to a different degree as a function of their energy and may appear to be coming from the same direction. This is indeed possible, but the efficiency of such a mechanism for production of the observed doublets and triplets is low.

If the UHECR were heavier nuclei, their rigidity would be lower and they would be affected more by the GMF. Iron nuclei of total energy 10^{20} eV would behave almost like 10^{18} protons are shown to behave in Fig. 9.17.

An interesting question is the extension of the galactic field and the possible existence of a magnetic halo of our Galaxy. The data on such a halo are inconsistent, although the velocity of the galactic wind could be limited to about 20 km/s because of the low anisotropy of the galactic cosmic rays. The most extreme suggestion for UHECR scattering in a halo is made in Ref. [332]. It is based on the observation that the magnetic field in the local arm points inwards towards the galactic center as do the fields in many other spiral galaxies. If one postulates the existence of a halo with the structure of a Parker spiral (similarly to the solar system) and an extension to distances of order 1 Mpc, all antiprotons backtracked from Earth would bend in the direction of the North Galactic Pole. Conversely, cosmic rays injected in this system at positions close to the North Galactic Pole would appear to us as an almost isotropic flux. The UHECR can in this way be connected to M87, the closest powerful radiogalaxy in the Virgo supercluster and about 15° away from the North Galactic Pole.

Because of the much bigger pathlengths involved the proton scattering in the extragalactic magnetic fields could be important if their strength is near to, or higher than, 1 nG. There are several effects in addition to the deflection angle. The most important one that certainly exists at some level is the increased pathlength and associated time delay because of scattering in the random field. The time delay Δt is defined as the difference between the time of propagation in the magnetic field and the light propagation time. It is derived to be

$$\Delta t \approx 30 \left(\frac{ZB_{-9}}{E_{20}}\right)^2 \left(\frac{D}{\text{Mpc}}\right)^2 l_0 \text{ years} \tag{9.15}$$

without the account for the proton energy loss, which will lower the proton energy and increase its time delay. The increase in the pathlength is proportional to Δt and also significantly increases the energy loss.

But even without the account for energy loss, the time delay itself limits the distances from which cosmic rays can reach us within Hubble time. For 1 nG random fields and distances of 100 Mpc, only protons of energy above 5×10^{17} can propagate to us within 10^{10} yr. One can turn the argument around and calculate the distance to sources from which UHECR can reach us in Hubble time – the cosmic ray horizon. Figure 9.23 shows the results of a calculation which accounts for the energy loss of the protons in propagation [333].

The horizon R_{50} is defined as the distance at which the number of particles keeping more than 50% of their injection energy decreases by a factor of e. At energies above 10^{20} eV the horizon closely tracks the proton energy loss distance. At smaller energy, however, it is significantly shorter even for the lowest magnetic field assumed. In the $10B_{-9}$ case the horizon starts decreasing approximately at the position of the GZK cutoff, although none of the particles has time delays of the order of Hubble time. The decrease is due to increased energy loss on pair production and cosmological expansion. The Hubble time restriction will kick in at somewhat smaller energy than can be estimated with (9.15) and is impossible to calculate with the Monte Carlo technique used for Fig. 9.23.

The assumption of regular extragalactic magnetic fields leads to different effects depending on the relative positions of the source and the observer, and on the direction of the magnetic field. Figure 9.24 shows two examples for the change of the proton injection spectrum in opposite directions.

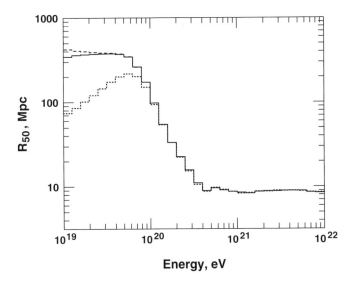

Fig. 9.23. UHE cosmic ray horizon as a function of energy and the strength of the random extragalactic magnetic field. See the text for the definition of the horizon R_{50}. The solid histogram is for B_{-9}, the dotted for $10B_{-9}$ and the dashed for $0.1B_{-9}$.

A cosmic ray source is positioned at the origin of the coordinates system in the middle of a magnetic wall of width 3 Mpc that coincides with the yz

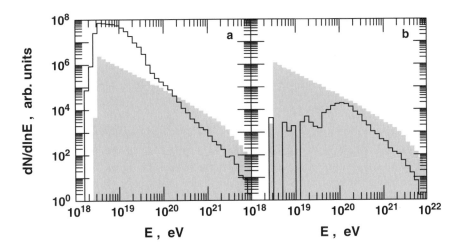

Fig. 9.24. Spectra of UHE protons at 20 Mpc from their source. The shaded area shows the injection spectrum. Panel (a) shows particles exiting the 20 Mpc sphere within the magnetic wall. Panel (b) shows particles exiting in the direction normal to the wall. See the text for a fuller description of the geometry.

plane. The magnetic field of the wall has a strength $B_{reg} = 10$ nG, points in the z direction and exponentially decreases outside the wall. It is accompanied by a random field of strength $B_{reg}/2$. The source injects isotropically UHE protons on an E^{-2} injection spectrum with an exponential cutoff, as shown with the shaded histogram in Fig. 9.24. The protons are followed in this environment with account of all energy losses until they cross a sphere of radius 20 Mpc. Their energy spectrum is studied as a function of the exit position.

Figure 9.24 shows the exit spectra in two 9 Mpc^2 patches. Patch (a) (front) is inside the magnetic wall at positive z around $y = 0, x = 0$ and patch (b) (side) is in direction perpendicular to the magnetic wall at $y = 0, z = 0$. The energy spectra above 3×10^{20} eV are almost identical, lower than the injection spectrum because of energy loss on photoproduction. At lower energy, however, the spectra are very different. The lower energy particles that exit through the front path propagate along the magnetic field lines. Particles that have been injected in almost any direction are magnetically contained in the wall, gyrate back and forth along the field lines and exit through the front patch. Thus the excess at 10^{19} eV in this patch, compared to injection, is about two orders of magnitude. These particles have on the average large Δt exceeding 10^8 years in comparison with the light propagation time of 6.5×10^7 years.

Particles that exit through the side patch have to move across the magnetic field lines. The highest energy particles are not much affected by the magnetic field, while at lower energy this movement is limited to a small number of protons that have been assisted by the random field.

Two observers located in these positions could never agree on the luminosity of the source and the shape of the injection spectrum at energies below 10^{20} eV. Observer A (located in patch (a)) would claim a high luminosity and a steep injection spectrum with spectral index close to -4. Observer B would insist on a luminosity lower by many orders of magnitude and a very flat E^{-1} spectrum. The figure underscores the need for much higher statistics above 10^{20} eV, which could be used for a good estimate of the luminosity, the injection spectrum shape, and the position of the source.

All these arguments have been developed for steady cosmic ray sources that continuously emit UHE cosmic rays. Many of the models, however, suggest time-dependent injection. The extreme case is again the GRB scenario. GRB would inject UHECR only during an extremely short period of time. The high energy end of the injection spectrum would reach the observer with the smallest time delay. The lower energy part of the spectrum would scatter for much longer. The end result is a time-dependent energy spectrum at the position of the observer.

9.5 Current status of the field

At the present time we do not understand even the character of the UHECR sources or their luminosity. There are several contradictions in the scarce world statistics of such events.

– The normalization of the flux at about 10^{19} eV varies by about 30–40% as derived by different experiments. This is not much above the systematic errors of the experiments but requires errors in the opposite direction to make the data compatible.
– While some experiments do not see obvious GZK features in the measured spectra, others find the spectrum consistent with GZK cutoff plus very few super-GZK events. The event statistics are, however, so low that the statistical significance of the differences in the spectral shape is not very high.
– Some experiments measure a change of the cosmic ray composition from heavy to light in the range above 10^{18} eV. Others do not see an obvious change in the UHECR chemical composition.

The mere existence of super-GZK events, which have been seen by all experiments, is a very interesting phenomenon. Its understanding is further complicated by the arrival direction distribution of UHECR, which shows both a large-scale isotropy and a small-scale clustering. These conclusions are also derived on the basis of sparse experimental statistics and are thus at a relatively low confidence level.

Some arguments have been recently developed about the nature of the UHE cosmic rays. The first came from the shape of the shower profile of the highest energy $(3 \times 10^{20}$ eV) shower detected by Fly's Eye and shown in Fig. 9.8. Although it is a single event, its shape is well measured and is not consistent with the expectations for γ-ray showers [334] under any assumptions for their development. The profile of this event could be generated by a proton shower, but is more consistent with showers initiated by heavier nuclei. Figure 9.25 shows a comparison of the observed shower profile to expectations from different primary cosmic rays.

The expected profile for γ-ray initiated showers in Fig. 9.25 includes gamma-ray interactions on the geomagnetic field and the change of the electromagnetic processes (LPM effect) at very high energy. It is important to note that, if the hadronic collisions are softer than in the interaction model used for this result, proton and Fe initiated showers will both move to smaller depth. The γ-ray shower profile can only move to bigger atmospheric depths if the primary particle did not interact on the geomagnetic field.

In 2002 the Haverah Park [335] and the AGASA [336] groups looked at their statistics to identify γ-ray initiated showers. The Haverah Park group concentrated on the zenith angle distribution of the detected showers which is different for hadronic and electromagnetic showers. The result was an upper limit of about 50% for the fraction of γ-rays in the cosmic ray flux above

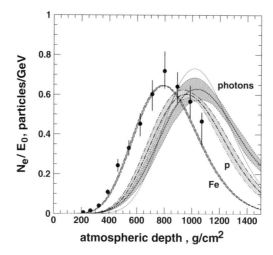

Fig. 9.25. The shower profile of the highest energy shower compared to the expectations for iron, proton, and γ–ray initiated showers. The bands show the expected widths of the respective shower profiles.

10^{19} eV. The AGASA group attempted to identify electromagnetic showers by their low fraction of muons. The limit above 10^{19} eV is 28%, and above $10^{19.5}$ eV is a weaker 67%. These results exclude models in which all UHECR are γ-rays but leave enough room for all possible source models, especially above 10^{20} eV, where a statistical analysis cannot be made.

9.6 Future detectors

It should be obvious from the discussion presented above that the study of the end of the cosmic ray spectrum needs a lot more experimental data. A vast increase of the statistics is not easy. The AGASA detector has an aperture of 125 km^2 ster and has observed only a handful of the most interesting events for almost twenty years of operation. HiRes is the first of the second-generation detectors with a significantly larger aperture that reaches 1,000 km^2.ster at 10^{20} eV, an improvement almost a factor of 10. There are several other projects that will contribute significantly to the event statistics of the future.

The Auger Observatory was first proposed in 1992 by James Cronin and Alan Watson who dreamed of and planned for a big increase of the UHECR statistics [337]. It will consist of northern and southern hemisphere sites, each with an area of 3,000 km^2. It takes the name of the French physicist who discovered the extensive air showers and in 1939 presented the evidence for the existence of 10^{15} eV cosmic rays [219]. The southern Auger observatory is now being built in the Mendoza province of Argentina. It will consist of

1,600 water tanks, each with an area of 10 m^2 and depth of 120 cm, that are positioned 1.5 km from each other. Each tank holds more than 11 tons of water. Charged particles emit Cherenkov light in the tanks, which is detected by three photomultipliers. Shower γ-rays convert into electron-positron pairs and also generate light. It is a hybrid array, a combination of surface array and 24 fluorescence telescopes that will be able to observe the shower profile of almost all events that trigger the array during clear nights.

The 'engineering' array, a requirement of the funding agencies, is already completed and has detected significant statistics of showers above 10^{18} eV. It consists of 40 surface detectors and two fluorescence telescopes. A fraction of the registered events have been seen by both components of the detector. The array will continue to collect data during most of the construction period.

The hybrid detector has the advantage of seeing simultaneously the shower profile and the lateral distribution of the shower particles on the surface. It thus determines the shower energy in two alternative ways and is quite sensitive to the nature of the primary cosmic ray particle. It will also be able to normalize the different ways of deriving the shower energy and answer the question why previous experiments have significant differences in energy assignment.

The choice of water tanks also has its advantages. Rather than measuring only charged particles, as a plastic scintillator usually does, they provide about three radiation lengths of water and convert all shower photons into electron-positron pairs. The measured quantity is the total shower energy, except for the fraction that has gone to neutrinos. Since the tanks are far apart muons and electrons have different time signatures and could be distinguished from each other. Muons arrive at the detector first because they do not scatter as much as electrons (or photons) do. Muons are characterized by a tight and high amplitude signal, while the electrons and photons have lower overlapping signals that last up to several microseconds.

The Southern Auger observatory expects to detect over 5,000 events above 10^{19} eV and about 100 events above 10^{20} eV per year if the AGASA cosmic ray spectrum is the correct one.

The Telescope Array is a project for an optical air shower detector of a different design from the HiRes [338]. A telescope array station consists of 40 telescopes that view independent regions of the sky. The field of view of a single telescope is 18° in azimuth and 15.5° in zenith angle. The telescopes are arranged in two 20-unit layers that view elevations from 3° to 34°.

The plan is to install the Telescope Array in Utah, USA, where individual stations will be separated by 30 to 40 km. A total of 10 stations are envisioned, three of which would complement the northern Auger observatory, which is not yet funded. The total effective aperture of the Telescope Array would be 5,000 km^2 ster for the typical 10% duty factor on clear moonless nights.

Individual units are designed so that they could be used as Cherenkov telescopes. Prototypes have already been installed in the Utah desert and have performed observations of TeV gamma-ray sources.

This original proposal was recently modified to also include a surface array of area 1,000 km^2. The surface array proposal is already funded.

EUSO is the acronym for the Extreme Universe Space Observatory [339], which is now funded by the European Space Agency for a possible deployment at the International Space Station. This is a new concept for observation of giant air showers from space which was suggested by the UHECR pioneer John Linsley in the early 1980s.

The idea is that an optical detector in space will observe a very large area and will have aperture higher than that of surface detectors by orders of magnitude. Such a detector should be able to observe the shower fluorescent light, as well as the shower Cherenkov light reflected from the surface – especially from the Earth's oceans. Initially the project was developed in Palermo, Italy, but now includes many European and US collaborators.

Since EUSO is going to the Space Station, it will fly at an altitude of about 400 km. It will be equipped with a wide Fresnel optics lens with an opening angle of 30° and a diameter of about 2.5 m, which concentrates the light onto a focal plane. The aim is to achieve a pixel resolution of 1 km^2 at Earth's surface, which will require more than 10^5 pixels for the total surface area of 150,000 km^2. It is difficult to estimate the total aperture of EUSO from general principles, but the aim is have it of the order of 10^5 km^2 ster. Most of the showers that will be observed by EUSO will be highly inclined showers, for which the shower track will be easier to reconstruct. Such showers will be created by UHECR and detected by EUSO at altitudes between 10 and 100 km. A typical shower observation would be a long shower trajectory of fluorescent light, followed by a splash of Cherenkov light when the shower hits the water surface. The Cherenkov light portion of the signal will depend strongly on the shower zenith angle.

The specific nanosecond signals of the shower front moving with the speed of light does not have any natural backgrounds. When the background light from the Earth is high, however, the signal has to be very strong to be detected and analyzed. One question that still remains to be answered is what the EUSO threshold energy will be. This depends crucially on the background light and the reflection from the Earth's surface that have not been previously studied at the level now needed. The collaboration has started a series of balloon flights that will measure both the background light and the reflectivity of the ocean water. The plan is also to monitor the light conditions from space and check them after every detector trigger. The aim of the group is to achieve an energy threshold of 5×10^{19} eV.

OWL (Orbiting Wide-angle Light-collectors) [340] project is an even more ambitious project for UHECR shower observations from space. OWL aims to monitor 3×10^6 km^2 ster of atmosphere which, with the typical 10% duty cycle, would collect about 3,000 events above 10^{20} eV per year.

The concept of this project is to perform a stereo fluorescent experiment from space. Two satellites flying on a medium altitude orbit (higher than EUSO) will be viewing a common volume of the atmosphere. When a cosmic ray shower develops in this volume the simultaneous observation will help fight the light background. It will also provide a better analysis of the events in terms of energy, interaction depth and arrival angle.

If the two-satellite experiments are performed they will indeed increase the UHECR statistics by orders of magnitude. ON a 5-year mission OWL should detect at least 10,000 cosmic ray showers of energy above 10^{20} eV. Some more optimistic assumptions would double and triple this number. This will lead to a much better knowledge of the nature of UHECR and the type and probably the location of their sources. The mystery of the highest energy cosmic rays will be finally solved.

10 High energy neutrino and gamma-ray astronomy

There are at least three arguments why we need a new type of astronomy – high energy neutrino astronomy. Low energy (MeV) neutrino astronomy exists. It detected the neutrinos from SN1987a and continuously monitors the emission of solar neutrinos in different energy ranges. Davis and Koshiba received the 2002 Nobel prize in physics for these observations.

The usual argument for extending the frequency range of telescopes and creating a new type of astronomy is that multiwavelength observations reveal much better the total energy output, as well as the dynamics of an astrophysical object. Such is the first argument in favor of neutrino astronomy. The emission of many luminous astrophysical systems is absorbed from intervening bodies or clouds, which are often related to the emission of the system itself. A typical example would be a star with very heavy stellar winds that shield the emission of the star in many frequencies. In this respect the detection of neutrinos would be a drastic improvement.

Let me explain what a drastic change means here: if the typical electromagnetic cross-section is Thomson's cross-section $\sigma_T = 6.65 \times 10^{-25}$ cm^2, neutrinos of energy 1 GeV have an interaction cross-section of 5×10^{-39} cm^2, smaller by 14 orders of magnitude. GeV neutrinos can propagate without interactions not only through the densest molecular clouds, but also through whole stars. The MeV solar neutrinos (that have even smaller interaction cross-sections) are a very good example. Solar neutrinos are generated in nuclear processes in the core of the Sun and are not absorbed by interactions in the matter of the star. The neutrino cross-section grows with energy but is still much lower than the electromagnetic cross-sections.

The second argument is that at high energy the Universe is no longer transparent to γ-rays. In Chap. 9 we discussed a bit the γ-ray interactions on the microwave background and on the isotropic radio background. Another prominent photon background is the infrared/optical background which is due to direct and re-scattered light emission from all stars. Figure 10.1 shows the γ-ray interaction length of all these three backgrounds with a solid line. The main contribution around 5×10^{14} eV is from interactions of the microwave background. The lower energy shoulder is from interactions on the IR/O background. It prevents the observations of 10^{13} eV γ-rays from distant objects. The high energy shoulder, discussed in the previous chapter is due to the radio background. If we want to extend the observations of the powerful astrophysical objects to very high energy, we cannot do it with γ-rays. Neutrino astronomy is the only solution.

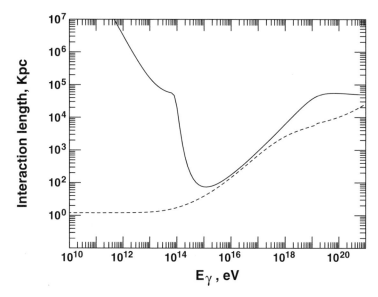

Fig. 10.1. Interaction length for γ-ray (solid) and electron (dashed line) interactions on the universal photon backgrounds. Only the major process of production of electron–positron pairs is plotted for gamma-rays. The electron interaction length is shown for inverse Compton effect.

The third argument concerns the character of the most energetic processes in powerful astrophysical systems. Gamma-rays can be produced in either electromagnetic or hadronic processes. To assume that all observed γ-rays are of electromagnetic origin means that only electrons are accelerated to high energy in astrophysical objects. Neutrinos can only be produced in hadronic processes. The detection of high energy astrophysical neutrinos will show what the role of hadronic processes is in the powerful astrophysical systems. Neutrinos are also a common product of the massive particles that could be remnants from the early stages of the evolution of the Universe.

In the same way the very large penetration ability of neutrinos makes them important for the observation of hidden sources and processes; it makes them very difficult to detect. If the ratio of the γ-ray to neutrino cross-sections is 10 orders of magnitude, this also means that the neutrino detector has to be 10^{10} times more massive to obtain the same statistics as a γ-ray telescope.

10.1 The neutrino cross-section at very high energy

The general behavior of the neutrino cross-section was described and the formulae for its calculation were given in Sect. 7.2. The only process that is important at high energy is the deep inelastic scattering. The charge current (CC) DIS cross-section is proportional to the neutrino energy up to Lab energies of a TeV and start bending over at the approach of m_W^2. The neutral current (NC) cross-section bends over at slightly higher energy corresponding to the higher Z_0 mass. Well above the squared masses of the vector bosons both cross-sections become proportional to $\ln E_\nu$. The cross-sections calculated with the GRV98 structure functions are plotted in Fig. 10.2.

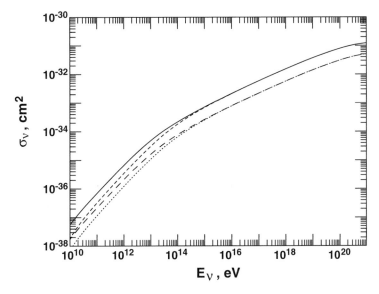

Fig. 10.2. Cross-sections for deep inelastic neutrino scattering. Neutrino CC cross-section is plotted with a solid line, the antineutrino with short dashes. The NC cross-section for neutrinos are plotted with long dashes, and for antineutrinos with a dotted line.

The absolute threshold for neutrino nucleon CC scattering is the mass of the respective charged lepton m_l. The cross-sections are depressed to more than 10 m_l. At 10 GeV, for example, the ν_τ cross-section is still much lower than the ν_2 and ν_μ cross-sections. Only at $E_\nu \gg m_l$ are the CC cross-sections of all neutrinos equal, as shown in Fig. 10.2.

At Lab energy of 1 GeV the antineutrino cross-section is lower than σ_ν by about a factor of 3. This difference decreases with E_ν and disappears at E_ν of about 10^6 GeV. The neutral current cross-section is always smaller than σ_{CC}.

The behavior of the very high energy neutrino cross-section is quite well known now, after the structure functions that define it were studied to low x and high Q^2 values. The exact σ_ν behavior at the highest end of Fig. 10.2 depend on the structure functions at values much lower than $x = 10^{-4} - 10^{-5}$ and are still somewhat uncertain, although not more than by a factor of 2.

The increase of the cross-section at very high energy is due to the contributions of sea quarks heavier than u and d. The exact content of such quarks is also somewhat uncertain. A good analysis of the contributions of all quarks and the uncertainties of the neutrino nucleon cross-section is given in Ref. [341].

10.2 Galactic gamma-ray and neutrino fluxes

The current picture of astrophysical neutrino production in galactic sources is that they are produced in pp inelastic interactions. Other meson decays also contribute at high energy, but the typical process is the production of neutral and charged secondary pions

$$p + p \rightarrow p(n) + m\pi^0 + 2m\pi^\pm .$$

The number of charged pions is roughly twice as large. At moderate energies the interacting proton loses on the average one-half of its energy and the other half is distributed between the secondary particles. Charged pions decay into a muons and muon neutrinos, and in the astrophysical environment muons also always decay into a muon neutrino, electron neutrino and an electron. So imagine an interaction where one secondary pion of every charge is produced.

π^0	π^+				π^-				p
$\frac{1}{6}$	$\frac{1}{6}$				$\frac{1}{6}$				$\frac{1}{2}$
$\gamma \;\; \gamma$	e^+	ν_e	ν_μ	$\bar{\nu}_\mu$	e^-	$\bar{\nu}_e$	$\bar{\nu}_\mu$	ν_μ	

The two γ-rays share on the average equally the π^0 energy and the four particles from the charged-pion – muon decay chain get approximately 1/4 of the the $\pi^\pm$ energy. So the total amount of energy that goes into electromagnetic particles (γ-rays and $e^\pm$) is about equal to energy in neutrinos, while the energy of the individual neutrinos is smaller that that of γ-rays. Folded with the steep cosmic ray spectrum $E^{-\alpha}$ the flux of neutrinos is smaller than that of γ-rays:

$$\phi_\nu = \phi_\gamma (1 - r_\pi)^\alpha , \tag{10.1}$$

where $r_\pi = (m_\mu/m_\pi)^2$.

The π^0 gamma-ray spectrum peaks at one half m_{π^0} – about 70 MeV. The neutrino spectra peak at about $m_{\pi^\pm}/4$. Particles of such energy arrive at Earth in fluxes much smaller than the ones produced by cosmic rays in the atmosphere. While for γ-rays it is possible to put a detector on a satellite, the required high mass makes that impossible for neutrino detectors.

The assumption that galactic neutrinos come from pp interactions requires the existence of dense matter targets at the source. The first suspects are then supernova remnants and binary systems powered by accretion.

More recent astronomical discoveries of microquasars, galactic objects with extended jets, may suggest neutrino production via photoproduction interactions, but this possibility has not been fully theoretically explored.

10.2.1 Galactic binary systems – Cygnus X-3 tales

The current interest in very high energy γ–ray and neutrino astronomy started about 20 years ago with a result that was never proven right or wrong. An excess of 10^{15} eV showers were detected by the Kiel group from the direction of the powerful binary system Cygnus X-3 [342]. The excess showers seemed to come with a periodicity approximately coinciding with the binary rotation period of this source. Soon afterward this result was confirmed by another air shower array with somewhat different periodicity. There were several detectors built to study the emission of Cyg X-3, and many theoretical papers explaining the dynamic of the source and the emission process. They involved the acceleration of protons close to the compact object (neutron star in this case) and the production of the signal in the vicinity of the companion star, either in the star itself or in the accretion disk or the accretion flow.

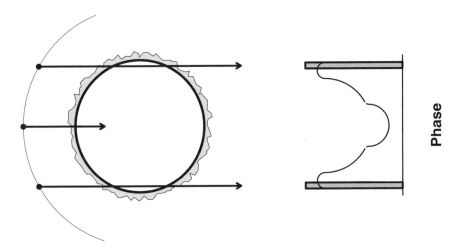

Fig. 10.3. The Vestrand–Eichler model of signal production in a binary system and the light curve for γ-ray (shaded) and neutrino fluxes.

The first and most influential papers [343] postulated that protons are accelerated at the neutron star continuously but they do not generate a signal if the observer is not in the line of sight defined by the two objects. When it is, as shown in Fig. 10.3, the γ-rays generated by the protons that hit the companion star become visible. Actually the generated 10^{15} eV γ-rays will survive only when the protons hit the very edge of the star – otherwise they would be absorbed. Two narrow peaks of γ-ray emission are indicated in the figure with shaded lines. This symmetry comes from the assumption that the observer is in the equatorial plane of the binary system. One can easily find a geometrical configuration that produces a single peak, even without including in the model the strong magnetic field needed for the proton acceleration.

Figure 10.3 also shows the the neutrino light curve which is expected from this model [344]. At the edges of the star the neutrino flux is almost as high as the γ-ray. Some low energy neutrinos, however, succeed in penetrating the companion star, whose density profile is reflected in the light curve.

The Cygnus X-3 saga finished in an unsatisfactory way. The source was very active in radio while the new detectors were constructed, but its activity decreased when they were ready for observations. None of the more sophisticated detectors confirmed the signals. It is still not obvious to me whether the source is in a quiet epoch, or the reported signals were flukes due to the small signal statistics, as most of my colleagues believe.

The main reason for this pessimistic view should be not the fact that the signal disappeared – we know that many astrophysical objects are only active during a short period of time – but the required energetics of the source. Binary systems are powered through accretion, which is limited to the maximum Eddington luminosity

$$L_{\mathrm{Edd}} \; = \; 1.3 \times 10^{38} M \; \mathrm{ergs\, s^{-1}} \tag{10.2}$$

where M is the mass of the object measured in $M_\odot$. The reported γ-ray signals from Cygnus X-3 would exceed L_{Edd} in any of the models for the emission. There are arguments suggesting the L_{Edd} is calculated on the basis of σ_T, which is higher than the hadronic cross-sections, and it can be exceeded in the case of proton acceleration and emission. This is true, but it is very unlikely that the total luminosity of a system would go only in the production of very high energy signals. We also know that the object was very active at all wavelengths.

In spite of probably being a statistical fluctuation, the Cygnus X-3 story was very helpful for the development of the ultra high energy astrophysics. The model described above is also typical for most models for hadronic production in binary systems. They consist of the same two important parts: an acceleration site and a production site. The acceleration site has to be luminous enough to explain the source luminosity and the production site has to contain enough matter to provide sufficient target for proton interactions.

10.2.2 Supernova remnants

There are two epochs in supernova remnant (SNR) evolution when one can expect γ-ray and neutrino emission. One of them is shortly after the supernova explosion, when the density of the expanding supernova envelope is very high and thus contains enough of a target for hadronic interactions. The basic theory for emission at this stage was developed by Berezinsky & Prilutski [345]. If one could think of a way to accelerate protons in the dense environment, and there are several ideas about that, one could think of several important stages that define the strength and the duration of the high energy γ-ray emission. The corresponding time frames are: τ_π when pions decay rather then interact, $\tau_{\gamma\gamma}$ when the supernova remnant cools down and the γ-rays from π^0 decay are not absorbed in production of electron–positron pairs on the SNR photon field, and $\tau_{\gamma p}$ when the γ-rays are not absorbed in bremsstrahlung. Since protons are contained in the magnetic fields of the expanding SNR, and γ-rays propagate in straight lines, there is room for proton interactions while γ-rays leave the young SNR without interaction. This is the time when the high energy γ-ray emission could start.

The emission will continue for about 2 to 10 years (depending on the mass distribution and expansion velocity of the SNR) until τ_a, when the proton energy loss on inelastic interactions becomes dominated by the adiabatic loss due to the SNR expansion. The γ-ray emission will fade for a long time, until the SNR reaches the Sedov phase, when most of the galactic cosmic rays are accelerated.

The emission in the early periods of the SNR evolution is not as important as the one during the Sedov phase for at least two reasons:

–   the emission is relatively short lived. With two or three galactic supernova explosions per century the chance for observation of such emission, even if it exists, is small, of the order of 10%.
–   the Sedov phase lasts for a long time, and there are always many SNR in this phase. The detection of such emission would be a confirmation of the our current understanding of the origin of the galactic cosmic rays.

The modern expectations of the gamma-ray emission of mature supernova remnants was developed by Drury, Aharonian & Völk [346]. The assumption is that cosmic rays at the source have a much flatter spectrum than the one observed at Earth. The authors consider the cases for power law differential acceleration spectra with indices 2.1 to 2.3. The γ-ray fluxes at Earth derived for the flat ($\gamma = 1.1$) spectrum is

$$F(> E_\gamma) = \simeq 10^{-10}\theta \left(\frac{E_\gamma}{\text{TeV}}\right)^{-1.1} \left(\frac{E_{SN}}{10^{51}\,\text{erg}}\right) \left(\frac{d}{1\,\text{Kpc}}\right) n_1 \quad \text{cm}^{-2}\,\text{s}^{-1}\ .$$

(10.3)

The unknown parameter θ is the fraction of the total SNR kinetic energy E_{SN} that is converted to accelerated cosmic rays, d is the distance to the

SNR and n_1 is the matter density in the vicinity of the blast shock in terms of 1 atom per cm^3. If only very high energy γ-rays are considered the flat spectrum is the most energetically efficient. The γ-ray flux would inevitably be accompanied by neutrino fluxes scaled down as in (10.1). The critical parameters are obviously θ and n_1. While θ could now be estimated only on the basis of theoretical predictions and in the future from the observation of SNR γ-ray fluxes, n_1 can be estimated from the SNR emission at other wavelengths. The observation of γ-ray fluxes would be helped significantly if the SNR was in the vicinity of a dense molecular cloud with $n_1 > 100$ or if the SNR expands in the dense envelope of the pre-supernova star wind as in Ref. [61].

As an example of the expectations from a concrete SNR Ref. [346] applies the calculation to the Tycho (1572) supernova remnant which should be close to the Sedov phase. One can take the average supernova energy and density from different estimates $E_{rmSN} = 4.5 \pm 2.5 \times 10^{50}$ ergs, $n_1 = 0.7 \pm 0.4$ and estimate the γ-ray flux for $\theta = 0.2$ and $d = 2.25 \pm 0.25$ kpc. The expected flux is

$$F(> E_\gamma) = \simeq 1.2 \times 10^{-12} \left(\frac{E_\gamma}{\text{TeV}} \right)^{-1.1} \text{cm}^{-2}\,\text{s}^{-1} . \tag{10.4}$$

The detection of such a flux is within the capabilities of the new generation of γ-ray Cherenkov telescopes that are being built now.

10.2.3 Observational constraints

The EGRET instrument on the Compton gamma-ray observatory detected 271 astrophysical sources of GeV γ-ray radiation. Some 93 sources were identified as extragalactic and 5 as galactic pulsars, and 170 sources were not identified. These unidentified sources cluster around the galactic plane – about 9 sources per degree of galactic latitude within $2°$ of the galactic plane where most of the SNR reside. Figure 10.4 shows the direction of these 170 sources and the position of known SNR.

Subsequent research matched the direction of the γ-ray signals with the position of galactic SNR and found four coinciding directions [347]. It is not obvious that the γ-ray fluxes are indeed produced by the supernova remnants – in two of the four cases the suspicion is that there is a different source in the same direction as the supernova remnant.

The emission from two of the sources, γ-Cygni and IC443 were analysed in Ref. [94] using the production mechanisms described in Sect. 4.4, including π^0 production and decay, inverse Compton scattering on the microwave and galactic IR/O background and on the photon field in the specific sources and bremsstrahlung of the accelerated electrons. Since the important parameters are either unknown or known with large uncertainty, the problem is not very well defined and the emission of IC443 could be fitted with different production mechanisms.

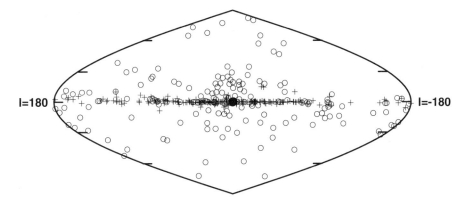

Fig. 10.4. Position of the EGRET unidentified sources in galactic coordinates. The galactic center is in the middle marked with a dot. The crosses show the positions of known supernova remnants from D.A. Green's catalog.

Figure 10.5 shows four different fits that explain the observed emission equally well. One of the problems with the fit is that the detected emission from IC443 extends only two orders of magnitude above 40 MeV and does not reach a high enough energy where the π^0 decay products are expected to dominate. There is a slight curvature in the data that seems to coincide with the position of the π^0 decay γ-ray peak. It also seems that the emission below 100 MeV has a different shape, although the accuracy of the EGRET detector in this range is small. There are only eight data points plus two upper limits for the flux.

The number of the possible fitting parameters is six, assuming that the radiation field in IC443 is well known: slopes of the spectra of accelerated protons and electrons (which do not have to be equal), maximum acceleration energies for protons and electrons, the ratio of accelerated electrons and protons, and the matter density at the SNR. Since the number of fitting parameters almost matches the number of data points, the fits can only be successfully performed with some of the parameters fixed, and others fitted.

The general conclusions from this work is that the observed emission could be of π^0 origin only if the matter density in IC443 is high, n_1 above 100. In case of such high matter density the π^0 decay and the electron bremsstrahlung dominate over the inverse Compton scattering. In cases of high matter density the ratio of the accelerated electrons to protons is similar to those in the solar system – 1/100. The low energy end of the detected emission is described best with electron bremsstrahlung. Because of the low energy range of the observed emission the fits are not sensitive to the maximum acceleration energy it could be as low as 100 GeV, both for protons and for electrons.

Scenarios such as the one shown in Fig. 10.5d are, however, also possible. In this case of low matter density the sub-100 MeV γ-rays are generated by inverse Compton scattering and both the electron and proton spectra cut off

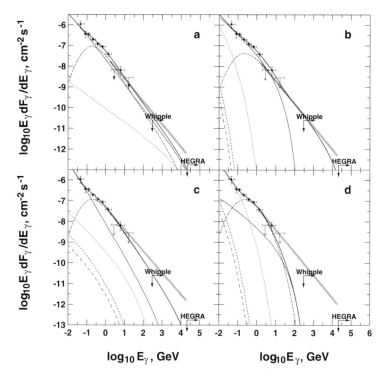

Fig. 10.5. Four fits of the gamma-ray emission from SNR IC443 from Ref. [94]. The data points are from Ref. [347]. The thin solid line indicates production by bremsstrahlung, the dotted line, by the inverse Compton effect, and the long and short dashed line, in hadronic interactions. An extension of the hadronic gamma-rays above 2 GeV with a flat 2.1 spectrum is shown with thick gray line. The upper limits of the emission from the Whipple and HEGRA TeV Cherenkov telescopes are also shown.

below 1 TeV. One could also predict γ-ray fluxes above 100 GeV generated only by inverse Compton scattering in the case of low matter density and high electron to proton ratio.

With the exception of this last fit the three others predict TeV γ-ray fluxes above or at the limit set by the Whipple Cherenkov gamma-ray telescope [348]. There are alternative explanations: either the observed γ–ray flux is not produced by π^0 decay and is of purely electromagnetic origin, or the spectrum of the accelerated cosmic rays cuts off at low energy or is steeper than $E^{2.2}$. Both these assumptions challenge the standard theory for cosmic ray acceleration and propagation.

Reference [349] performs an analysis of the radio and γ-ray emission of supernova remnants, demonstrates that the radiation of IC443 could be of purely electromagnetic origin and discusses the relation between TeV γ-ray production and cosmic ray acceleration. The conclusion is that γ-ray generating supernova remnants may not be the same objects that accelerate the galactic cosmic rays to an energy of 10^{15} eV. The SNR population that generates γ-rays may belong to a subtype that expands in high interstellar density that limits the maximum acceleration energy to well below the 'knee'. The authors claim an anticorrelation between the SNR luminosity in the GeV range (as detected by EGRET) and that in TeV gamma-rays.

TeV γ-rays have been detected from the Crab nebula and the supernova remnants SN1006, Cas A, and most recently from RX J1713.7-34.6. The Crab nebula is a standard candle in TeV γ-ray astronomy; it has a steady flux which is used to measure the fluxes of other sources. The radiation from the Crab nebula is shown to be of purely electromagnetic origin, as is that from SN1006. The origin of the fluxes from the other SNR is still being debated, but there are certainly no clear indications that they are of hadronic origin. We cannot thus predict detectable neutrino fluxes from the same sources with any certainty.

Another possibility is that the measured $E^{-0.6}$ dependence of the secondary to primary cosmic rays is only valid at very low energy and at high energy it is $E^{1/3}$ as predicted by the Kolmogorov spectrum of the plasma turbulence in the Galaxy [350]. In such a case the cosmic ray acceleration at SNR generates $E^{-2.4}$ energy spectrum, which is not inconsistent with the fits shown above. The TeV γ-ray fluxes with hadronic origin will be than lower than the sensitivity of the current TeV telescopes, but could be detectable in the near future. The GeV γ-ray flux from the central region of the Galaxy includes unresolved supernova remnants and future neutrino telescopes will detect the corresponding diffuse neutrino emission.

10.3 Extragalactic sources

The matter density in the luminous extragalactic sources is low compared to the photon field density in many regions. This determines the biggest difference in our current understanding – extragalactic hadronic γ-ray and neutrino signals should be produced in photoproduction interactions. The ratio of the photoproduction cross-section to σ_{pp} is roughly 1:100. Even in the Galaxy the ratio of the photon and nucleon density is higher than 400:1. The problem is that the energy threshold for pp is of the order of m_p while the energy threshold for proton interactions on the universal microwave background exceeds $10^{10} m_p$. The hope is that these luminous astrophysical systems generate a much more energetic photon background, as seen at optical and X-ray frequencies. For interactions on the 1 eV optical radiation the threshold decreases by three orders of magnitude, and in interactions on the X-ray background the threshold gains another three orders decreasing to $10^4 m_p$.

10.3.1 Observations of GeV and TeV gamma-rays from extragalactic sources

One of the greatest and not totally expected successes of the Compton GRO was the detection of numerous extragalactic sources of GeV γ-rays. Sixty six sources were definitely identified with extragalactic blazars and 27 more have a somewhat lower level of identification. Blazars are identified as highly variable active galactic nuclei. All EGRET gamma-ray sources are radio-loud, flat-spectrum radio sources. Some of them (3C279 and 3C273 for example) are superluminal objects. The emission in such objects is supposed to come from a jet oriented at small angle with respect to the line-of-sight to the system. When sources were observed in more than one viewing period, all extragalactic sources showed a high degree of variability. Figure 10.6 shows the directions of all EGRET extragalactic sources.

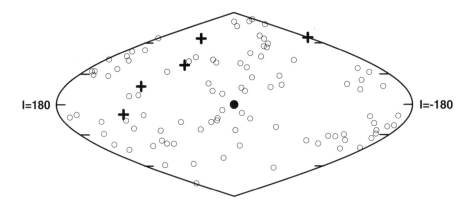

Fig. 10.6. Positions of the EGRET GeV gamma-ray sources identified as extragalactic (circles). Confirmed TeV sources are shown with heavy crosses. The position of the galactic center is marked with a dot.

The EGRET blazars are subdivided into two general classes – flat-spectrum radio quasars (FSRQ) and BL Lac objects. The sources are at redshifts $z = 0.03$ to 2.28 (distances from 120 Mpc to 9.12 Gpc). The γ-ray spectrum has been fitted with power laws for all individual objects. There is significant variations in the shape of the spectrum, but the average power law index is $\sim$2.2. 76% of the FSRQ are detected to be variable on the scale of several days to months. There is a general trend for the γ-ray spectrum to become harder during the flaring of the objects.

BL Lac objects are also variable, but to a smaller extent. This may be an indication that they less often have periods of increased activity.

The general feature of the blazar gamma-ray emission is that their luminosity in the GeV γ-ray band exceeds the emission in the infrared/optical band. Some details and references to the publications concerning individual objects are given in Ref. [351].

Air-Cherenkov gamma-ray telescopes aim at the detection of gamma-rays above 100 GeV. There is thus a gap in the energy spectrum, extending from 10 to about 100 GeV. The sensitivity of the air-Cherenkov telescopes is not much lower than that of a typical satellite γ-ray detector. This is determined by the shower Cherenkov light footprint (that reflects the lateral distribution of the shower electrons at shower maximum) is roughly 10^4 m^2 compared to the typical linear dimension of 1 meter for a satellite detector.

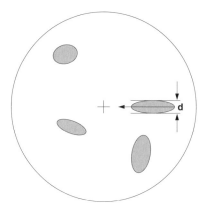

Fig. 10.7. Illustration of the principles of the imaging technique used by the air-Cherenkov TeV gamma-ray telescopes.

The air-Cherenkov technique has been tested since the early 1960s, starting at the Crimean Astrophysical Observatory, without much success. The detection of γ-ray showers has to compete with ordinary hadronic showers that have higher flux. The only way to push the background down used to be the improvement the angular resolution of the telescope. A revolution of the Cherenkov telescope technique started in 1990s after A.M. Hillas outlined the differences in the development of γ-ray and hadronic showers and their Cherenkov image.

The main difference is that the secondary particles in hadronic interactions have average transverse moments $p_\perp$ of more than 300 MeV/c. The transverse momenta in electromagnetic interactions are of the order of m_e, i.e. about 1,000 times lower. For this reason hadronic showers are ragged and irregular, while electromagnetic showers are smooth. Another difference comes from the pointing of the air-Cherenkov telescope at a potential γ-ray source. The images of showers coming from the direction of the source point at the center of the mirror, as shown in Fig. 10.7. From that illustration one can see how one of the showers is thinner than the rest (smaller d) and points at the center of the mirror. It is likely to be a signal shower while the other three showers are most likely background.

The imaging technique which is now used by many telescopes includes many more parameters that are extracted from Monte Carlo calculations of the shower development and the detector performance. The background is suppressed by orders of magnitude, the observation time is shorter and the detection significance is increased. The imaging is realized by a high-resolution camera that views the focal plane of the large air-Cherenkov telescope reflector dish – 10 meter diameter in the case of the Whipple telescope. The camera consists of a large number of phototubes that observe the field of view (490 at Whipple). The smaller the pixel size of the camera (the larger the number of photomultipliers), the better is the imaging.

There are at least five active galactic nuclei that are sources of γ-rays above 300 GeV. The list of these AGN is given in Table 10.1.

Table 10.1. Galactic coordinates, names and redshifts of the confirmed extragalactic TeV γ-ray sources.

l, deg. b, deg.		Name	Redshift
166.1	38.2	Mrk 421	0.030
217.1	42.7	H1426+428	0.129
253.5	39.8	Mrk 501	0.034
300.0	65.2	1ES1959+650	0.047
356.8	51.7	1ES2344+514	0.044

Mrk 421 was detected by EGRET as a low-level source with γ-ray flux of $1.4\pm0.4\times10^{-7}$ cm^{-2}s^{-1} above 100 MeV. Mrk 501 was not an EGRET source – only an upper limit of its emission exists (1.5×10^{-7} cm^{-2} s^{-1}). The average fluxes detected by the Whipple observatory above 300 GeV are respectively 4.0×10^{-11} cm^{-2} s^{-1} and 8.1×10^{-12} cm^{-2} s^{-1}. Both sources, however, had powerful outbursts that increased their γ-ray emission, sometimes by an order of magnitude. One of the best examples is the dramatic flare of Mrk 421 in 1996 when the flux increased by a factor of 2 every 15 minutes.

It is not yet obvious what is the intrinsic source energy spectrum and the maximum energy of the extragalactic gamma-ray sources. Five different detectors observed the several months' outburst of the source Mrk 501 in 1997. The spectra recorded by different telescopes are in a reasonable agreement and reach 20 TeV. The mean free path of 20 TeV γ-rays on the IR/O background is expected to be about 90 Mpc for $H_0 = 75$/km/s/Mpc as shown in Fig. 10.1. The distance to Mrk 501 ($z = 0.034$) is 136 Mpc for the same H_0 value and the detected flux of 20 TeV γ-rays should be a factor of 5 less than that of the source one because of absorption on propagation. The maximum acceleration energy at the source is even more difficult to establish because there could be absorption due to $\gamma\gamma$ interactions at the source. The spectrum of Mrk 501 seems to have a break and becomes steeper above 5 TeV. This is consistent with absorption on IR/O as well as with a possible change of the γ-ray production mechanism. The question about the density of the IR/O background was complicated even more after the discovery of the source H1426+428 at a redshift of 0.129, i.e. a distance of 500 Mpc. This source shows a steep energy spectrum with a power law index of about 3.5. See Ref. [352] for a recent review of the observation of TeV γ-rays from active galactic nuclei.

10.3.2 Models of gamma-ray production in AGN

To understand the gamma-ray production models one has to first envision the structure of the source itself. A cartoon of the generic AGN model is shown in Fig. 10.8. The source of the AGN power is a massive black hole of 10^8 $M_\odot$ or more. It is surrounded by an accretion disk, formed by the matter accreting on to the black hole. The accretion disk is heated and is the main source of the AGN optical luminosity. The inner parts of the accretion disk are usually brighter at blue wavelengths, powered by the thermalized emission of the central engine.

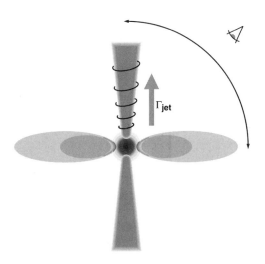

Fig. 10.8. Cartoon of the structure of AGN.

Some 10% of all identified AGN show one or two jets starting in the vicinity of the black hole and extending far into the extragalactic medium – sometimes at 100 Kpc from the central engine. The unification AGN model states that all AGN have similar structure and the observation of the AGN jet depends on our viewing angle. The jet consists of plasma moving away from the black hole with relativistic velocity. The plasma is contained by strong magnetic fields, possibly related to the spinning accretion disk. Some of the observed jets (in 3C279, for example) show super-relativistic motion, which means that we look inside the opening angle of the jet.

An important consequence of the high velocity of the jet plasma is the boost of the emission in the jet frame. If a source co-moving with the Jet plasma emits flux F, an outside observer looking towards the jet sees flux $\delta^3 F$. Here δ is the Doppler factor of the jet,

$$\delta = \frac{1}{\Gamma(1 - \beta \cos \theta)} ,\qquad (10.5)$$

where θ is the angle between the jet axis and the line of sight and β is the flow velocity measured in units of the speed of light. Γ is the Lorentz factor of the plasma. Lorentz factors Γ up to 20 have been observed in AGN jets. Thus for $\Gamma = 10$ an observer looking along the jet axis could see a flux enhancement by a factor of 10^3. In addition the energy of the particles is also boosted by a factor δ so that the total luminosity of the source would appear higher by a factor of 10^4. The observed superluminal motion with β_{obs} is linked to the actual velocity β of the jet plasma

$$\beta_{obs} = \frac{\beta \sin \theta}{1 - \beta \cos \theta} \qquad (10.6)$$

and has a maximum value of $\beta \Gamma$.

The prevailing majority of the γ-ray production models in AGN jets are purely electromagnetic and do not involve hadronic interactions. The synchrotron self Compton (SSC) model [353] assumes the acceleration of electrons in the jet frame. Because of the magnetic field, necessary for the support of the jet, these electrons suffer synchrotron radiation and generate a high density of synchrotron photons. The same electrons have inverse Compton scattering on the synchrotron radiation field and boost these photons to energy comparable to their own. Thus the radiation from the jet has a double-peaked energy distribution – one peak covers the synchrotron photons and the second one covers the products of the inverse Compton scattering. Since both components of the emission are produced in the jet frame, they are boosted to higher energy by a factor δ. The exact positions of the two peaks depend on the maximum electron energy, the radius of the emitting region and the strength of the magnetic field. SSC predicts a strong correlation between the emission in the two peaks – if more electrons are accelerated in the jet, this increases both the number of electrons and the density of the synchrotron photons, their target of inverse Compton scattering. The increase at the TeV peak can reach the square of the increase in the KeV region.

The SSC model has an intrinsic limit of the maximum γ-ray energy because of the decrease of the inverse Compton cross-section in the Klein–Nishina regime. The SSC model fits well not only the multi-wavelength emission of blazars, but also the emission of the Crab nebula. Figure 10.9 gives an idea of the typical SSC photon energy spectrum. More realistic models that use the observed multi-wavelength radiation from particular objects have more complicated spectra, but still preserving the two-peak structure.

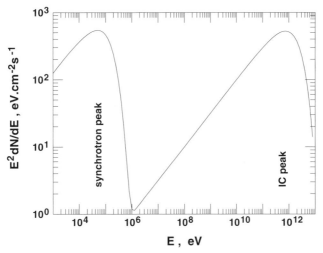

Fig. 10.9. Two-peaked photon spectrum generated by the Synchrotron Self Compton model of γ-ray production.

The external radiation Compton (ERC) model [354] differs from the SSC in the location of the seed radiation for inverse Compton scattering. Instead of being synchrotron radiation internal to the jet, the seed radiation is the optical emission of the accretion disk, re-scattered by the dust that surrounds the jet. This class of models may be more applicable to the FSRQ blazars that show strong atomic lines in their optical and UV spectra. On the other hand, the BL Lac type objects do not have strong lines and the existence of radiation external to the jet is not proven. BL Lac objects may be on a later stage of the AGN evolution when the black hole is already very massive and the amount of matter outside of the black hole has declined. It seems that the SSC model is more suitable for BL Lac objects and the ERC model is at least as good for FSRQ objects.

The only models that can generate neutrinos as well as gamma-rays are the ones that include hadronic processes [355]. These models also have to generate double-peaked γ-ray spectra to fit the observations. This can be achieved in two ways: electrons can be accelerated together with the protons and can then supply synchrotron radiation as in the SSC model. In addition, a fraction of the proton energy goes into secondary neutral particles that decay and the decay products cascade down to add to the radiation field. Finally the seed radiation could be generated by synchrotron radiation directly emitted by the protons and all charged hadrons. This, however, requires a very strong magnetic field.

The accelerated protons have photoproduction interactions on the photon field in which both γ-ray and neutrino fluxes are produced. The products of the photoproduction radiation supply the TeV peak while the KeV peak could be a product of all electromagnetic loss processes. The biggest difference in the γ-ray spectral shape is that the high energy radiation is not cut off in the multi-TeV region as it is in the inverse Compton scattering process. The γ-ray and correspondingly the neutrino spectra can extend to very high energy, depending on the maximum proton acceleration energy. Since hadronic models concentrate mostly on the high energy radiation, they are usually not as exact in the description of the KeV and lower energy flux, although there are no reasons for which they cannot be fine-tuned to correctly describe them.

10.3.3 Models of neutrino production in AGN

There are two possible locations for neutrino production in AGN: in the region close to the central engine and in the AGN jets. Gamma-rays can also be produced close to the black hole, but this is academic, because they would be immediately absorbed in the dense radiation environment. If neutrinos are produced they would survive and easily leave the AGN. The first calculation of neutrino production in AGN was published in Ref. [356]. While only about 10% of all AGN have visible jets, all of them have a massive central black hole. For this reason we shall call such models generic AGN models - see Fig. 10.10.

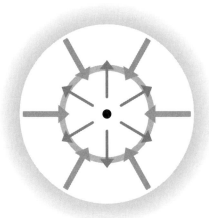

Fig. 10.10. The generic AGN model for neutrino production close to the central engine. The black dot in the middle is the Schwartzschild radius of the black hole. The gray circle where the in-falling matter meets the radiation pressure indicates the position of the shock.

The basis of the generic AGN models is the assumption that close to the black hole a shock would be produced where the ram pressure of the accreting matter is compensated by the radiation pressure from the AGN core. Let us assume that the distance from the shock to the black hole is R_1 or $x_1 = R_1/R_S$ in units of the Schwartzschild radius $R_s = 2GM/c^2$. R_S is the distance from the black hole within which photons cannot escape it. The scaled shock radius x_1 is assumed to have values between 10 and 100. From the relation between the source luminosity and the black hole mass (assuming Eddington luminosity) one can derive the radiation energy density and the proton density at the shock. U_{rad} is of the order of $(10^6/L_{45})(30/x_1)^2$ erg/s ($L_{45} = 10^{45}$ erg/s). The magnetic field at the shock is assumed in equipartition with the photon field and is $B \sim 7,000(30/x_1)/\sqrt{L_{45}}$ G. From the accretion rate needed to support the source luminosity the proton density comes out of the order of $10^8 x_1^{1/2}$ cm^{-3}. The photon density can be estimated from U_{rad} and the AGN emission spectrum in the optical to X–ray region. For an average photon energy of 5 eV the ratio between the proton and photon density $n_p/n_{ph} \sim 10^{-13} x_1^{3/2}$. Because the photoproduction to pp interaction cross-section is 0.01, the dominant energy loss of the protons will be on photoproduction [65].

The next step is the calculation of the maximum proton energy at acceleration at the shock. Since the magnetic field is large, the maximum energy would be determined by the energy loss on photoproduction, which depends strongly on x_1. The further away the shock is from the black hole, the smaller the photon density is, and the higher the maximum acceleration can be. For the chosen x_1 range, however, the acceleration proceeds to sufficiently high energy to generate neutrinos of energy exceeding 10^{18} eV. Neutrinos are the only particles that can leave the AGN. Gamma-rays are absorbed and thermalized. Protons are contained magnetically and only sufficiently high energy neutrons generated in the photoproduction interactions can also leave the AGN. As all photon energy is thermalized, only the neutrino flux could testify for the hadronic origin of the radiation. The model can be proven right only by the detection of the neutrinos emitted by the AGN core. It is not obvious how the neutrino luminosity of such a source has to be normalized. Following Ref. [356] it is more or less arbitrarily normalized to a fraction (30%) of the diffuse X-ray background assuming that it has the same origin.

The other natural location for proton acceleration and neutrino production are the AGN jets. The main process is again photoproduction either on internal synchrotron photons, or on the thermal UV photon background of the accretion disk. The matter density in the jet is very low and the proton energy loss on pp interactions (and the neutrino production from this process) is small. The high energy neutrino spectra will have the shape of those of the accelerated protons, usually assumed with power law index of 2. At energies below the proton photoproduction threshold the neutrino spectrum is a flat decay spectrum with $\alpha = 0$. Hadronic jet models are criticized for not being able to explain the short period of variability of the TeV γ-ray sources. The reason is that protons have to be accelerated to very high energy and this takes more time than the variability scale allows. In addition, very high energy protons are difficult to contain inside the jet. The containment of 10^{20} eV protons requires magnetic fields of 1 G at distances of 0.1 pc from the the central object. This is at the limit of the magnetic field calculated from the AGN luminosity assuming equipartition between the photon and the magnetic fields.

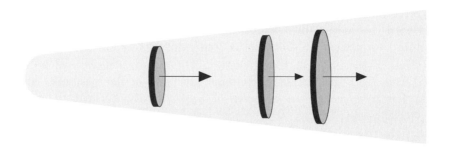

Fig. 10.11. Cartoon of the acceleration in an AGN jet scenario where the plasma is injected in the jet in separate lumps that have different velocities, indicated by the length of the arrow attached to each disk.

One possibility for generating a variable proton blazars model is to think of a suitable proton acceleration model. It is well known from astronomical observations that the jets are not static objects. One could imagine that the plasma is injected in the jet as separate blobs, or disks as shown in Fig. 10.11. These separate lumps have different velocities and the faster ones reach and override the slower lumps. Ideal conditions for shock creation and proton acceleration would exist when the plasma lumps approach each other.

The neutrino fluxes from photoproduction AGN jet models peak between energies of 10^{16} and 10^{18} eV. At lower energy there could be a contribution from pp scattering if the jet hits some higher density clouds. The neutrino emission from jets is boosted in energy and beamed in a cone with opening angle $1/\Gamma$ in reverse proportion to the jet Lorentz factor.

10.3.4 Neutrino production in GRB

Gamma-ray bursts (GRB) are the most luminous violent astrophysical phenomena known. They are understood as the expansion of a relativistic fireball [357] that is caused by the merger of two neutron stars or by the explosion of a supermassive star. The GRB emission is almost certainly beamed. These are objects at cosmological distances, many of them at redshift exceeding 1.

The luminosity required for such distances is 10^{51} erg/s during the burst if the emission were isotropic.

Especially if the the source of the GRB energy is a hypernova, there will be an associated flux of thermal neutrinos, similar to those detected from SN1987A. High energy neutrinos can only be associated with proton acceleration and interactions in the GRB [358]. The physics of the neutrino production is very close to that in AGN jets. The main differences are the high average Lorentz factor of the GRB ($\Gamma = 300$) jet and the short duration of the emission. The photon fluxes detected from GRB have a variety of shapes, but they are all distributed around a power law spectrum with a break at about $\epsilon_b = 1$ MeV. The radiation below the break has a differential energy spectrum with a power law index of 1, and that above the break is steeper by one power of the energy. Protons of energy above $\Gamma^2 E_{th}/\epsilon_b$, where E_{th} is the photoproduction threshold, would interact with the flatter part of the photon spectrum, while lower energy protons will only interact on the steeper high energy end.

This leads to a specific shape of the GRB high energy neutrino spectra that has two breaks – one at 10^5 GeV where the neutrino flux shape changes from E_ν^{-1} to E_ν^{-2}, and a second one at 10^7 GeV where it becomes steeper by another power of the energy. These power law indices correspond to a flat E^{-2} proton acceleration spectrum. The normalization of the GRB neutrino flux requires assumptions for the proton luminosity of the GRB. Only infrequent nearby GRB at distances less than 100 Mpc would produce detectable high energy neutrinos in coincidence with the photon signal, but the neutrino flux of all unidentified GRB will be detectable by the future neutrino telescopes.

10.3.5 Diffuse extragalactic neutrinos

Even if we cannot observe source neutrinos there is a chance to see the diffuse neutrino flux from unidentified sources, none of which is powerful enough to be observed individually. A simple example is to imagine 50 sources in the Universe, each of which generates only one event in a detector. We can never be certain that this neutrino event came from an astrophysical source, but could define the diffuse flux of 50 events if they had a distinguished signature. Such a signature is their energy spectrum.

The neutrinos generated by cosmic rays in the atmosphere – the atmospheric neutrinos – have a steep energy spectrum. It is partially due to the steep galactic cosmic ray spectrum, but also to the small dimensions of the atmosphere. High energy pions start interacting before they decay and high energy muons hit the ground and lose their energy. In astrophysical environments secondary interactions are not very likely and all unstable particles decay. Thus the diffuse neutrino background is expected to have indeed a much flatter energy spectrum.

Since the extragalactic neutrino sources are based on photoproduction, the expected neutrino signal peaks at very high energy. The rough relation between neutrino and proton energy is $E_p/E_\nu \sim 20$. For proton interactions on the microwave background this leads to neutrino energy exceeding 10^{18} eV. Neutrino fluxes from extragalactic sources should extend to lower energy because of the more energetic photon fields of the sources.

A reliable estimate of the diffuse neutrino flux is not easy because one has to account not only for the processes that happen in individual sources, but also for the luminosity spectrum of these sources and their cosmological evolution. One of the big contribution of Ref. [356] is the estimate of diffuse neutrino fluxes from generic AGN. The diffuse flux is calculated as

$$\frac{dF}{dE} = \frac{c}{H_0}\frac{1}{ER_0^3}\int dL_x \int dz \rho(L_x, z)\frac{dt}{dz}\frac{dL}{dE}(E_z, L_x) \qquad (10.7)$$

Here H_0 is the Hubble constant and ρ is the luminosity evolution of the sources with redshift z. The matrix element dt/dz is given by (9.10) and is $(1+z)^{-5/2}$ for the Einstein–de Sitter Universe.

The integration in redshift has to be carried to significant redshifts because the sources that can contribute to the diffuse background could be at large distance. Neutrinos do not lose energy on propagation and in this respect the situation is very different from UHECR that have to be produced in cosmologically nearby processes. The luminosity evolution of the neutrino sources can also change the strength and, to a certain extent, the shape of the spectrum.

10.3.6 Summary of the expected neutrino fluxes

The field of neutrino astronomy is young and it is difficult to judge all predicted neutrino emission from astrophysical sources. The calculations are based on important assumptions that may or may not be true. In Fig. 10.12 we show several predictions for neutrino emission from different types of sources compared to the flux of atmospheric neutrinos within 1°.

These predictions are obtained in different ways. The prediction for SNR IC443 comes from the most favorable fit of Ref. [94] where the proton acceleration is extended to the knee of the cosmic ray spectrum. The prediction for Mrk 501 is based on a proton acceleration model for the observed γ-ray emission. It is normalized to the emission during the outburst and not to the significantly lower average flux. The dashed lines (3) show the predictions for four different models of neutrino emission of 3C273 as described in Ref. [65]. It consists of two parts – the high energy part (above 10^{15} eV) is due to photoproduction interactions and the lower energy part comes mostly from pp interactions. The model variations at high energy come from the maximum proton acceleration energy. Model (4) shows the prediction of neutrino flux from the 3C279 jet of K. Mannheim. It includes a boost of 10 in the neutrino energy.

Figure 10.13 shows some of the predictions for diffuse high energy neutrino fluxes. Note that all predictions and the atmospheric fluxes are given per

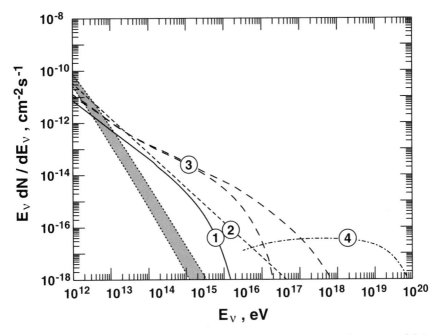

Fig. 10.12. Predictions of neutrino fluxes from different types of sources. (1) is the neutrino flux that would correspond to the gamma-rays of SNR IC443. (2) is the neutrino emission that would correspond to hadronic origin of the Mrk 501 gamma-ray outburst. (3) is the range of neutrino emission from the core of 3C273, and 4) is a prediction for the neutrino emission of 3C279. The shaded are shows the atmospheric neutrino flux within 1° – from high (horizontal) to vertical.

steradian. Before the discussion of this figure one has to note that the atmospheric neutrino flux is uncertain at energy above 10^{13} eV. The band shown would correspond to the lowest fluxes, that could increase after the neutrinos from forward charm and heavier flavors are added to it. The production cross-sections for forward charm are not well known and the estimates are uncertain.

The prediction for the neutrino flux from the central galaxy is based on the assumption that the flux above 1 GeV detected by EGRET is of π^0 origin. It is safe to claim that this prediction is quite certain and can only be wrong by not more than about 30%. Flux (2) is due to the models of Szabo & Protheroe, which are folded with the cosmological evolution of AGN. The cosmological evolution model used accounts separately for the luminosity evolution and for the source density evolution and cannot be easily compared to the simple formulae given in this book. Flux (3) is the isotropic AGN neutrino flux from Ref. [359]. Curve (4) is for the prediction of the diffuse neutrino fluxes from gamma-ray bursts. The shape of the flux follows the discussion in Sect. 10.3 and does not account for the most likely fluctuating parameters in GRB that would smooth the sharp transitions at 10^{14} and 10^{16} eV.

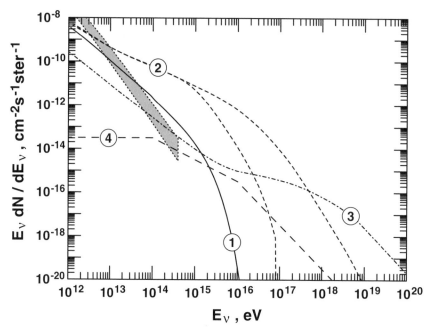

Fig. 10.13. Predictions for diffuse neutrino fluxes. The shaded area shows the horizontal (higher) and vertical fluxes of atmospheric neutrinos. Curve (1) is for the central region of the Galaxy, (2) corresponds to the curves (3) from Fig. 10.12, (3) is the prediction of Ref. [359] and (4) is the prediction of the GRB neutrinos of Ref. [358].

10.4 Neutrinos and UHE cosmic rays

We have described as potential high energy neutrino sources almost the same types of objects that may accelerate the ultra high energy extragalactic cosmic rays. The most obvious case is for gamma-ray bursts, where the models for acceleration and neutrino production are identical.

This relation has been used as a mean to estimate what the expected flux of very high energy neutrinos could be. In somewhat stricter terms it was used in Ref. [360] to derive the upper limit for the diffuse flux of high energy cosmic neutrinos. The limit is obtained in the following way.

The cosmic ray injection flux is assumed to be on a E^{-2} power law spectrum in the energy range from 10^{19} to 10^{21} eV. A normalization to the measured UHECR flux at 10^{19} leads to a total luminosity of the UHECR sources above that energy of 4.5×10^{44} erg Mpc^{-3} yr^{-1}. The following step is to relate the neutrino flux to the cosmic ray flux at injection as

$$E_\nu^2 \frac{dN_\nu}{dE_\nu} \simeq \frac{t_H}{4} \varepsilon E_p^2 \frac{dN_p}{dE_p} ,$$ (10.8)

where ε is a factor (<1) that reflects the kinematics of neutrino production by photoproduction. This factor is assumed to be independent of energy – every proton generates three neutrinos, ν_e, ν_μ, $\bar{\nu}_\mu$, each of which carries 5% of the proton energy. The maximum neutrino flux I_{max} is achieved for $\varepsilon = 1$, i.e. when the total proton energy is exhausted in photoproduction interactions and respective neutrino production. The maximum neutrino flux is

$$I_{max} = \frac{t_H}{4}\xi_z \frac{c}{4\pi} E_p^2 \frac{d\dot{N}_p}{dE_p} , \tag{10.9}$$

where ξ_z is a constant summarizing the product of the cosmological evolution of the UHECR sources and the relation of redshift to injection time. The maximum diffuse neutrino flux is then

$$E_\nu^2 \Phi(\nu_\mu + \bar{\nu}_\mu) = \varepsilon I_{max} = 1.5 \times 10^{-8}\xi_z \,\text{GeV cm}^{-2}\,\text{s}^{-1}\,\text{ster}^{-1} \tag{10.10}$$

and depends only on the source cosmological evolution. For evolution proportional to $(1 + z)^3$ ξ_z has a value of about 3.

It is important to note that (10.10) gives the upper limit of the neutrino flux rather than an estimate of its strength. This limit is applicable for sources with optical depth less than 1 for UHE protons. Sources such as the generic AGN model for cosmic ray acceleration and neutrino production are not accounted for, since they do not contribute to the UHECR flux reaching us. Models for neutrino production in jets (or GRB) are limited by it.

The upper limit of Waxman & Bahcall is not universally accepted. Another, more detailed, model for the upper limit of the neutrino flux was derived in Ref. [361] The authors of this paper develop the following points: (a) Waxman & Bahcall do not account correctly for the difference of the proton and neutrino energy loss horizon. Cosmic rays have $\lambda_p^{-1} = \lambda_z^{-1} + \lambda_{BH}^{-1} + \lambda_{phot}^{-1}$, where λ_{BH} stands for the pair production loss and λ_{phot} for the photoproduction loss. Neutrinos, however, have only loss from cosmological expansion, i.e. $\lambda_\nu = \lambda_z$. This relaxes the limit above neutrino energy of 10^{19} eV, where the proton energy loss is indeed very large. The main reason for the relaxation of the limit at lower energy is the uncertainty in the cosmic ray injection spectrum. Any spectrum steeper than E_p^{-2} would generate a higher limit. Some other more detailed considerations, such as the adiabatic loss of the cosmic ray protons related to the decrease of the jet plasma velocity with time, can also contribute to changes in the limit. The limit is derived for sources that absorb all the energy of protons, but are optically thin for neutrons.

Figure 10.14 shows the limits on the diffuse neutrino flux derived in these two ways. Ref. [361] also defines a limit for sources that are optically thick for neutrons. It is significantly higher straight line at 1.7×10^{-6} GeV cm^{-2} s^{-1} ster^{-1}. The MPR line for sources optically thin for neutrons in Fig. 10.14 approaches that limit at neutrino energy of about 10^5 GeV.

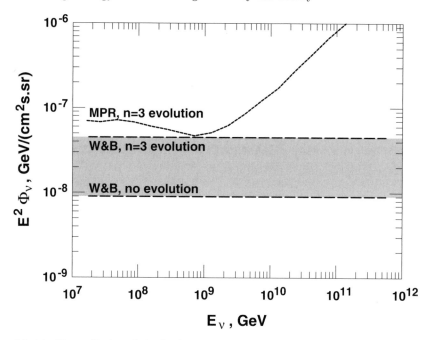

Fig. 10.14. Upper limits of the high energy diffuse neutrino flux derived by Waxman & Bahcall (W&B) and Mannheim, Protheroe & Rachen (MPR).

10.4.1 Neutrinos from propagation of ultra high energy cosmic rays

Neutrinos are not only produced in and at the cosmic ray sources. When high energy protons (or other cosmic ray nuclei) lose energy on photoproduction in propagation, the end product of the pion and other mesons decay chain are neutrinos. Because of the sharp threshold for photoproduction interactions, the neutrino yield of protons of energy less than 10^{20} eV is a strong function of the proton energy. A proton of energy $10^{19.5}$ eV generates 0.01 muon neutrinos and antineutrinos in propagation on 100 Mpc in the present Universe. A proton of energy 10^{20} eV generates 2.6 muon neutrinos and antineutrinos, mostly in the energy range from 10^{16} to 10^{19} eV. Higher energy protons generate many more neutrinos, because the secondary protons from the photoproduction interaction remain above the interaction threshold for a long time. The total generated flux thus depends on the acceleration spectrum of the ultra high energy cosmic rays.

Another source of neutrino production are neutron decays. Secondary neutrons are the product of proton photoproduction interactions in one-third of the interactions. The neutron and proton photoproduction cross-sections are almost identical, but the neutron decay length is smaller than the neutron interaction length below 4×10^{20} eV. Such neutrons rarely interact and decay, generating a $\bar{\nu}_e$. Magnetic fields of any size would scatter the protons and increase their pathlength over the neutron pathlength. For this reason, except at energies below 10^{16} eV the neutrino flux from cosmic ray propagation contains very few electron antineutrinos.

The existence of ultra high energy neutrinos from cosmic ray propagation was suggested by Berezinsky & Zatsepin [362], and independently by Stecker [363] soon after the realization that the GZK cutoff should exist. At that time the photoproduction was not as well studied as it is now. There have been many other calculations of this diffuse neutrino flux since then. Figure 10.15 shows the spectra of $(\nu_e + \bar\nu_e)$ and of $(\nu_\mu + \bar\nu_\mu)$ neutrinos generated in proton propagation in the assumption of uniform homogeneous source distribution of the sources of UHE cosmic rays and E^{-2} acceleration spectrum from a recent paper [364].

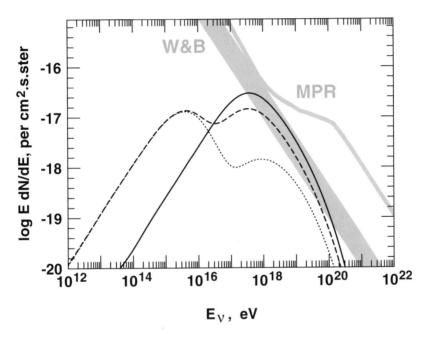

Fig. 10.15. Fluxes of neutrinos from proton propagation compared to the neutrino flux limits of W&B and MPR. The solid line is for the sum of muon neutrinos and antineutrinos and the dashed line is for the sum of electron neutrinos and antineutrinos. The dotted line shows separately the flux of $\bar\nu_e$.

The muon neutrino spectrum peaks (after multiplication by E_ν) at $10^{17.5}$ eV and has the characteristic flat decay spectrum well below the peak. With source evolution $(1 + z)^3$ the muon neutrino flux approaches the W&B limit calculated for the same cosmological evolution and source luminosity above 10^{18} eV. Proton propagation thus contributes to the UHE diffuse neutrino fluxes an amount that is equal to the amount contributed by cosmic ray sources. This is not strange, because in both cases protons have approximately the same number of interactions. The main difference is that the only essential interaction target in propagation is the microwave background, while higher energy photon fields exist in sources and lower energy neutrinos can also be produced there. The spectra of UHE neutrinos from proton propagation are well below the MPR limit which is much more relaxed at ultra high energy.

Electron neutrino spectra have a more interesting double-peak spectrum. The higher energy peak, containing mostly ν_e, coincides with the muon neutrino one. The lower energy spectrum peaking at $10^{15.5}$ eV contains only $\bar{\nu}_e$ from neutron decay. The antineutrino spectrum is shown with a dotted line in Fig. 10.15.

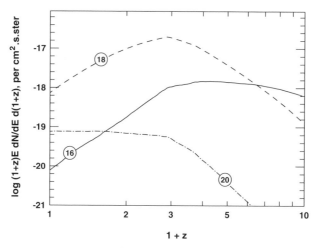

Fig. 10.16. Contribution of different cosmological epochs to the flux of neutrinos of different energy. Log_{10} of the neutrino energy is indicated by each curve. The normalization coincides with the one in Fig. 10.15.

We stated above that neutrino production is strongly dependent on the cosmic ray acceleration spectrum. There is, however, an interplay between spectral shape and source evolution when the final result is obtained. At larger redshifts the photon density is higher by $(1 + z)^3$ and both the microwave and proton energy are higher by $(1 + z)$. As a result the photoproduction threshold decreases, and the number of protons above threshold increases. Quantitatively, if one defines a neutrino yield function $Y(E_p, E_\nu, z) = E_\nu \, dN_\nu/dE_p \, dE_\nu$ for the production of neutrinos of energy E_ν by protons of energy E_p at redshift z, it will scale with redshift from the yield function at 0 redshift as

$$Y(E_p, E_\nu, z) = Y\left[(1+z)E_p, (1+z)^2 E_\nu, 0\right] . \qquad (10.11)$$

For a steeper proton acceleration spectrum the number of subthreshold protons that are above threshold at redshift z increases more rapidly and compensates for the smaller number of protons that interact at present. As a result the neutrino spectra generated by steeper proton spectra move to lower energy compared to the spectra of Fig. 10.15, but keep the total flux normalization almost constant. The number of high energy neutrinos decreases but the number of lower energy ones increases.

To illustrate fully the importance of cosmological evolution we show in Fig. 10.16 the contribution of different cosmological epochs to the fluxes of Fig. 10.15 for $E_\nu = 10^{16}$, 10^{18}, and 10^{20} eV.

The curve for 10^{18} eV neutrinos peaks at redshift of about 3 and is symmetric in $\log(1 + z)$. The higher energy 10^{20} eV neutrinos are mostly produced at present. A part of the decrease is the redshift in E_ν with z. For the lower energy neutrinos the trend is the opposite – the highest contribution is from high redshifts.

10.5 Detection of high energy astrophysical neutrinos

The standard method for detection of high energy astrophysical neutrinos is upward-going neutrino-induced muons as discussed in Sect. 7.2.1. Chapter 7 also gives the neutrino cross-section for the CC deep inelastic scattering that is applicable to very high energy. All detectors listed in that chapter have generated sky maps of the detected neutrinos and have set limits for neutrinos from possible astrophysical sources. These limits are significantly higher than the estimates given in the previous section.

At high energy the neutrino–muon angle in deep inelastic scattering is small and muons define well the direction of the parent neutrinos. The neutrino cross-section grows with energy and at some point the Earth starts becoming opaque for neutrinos. This happens when the neutrino mean free path $(\sigma_\nu N_A)^{-1}$ exceeds the thickness of the Earth in g/cm^2. Figure 10.17 shows how the Earth becomes gradually opaque as a function of the neutrino energy and the zenith angle of the neutrino.

When the neutrino mean free path becomes 10^{10} g/cm^2 neutrinos cannot penetrate along the diameter of the Earth. They interact and the generated muon or other particles have big energy loss and are absorbed in the Earth in the same way as electrons are at lower energy. This happens at E_ν exceeding 5×10^4 GeV. With energy the opaque cone grows and at neutrino energy about 5×10^9 GeV the whole lower hemisphere is opaque. There can be still be a small number of muons that penetrate the detector almost horizontally.

More generally speaking, the upward-going muons are not a good detection method for neutrinos of energy exceeding 10^6 GeV. One has to apply different detection methods. The obvious one is to detect cascades that are generated by the neutrinos. Cascades are generated in charge current interactions of electron neutrinos, where the total neutrino energy is released in

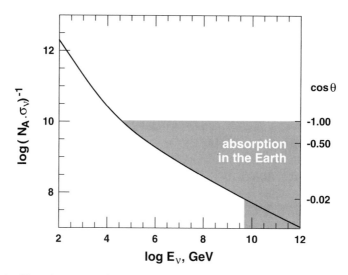

Fig. 10.17. Neutrino mean free path as a function of the energy. The right hand side scale shows the absorption in the Earth as a function of the neutrino zenith angle.

this form. Cascades are also generated in neutral current interactions, where most of the energy that does not go into the secondary neutrino can produce a cascade in the detector. Very high energy neutrino-induced muons that penetrate the detector from above can also be used in a statistical manner, because such muons are rarely generated in the atmosphere.

The existence of neutrino oscillations also changes the detection strategy. Tau neutrinos are very rarely generated in the cosmic ray sources and almost never in cosmic ray propagation. The production ratio of $(\nu_e + \bar{\nu}_e)$ to $(\nu_\mu + \bar{\nu}_\mu)$ is $(1 + \epsilon) : 4$, because the production of $\bar{\nu}_e$ in astrophysical environments is not strong, except by neutron decay. Because even the highest energy muon neutrinos easily oscillate into tau neutrinos on the vast astrophysical pathlength, on arrival at the Earth, the ratio between the three neutrino flavors is 1:1:1. As we will see later, the admixture of ν_τ significantly increases the neutrino survival probability as well as the neutrino-induced shower rate.

10.5.1 Detection of neutrino-induced showers

The detection rate of neutrino-induced showers in a large water Cherenkov detector by neutrinos of certain type and energy E_ν is the product of the neutrino flux Φ_ν and cross-section σ_ν with the Avogadro number N_A. For a neutrino cross-section of 10^{-35} cm^2 and a flux of 10^{-15} neutrinos.cm^{-2} s^{-1}, the rate is 6×10^{-27} s^{-1} g^{-1}, or 1.8×10^{-10} events per Kt yr. This simple example demonstrates that high energy neutrino telescopes have to be significantly bigger than the detectors of atmospheric neutrinos. The total rate of showers above an energy threshold, E_{thr}, is

$$\text{Rate} = N_A \int_{E_{thr}}^{\infty} \int_{y_{min}}^{1} \frac{d\sigma(E_\nu)}{dy} \, dy \, \Phi(E_\nu) \left(1 - \frac{\Omega(E_\nu)}{4\pi}\right) dE_\nu \; \text{g}^{-1} \text{s}^{-1} \, ,$$

(10.12)

where $\Omega(E_\nu)$ is the solid angle of the opaque cone. The detection technique could be exactly the same as for upward-going neutrino-induced muons in a water-Cherenkov detector. The charged particles from the cascade (mostly electrons) generate Cherenkov light. The arrival time of the Cherenkov light at the individual photomultipliers is used to determine the shower geometry and thus the arrival direction of the primary neutrino, and the total amount of light gives the total number of particles and thus the energy of the neutrino. It is much easier to analyze the event if the event is contained inside the detector.

With the increasing neutrino energy the cross-section grows and the event rate increases in spite of the absorption in the Earth. Figure 10.18 shows the shower rate from electron neutrinos and antineutrinos presented in Fig. 10.15.

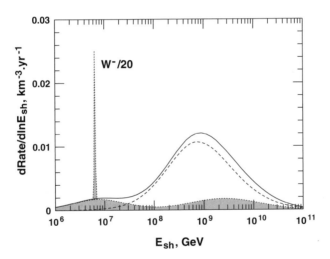

Fig. 10.18. Shower rate generated by electron neutrinos and antineutrinos (dotted, shaded) as a function of the neutrino energy. Neutrino-induced showers are shown with a dashed line, and the total with a solid line.

Although the peak in the $\bar{\nu}_e$ spectrum is of the same magnitude as the ν_e peak at higher energy, the number of showers generated by $\bar{\nu}_e$ is significantly smaller. This is the result of the increasing neutrino cross-section. The narrow sharp peak at energy 6×10^6 GeV (which is scaled down by a factor of 20 to fit in the graph) is from the *Glashow resonance* [365] – the resonant production of W^- bosons in $\bar{\nu}_e e$ interactions. The resonant cross-section is

$$\sigma(\bar{\nu}_e, e) = \frac{G_F^2 s}{3\pi} \frac{M_W^4}{(s - M_W^2)^2 + \Gamma_W^2 M_W^2} , \qquad (10.13)$$

where Γ_W is the width of the resonance. The peak of the resonance is at neutrino energy $E_\nu = s/(2m_e) \sim 6.4 \times 10^6$ GeV, where the cross-section is about 4.7×10^{-31} cm^2. The peak cross-section is high, but the total rate from the resonance is not because its width is only 2.1 GeV.

Most of the showers are produced by electron neutrinos in the energy range 10^8 to 10^{10} GeV. The total rate is not high: the integral is only about one shower in 20 years in a km^3 detector. Other neutrino flavors contribute twice that and neutral current processes contribute as much for a total of one event every 5 years. One cubic kilometer is a very large volume to instrument, even if not very densely. The hope is then to find a way to observe showers that are not contained in the detector volume.

One can, for example, detect the air showers initiated by 10^{10} eV astrophysical neutrinos and analyze them in the same way as ordinary air showers. The volume of air that the Auger observatory can inspect for neutrino-induced showers is equivalent to 30 km^3 of water [366] at neutrino energy above 10^{19} eV. The EUSO and OWL detectors will be sensitive to much larger volumes using the air fluorescence technique.

The detection is not direct; it can only work on a statistically large sample. The vertical thickness of the atmosphere is not large enough to cause many neutrino interactions. In horizontal directions, however, the atmosphere is 40 times thicker. From Fig. 10.17 the 10^{19} eV neutrino mean free path is about 5×10^7 g/cm^2. About 1/1000 of the neutrinos of that energy would interact and initiate showers. Ordinary hadronic showers are fully absorbed before reaching such atmospheric depths, but neutrino showers will on the average start much later and will reach the shower array as relatively young showers. So giant air shower arrays will look for young horizontal showers and identify them as generated by ultra high energy neutrinos. The challenge there is the ability of the air shower array to analyze horizontal events. One of the reasons for the selection of the water-Cherenkov tanks as the ground detector for the Auger observatory is that these tanks do not change their surface area with angle as much as thin scintillator detectors do. The orbiting air shower fluorescent detectors EUSO and OWL will also look for almost horizontal, late developing showers. They will probably have an even higher energy threshold but will be able to observe a much larger part of the atmosphere.

Another very exciting idea is the detection of the shower at radio frequencies. This is an old idea, suggested in the 1960s by Askaryan [367], which is now being implemented. The physics is that in every electromagnetic shower there is some degree of charge asymmetry. The asymmetry is due to the small differences in the scattering of electrons and positrons, to the Compton scattering of the shower photons on atomic electrons and on the positron annihilation. A charge that moves with the speed of light generates electromagnetic waves.

At low frequencies, when the wavelength is longer than the shower dimension, the emission is proportional to the charge. When the wavelength is shorter than the shower dimension the emission is proportional to its square. The charge asymmetry itself is proportional to the primary energy, so above some threshold energy showers can be detected from a large distance, depending on the transparency of the medium where they develop into radio waves.

Cold ice turns out to be a suitable medium. A shower initiated by a 10^{16} eV neutrino could be detected at distances of a few kilometers. One could then cover large distances of Antarctica with antennas positioned one kilometer apart from each other and observe the ice to the thickness of a kilometer. The direction of the neutrino-induced shower can be derived from the time when the radio signal reaches individual detectors. Ten thousand relatively inexpensive antennas on the surface of the ice can observe 10^4 km^3 of volume for high energy neutrino interactions. Antarctica seems a good place for radio detection, also because the noise at radio frequencies is smaller than at any other location.

The idea has been tested at accelerators, where a beam of electrons was dumped into sand. The antennas around the beam dump were able to estimate correctly the direction of the beam and the total amount of energy.

Another suggestion of Askaryan was that of acoustic detection. A particle cascade deposits its energy in a relatively small volume of matter. The length of 10^{10} GeV shower in water is of the order of 10 meters. The energy is thermalized and converted to heat in 30 nanoseconds. This creates a shock and an acoustic boom similar to that of an aircraft exceeding Mach 1. The radiation is coherent, and thus proportional to the square of the primary energy. One can then envision a neutrino telescope consisting of hydrophones – devices much less expensive and complicated than photomultipliers.

10.5.2 Neutrino oscillations and neutrino telescopes

Tau neutrinos change the strategy of neutrino detection because of the very short lifetime of the tau meson. The decay time of τ is 2.9×10^{-13} s. The decay length of a 10^{15} eV τ meson is $8.7 \times 10^{-3} E_\tau/m_\tau$ cm or about 50 meters. In addition, the electromagnetic energy loss of the τ is significantly smaller than that of a muon – the bremsstrahlung is very strongly suppressed and the pair-production loss is suppressed by the ratio of the τ/μ masses. So τ mesons lose only about 2×10^{-7} of their energy per g/cm^2. At a pathlength of 50 meters in standard rock of density 2.65 g/cm^3 a τ meson loses only 0.25% of its energy.

The lepton number is conserved in τ decay and one of the products of the decay is a ν_τ. The neutrino energy is decreased by a factor of 10 or so, but the neutrino is not absorbed. The sequence of ν_τ interactions and τ decay continues until the neutrino energy falls below 10^5 GeV and it does not interact any more even if it propagates through the center of the Earth [368]. For different propagation angles the ν_τ would survive with higher energy.

The fraction of the ultra high energy astrophysical neutrinos that have oscillated into τ neutrinos survives propagation through the Earth and appears at lower energy. The energy spectrum of the surviving τ neutrinos will be very different from their production spectrum. At energies between 10^{14} and 10^{15} eV there will be a pile-up after which the spectrum will be cut off.

Actually there will be some non-absorption of all neutrino types because of neutral current interactions $\nu + N \longrightarrow \nu + X$. The energy of the neutrino is decreased only slightly, by 30% or so at high energy. The NC cross-section is, however, significantly lower than the charge current cross-section (by a factor of 2.5 to 3) and the effect is not large because the interaction probability exponentiates this difference.

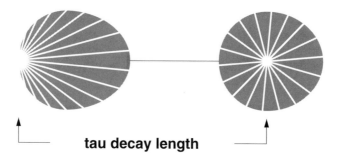

tau decay length

Fig. 10.19. Double-bang signature of τ neutrino interactions. The left-hand shower is generated by the ν_τ interaction and the right-hand one by the tau meson decay. The two showers are connected with the weak Cherenkov track of the τ meson.

The short lifetime of the τ lepton also creates a unique signature of the ν_τ interaction – the double bang [369], which is illustrated in Fig. 10.19. At the position of the ν_τ interaction there will be a shower initiated by the secondary particles created in the interaction. Because of its very low energy loss, the propagating τ meson will have a much weaker Cherenkov light track than a μ meson. Then a second shower will be created by the τ decay. The distance between the two showers depends on the τ energy and is on the average about 50 meters for 10^{15} eV ν_τ and the whole signature could be contained in a neutrino telescope.

At the energy of 10^{19} eV, where giant air shower arrays could be sensitive to neutrino-induced showers, the distance grows by four orders of magnitude and becomes on the average 500 km. It will be difficult for a ground array such as Auger to see the double bang, but satellite air shower detectors such as EUSO and OWL would be be able to observe the double bang ν_τ signature if they are sensitive enough. The event rate for the double bang is not very large because the whole elongated structure has to fit in the observed volume, but it is indeed unique and has to be looked for.

10.5.3 Neutrino telescopes

As we have already mentioned above, all underground neutrino detectors served the role of first-generation neutrino telescopes. They did map neutrino arrival directions and set limits on the magnitude of the high energy astrophysical neutrino fluxes. We will not discuss these activities, because it is obvious from the fluxes shown for different types of sources that much bigger detectors are needed for the detection of astrophysical neutrinos. All projects listed below are international collaborations.

The first proposal for a second-generation neutrino telescope was that of the Deep Underwater Muon and Neutrino Detector (DUMAND) that was submitted for funding in the late 1970s. The general design follows the suggestion of Markov that ocean water could be instrumented well enough to become a muon and neutrino detector. The first DUMAND proposal already incorporated all major features of the more modern neutrino telescope designs and probably underestimated some of the problems with the deployment and operation of the telescope.

Photomultipliers are to be arranged on strings that are anchored to the floor of the ocean. The power for the photomultipliers and the information goes on cables from the individual string to a junction box, which is connected to the shore with a 30–40 km long cable. The strings are shielded in optical modules (OM) that have to be watertight and are arranged in a cylindrical shape with distances of about 50 meters between strings. The optical modules have to go deep enough to be shielded from cosmic rays and from background light. It is very important that the modules know about each other position and in almost all designs they are equipped with light sources that are constantly monitored by nearby modules. This provides information not only about the module positions, but also about the optical properties of the medium. Figure 10.20 sketches the detection technique with a deep underwater neutrino telescope.

DUMAND II was funded between 1992 and 1995 and succeeded in deploying the junction box and a string. It took and analyzed muon data, but the funding was discontinued and the work did not progress any further. DUMAND II was to be built 30 km from the cost of Hawaii at depth of more than 4.5 km, which provides a very good shielding of the telescope from the atmospheric muons. The water at this depth is very clear and the bioluminescence is manageable.

Baikal NT-200 is the first neutrino telescope [370] that became operational. It was constructed in the Siberian Lake Baikal and was designed to consist of 200 photomultipliers that are mounted on several strings with more complicated shape. The lake freezes solidly every winter and this is when the experiment was constructed and maintained. The water in the lake is not as

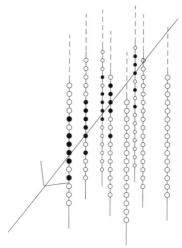

Fig. 10.20. Sketch of the detection technique with a deep underwater neutrino telescope. Upward-going neutrino-induced muon goes through the detector and generates Cherenkov light in a wide cone. The track is reconstructed from the timing of the triggered photomultipliers, shown here with full circles.

clear as in the deep ocean and the OMs have to be cleaned every season. Baikal is the first neutrino telescope that detected neutrinos, although they were most likely of atmospheric origin.

The idea of designing an underwater neutrino telescope is now continuing in the Mediterranean, where three different projects exist: NESTOR, NEMO and ANTARES. NESTOR [371] is to be situated at the southwestern tip of Greece at a depth of about 3.5 km and under one the best deep ocean conditions in the Mediterranean. The optical modules are not organized in strings, rather in towers, with different stories of a tower sliding on top of each other. There are many measurements and studies of the water conditions and the deployment process that have to be made before the actual deployment and NESTOR is still in this stage at the time of writing.

The biggest and most advanced Mediterranean project is ANTARES [372], which is based in the French Mediterranean close to Marseille. The telescope is to consist of 12 vertical strings that cover area of about 0.1 km². The strings are of total length 400 m and 300 meters of them is instrumented with about 1,000 optical modules with very high pressure resistance. The modules are in group of three every 16 m along the string. They have the center of their field of view 45° below the horizon. This means that the three PMTs together view the total lower hemisphere because the Cherenkov light cone in water is close to 45°. The ocean at the ANTARES location is shallower (2.5 km), and the atmospheric background will be higher, but the overlapping phototubes would make the filtering of the background events easier.

Like NESTOR, ANTARES is designed as a relatively low-threshold detector, whose efficiency should start approaching unity at 20 GeV.

All strings are connected to a junction box that is connected by a 40 km long electro-optical cable to the shore station. ANTARES is in a good position to enjoy the support of the excellent knowledge and experience of deep underwater work that exists in France. Test string deployments have been successfully performed and the first permanent string deployments should start in 2003.

NEMO is a relatively new project that started developing in Catania, Sicily [373]. The project is in the stage of environment studies for the best site in the Mediterranean for a big neutrino telescope, and of the design of separate aspects of a neutrino telescope. The NEMO group is in the planning phase for a bigger, 1 km^3 neutrino telescope that will succeed devices such as NESTOR or ANTARES.

AMANDA [374] (Antarctic Muon and Neutrino Detector Array) employs the same detection principles as all other neutrino telescopes. Details are, however, very different. AMANDA is located at the South Pole and works in the ice cover of Antarctica rather then in sea water. Deep ice develops a special structure that makes it extremely transparent. Holes are drilled by pumping hot water into the ice and strings are lowered to depths exceeding 1.5 km from the surface. Strings carry the power to the individual optical modules and the signals from them up to the surface.

AMANDA was built in several steps. In the first attempt the strings were in ice that was much too shallow and did not have the right optical properties. Then four strings (B4) were put under 1,500 m of ice that demonstrated the viability of an under ice neutrino telescope. B4 was complemented with six more strings to B10. Currently AMANDA II exists with a cross-section area of 3×10^4 m^2 and a depth of about 500 meters – not all strings are of the same length. AMANDA II consists of 19 strings with about 700 optical modules. It is the biggest operating neutrino telescope that has already published results on high energy atmospheric neutrinos and limits on neutrino fluxes of non-trivial origin. Not all results from AMANDA are analyzed because running an experiment at the South Pole requires very sophisticated logistics, not similar to a space probe, which often introduces delays.

The first neutrino telescope which would be able not only to detect, but also to study high energy astrophysical neutrino signals, ICECUBE [375], is now funded and in the stage of final design. It was proposed by a big international collaboration, which has as its nucleus the AMANDA collaboration. The project calls for the deployment of 80 strings with 60 optical modules each deployed at depths of 1,400 to 2,400 meters of ice. The total fiducial volume of the telescope is 1 km^3. The arrangement of the holes is in a triangular scheme, as shown in Fig. 10.21. The optical modules will be smart digital optical modules (DOM), they will digitize the PMT signals, contact their neighbors, and test the quality of the signals before transmitting them

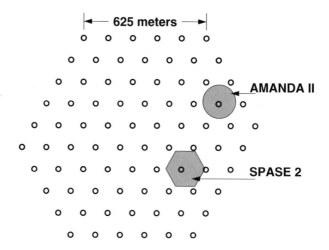

Fig. 10.21. Plan of the arrangement of the ICECUBE strings. Each string is accompanied by two ice-Cherenkov tanks that form the ICETOP air shower array. The shaded areas show the location of the AMANDA II neutrino telescope and the SPASE 2 air shower array.

to the surface. Every DOM will in practice be a powerful CPU, that will transmit the waveform received from the PMT in digital form to the main data acquisition system, if there is an array trigger.

Every string will be accompanied by two ice tanks at the surface that form ICETOP, an associated air shower array. ICETOP will help calibrate ICECUBE by detecting high energy muons from air showers. It will also give anticoincidence signals when a high energy cascade is spotted in ICECUBE and perform standard air shower physics.

ICECUBE would enjoy one of the advantages of every telescope that is located at the poles of the Earth: each astrophysical object is always observed at the same elevation. This helps the calibration of the device and gives equal exposure per unit solid angle. ICECUBE will be able to observe all sources at northern declination, as shown in Fig. 10.22 in galactic coordinates.

If the Mediterranean efforts lead to another km^3 neutrino telescope (as anticipated), the whole sky will be observable in high energy neutrinos. Because of the rotation of the Earth most of the sky is observed only part of the time, except for the dark area in Fig. 10.22 which is seen 24 hours a day. The galactic center is seen approximately half of the time. A large fraction of the fields of view of telescopes in these two locations coincide. This is also important because in such a novel type of observation detected signals have to be confirmed. It is not yet certain, but the expectations are that in several years this observing scheme will work and we shall know much more about the role of high energy processes in the dynamics of luminous objects.

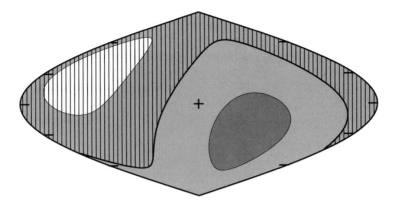

Fig. 10.22. The field of view of ICECUBE covers the Northern hemisphere shown here in galactic coordinates. A northern hemisphere neutrino telescope (in Mediterranean location) would observe the whole sky except for the white area. The dark area will be observed full time. The cross marks the position of the Galactic center.

There are serious new developments also in the radio technique for detection of neutrino-induced cascades. RICE [376] is a project that aims at the development of a large area (10^4 km^2) neutrino detector in the Antarctic ice. The tests that RICE has performed since 1998 include placing radio antennas at depths of about 100 meters in some of the AMANDA holes. The overburden shields to a certain extent the detector from atmospheric radio noise. The experimental group has proved that this technique could be used for detection of high energy cascades, but the prototype is not big enough to detect signals. On this basis RICE can place a limit on the flux of UHE neutrinos, which is still well above the theoretical expectations.

Another positive development is the ANITA [377] project that is already approved. ANITA will launch circumpolar balloon flights to look for neutrino interactions in the Antarctic ice. The balloons will fly at an altitude of about 40 km and will be able to monitor the ice for radio signals up to 680 km away. Antarctic ice is transparent to radio at frequencies up to 1 GHz, where the shower radio emission is coherent and proportional to E_ν^2. A 12-day flight, expected from a circumpolar balloon, would provide an exposure of almost 2×10^5 km^2 yrsr and should be able to detect UHE neutrinos. One important parameter that is not yet known is the threshold energy for the cascades that should depend on the distance to the balloon. The elevation resolution of ANITA is about 3°, which should make the determination of the distance to the cascade fairly good. ANITA flights are scheduled for 2006.

The field is developing rapidly and the risk that these pages will be outdated soon is large. New methods and ideas are developed and sometimes implemented quickly.

A Appendix

A.1 Physical constants

Quantity	Symbol	Value
speed of light in vacuum*	c	2.998×10^{10} cm/s
reduced Planck constant	$\hbar \equiv h/2\pi$	6.582×10^{-22} MeV s
electron charge	e	1.602×10^{-19} C
electron mass	m_e	0.511 MeV/c^2
		9.109×10^{-28} g
proton mass	m_p	938.3 MeV/c^2
		1.007×10^{-24} g
$W^{\pm}$ mass	m_W	80.42 GeV/c^2
Z^0 mass	m_Z	91.19 GeV/c^2
fine structure constant	$\alpha = e^2/4\pi\epsilon_0\hbar c$	$1/137$
classical electron radius	$r_e = e^2/4\pi\epsilon_0 m_e c^2$	2.818×10^{-13} cm
gravitational constant	G_N	6.710×10^{-39} $\hbar c/(\text{GeV}/c^2)^2$
Avogadro number	N_A	6.022×10^{23} g^{-1}
Boltzmann constant	k	8.617×10^{-5} eV/K
Fermi coupling constant	$G_F/(\hbar c)^3$	1.166×10^{-5} GeV^{-2}
Thomson cross-section	$\sigma_T = 8\pi r_e^2/3$	6.652×10^{-25} cm^2

* The definition of a meter is the distance that light travels in a vacuum in 1/299 792 458 of a second.

A.2 Astrophysical constants

Quantity	Symbol	Value
astronomical unit*	AU	1.496×10^{13} cm
parsec+	pc	3.086×10^{18} cm
light year	ly	0.3066 pc
solar luminosity	$L_\odot$	3.846×10^{33} erg/s
solar radius	$R_\odot$	6.961×10^{10} cm
solar mass	$M_\odot$	1.989×10^{33} g
Earth mass	$M_\oplus$	5.975×10^{27} g
1 eV equals	wavelength	1.240×10^{-4} cm
	frequency	2.418×10^{14} Hz
	temperature	1.161×10^{4} K
	energy	1.602×10^{-12} erg

* average distance between the Sun and the Earth.
+ distance at which 1 AU is seen as 1 arc sec.

A.3 Properties of particles discussed in the book

Leptons

electron $e^{\pm}$, $m_e = 0.511$ MeV, stable
electron neutrino ν_e, $m_{\nu_e} < 3$ eV

muon $\mu^{\pm}$, $m_\mu = 105.7$ MeV, mean life τ_μ 2.197×10^{-6} s
decay mode $\mu^{\pm} \longrightarrow e^{\pm} \nu_e (\bar{\nu}_e) \bar{\nu}_\mu (\nu_\mu)$
muon neutrino ν_μ, $m_{\nu_\mu} < 0.19$ MeV

tau $\tau^{\pm}$, $m_\tau = 1{,}777$ MeV, mean life τ_τ 2.906×10^{-13} s
multiple decay modes
tau neutrino ν_τ, $m_{\nu_\tau} < 18$ MeV

Hadrons

Mesons

charged pion $\pi^{\pm}$, quark content $(u\bar{d}, d\bar{u})$, $m_{\pi^{\pm}}$ 139.57 MeV,
mean life $\tau_{\pi^{\pm}}$ 2.60×10^{-8} s
decay mode $\pi^{\pm} \longrightarrow \mu^{\pm} \nu_\mu (\bar{\nu}_\mu)$

neutral pion π^0, quark content $(u\bar{u} - d\bar{d})/\sqrt{2}$, m_{π^0} 134.98 MeV, mean life τ_{π^0} 8.4×10^{-17} s
decay mode $\pi^0 \longrightarrow \gamma\gamma$

charged kaon $K^\pm$, quark content $(u\bar{s}, \bar{u}s)$, $m_{K^\pm}$ 493.68 MeV,
mean life $\tau_{K^\pm}$ 1.24×10^{-8} s
main decay modes:
$K^\pm \longrightarrow \mu^\pm \nu_\mu (\bar{\nu}_\mu)$ 63.4%
$K^\pm \longrightarrow \pi^\pm \pi^0$ 21.1%

neutral kaon $K^0 = 1/2 K_S^0 + 1/2 K_L^0$, quark content $d\bar{s}$, m_{K^0} 497.67 MeV
K_S^0 – mean life $\tau_{K_S^0}$ 8.94×10^{-11} s
main decay modes:
$K_S^0 \longrightarrow \pi^+ \pi^-$ 68.6%
$K_S^0 \longrightarrow \pi^0 \pi^0$ 31.4%
K_L^0 – mean life $\tau_{K_L^0}$ 5.17×10^{-8} s
main decay modes:
$K_L^0 \longrightarrow \pi^\pm e^\mp \nu_e$ 38.8%
$K_L^0 \longrightarrow \pi^\pm \mu^\mp \nu_\mu$ 27.2%
$K_L^0 \longrightarrow \pi^0 \pi^0 \pi^0$ 21.1%
$K_L^0 \longrightarrow \pi^+ \pi^- \pi^0$ 12.6%

Baryons

proton p, quark content (uud), m_p 938.27 MeV,
stable
neutron n, quark content (udd), m_n 939.57 MeV,
mean life τ_n 886±1 s
Delta resonance $\Delta(1232)$, quark content:
$\Delta^{++}(uuu)$, $\Delta^+(uud)$, $\Delta^0(udd)$, $Delta^-(ddd)$
m_Δ 1,232 MeV

References

1. B. Rossi, High Energy Particles (Prentice Hall, New York) 1952
2. V.L. Ginzburg & S.I. Syrovatskii, The Origin of Cosmic Rays (Pergamon Press, Oxford) 1964
3. S. Hayakawa, Cosmic Ray Physics: nuclear and astrophysical aspects (Wiley–Interscience, New York) 1969
4. A.M. Hillas, Cosmic Rays (Pergamon Press, Oxford) 1972
5. V.S. Berezinskii, S.V. Bulanov, V.A. Dogiel, V.L. Ginzburg (editor) and V.S. Ptuskin, Astrophysics of Cosmic Rays (North Holland, Amsterdam) 1990
6. T.K. Gaisser, Cosmic Rays and Particle Physics (Cambridge University Press, Cambridge) 1990
7. M.S. Longair, High Energy Astrophysics (Cambridge University Press, Cambridge) 1992
8. P.K.F. Grieder, Cosmic Rays at Earth (Elsevier, Amsterdam) 2001
9. R. Schlickeiser, Cosmic Ray Astrophysics (Springer-Verlag, Berlin, Heidelberg) 2002
10. J.N. Bahcall, Neutrino Astrophysics (Cambridge University Press, Cambridge) 1989
11. *Review of Particle Properties.* For the most recent edition see *http://pdg.lbl.gov.*
12. E. Rutherford, Phyl. Mag., **21**, 669 (1911)
13. H. Bethe, Ann. d. Phys., **5**, 325 (1930)
14. F. Bloch, Zs. Phys., **81**, 363 (1933)
15. R.M. Sternheimer, Phys. Rev., **103**, 511 (1956); R.M. Sternheimer et al., At. Data Nucl. Data Tables, **30**, 281 (1984)
16. H.A. Bethe & W. Heitler, Proc. Roy. Soc., A**146**, 83 (1934)
17. V. Flaminio et al. CERN–HERA 84-01
18. A.M. Hillas, in Proc 16th Int. Cosmic Ray Conf. (Kyoto), **8**, 7 (1979)
19. R.P. Feynman, Phys. Rev. Lett., **23**, 1415 (1969)
20. E. Yen, Phys. Rev., D**10**, 836 (1974)
21. E. Albini et al., Nuovo Cim., A**32**, 101 (1976)
22. W. Thome et al., Nucl. Phys., B**129**, 365 (1977)
23. M. Antinucci et al., Lett. Nuovo Cim., **6**, 121 (1973)
24. J. Rachen, PhD Thesis, Bonn University (1996)
25. For a review see J. Hufner, Phys. Rep., **125**, 129 (1985)
26. W.R. Webber, J.C. Kish & D.A. Shrier, Phys. Rev. C**41**, 574 (1990)
27. J. Engel et al., Phys. Rev., D**46**, 5013 (1992)
28. J.R. Letaw, R. Silberberg & C.H. Tsao, Ap. J. Suppl., **56**, 3691 (1984)
29. R. Silberberg and C.H. Tsao, Ap. J. Suppl., 25, 315 (1973), 25, 335 (1973)

30. W.R. Webber et al., Ap. J., **508**, 940 (1998)
31. W.R. Webber et al., Ap. J., **508**, 949 (1998)
32. B.T. Cleveland et al., Ap. J., **496**, 505 (1998)
33. W. Hampel et al., Phys. Lett., **B447**, 127 (1999)
34. J.N. Abdurashitov et al., Phys. Rev., C**60**:055801 (1999)
35. K.S. Hirata et al., Phys. Rev. Lett., **65**, 1297 (1990)
36. Y. Fukuda et al., Phys. Rev. Lett., **82**, 1810 (1999)
37. Y. Fukuda et al., Phys. Rev. Lett., **82**, 2644 (1999)
38. J.N. Bahcall, M.H. Pinsonneault & S. Basu, Ap. J., **555**, 990 (2001)
39. A.S. Brun, S. Turck-Chiese & P. Monel, Ap. J., **506**, 913 (1998)
40. J.W. Bieber et al., Nature, **348**, 407 (1990)
41. The SNO Collaboration, Phys. Rev. Lett., **87**, 071301 (2001)
42. K.S. Hirata et al., Phys. Rev. Lett., **58**, 1490 (1987)
43. R.M. Bionta et al., Phys. Rev. Lett., **58**, 1494 (1987)
44. E.N. Alekseev et al., JETP Lett., **45**, 589 (1987)
45. M. Aglietta et al., Europhys. Lett., **3**, 1315 (1987)
46. D.N. Schramm, Comm. Part. Nucl. Phys., **17**, 239 (1987)
47. A. Burrows & J.M. Latimer, Ap. J., **307**, 178 (1986)
48. S.E. Woosley, J.R. Wilson & R. Mayle, Ap. J., **309**, 19 (1986)
49. J.N. Bahcall et al., Nature, **327**, 682 (1987)
50. G.T. Zatsepin, ZhETP Lett., **8**, 205 (1968)
51. E. Fermi, Phys. Rev., **75**, 1169 (1949)
52. W.I. Axford, E. Lear & G. Skadron, Proc. 15th Int. Cosmic Ray Conf. (Plovdiv), **11**, 132 (1977)
53. A.R. Bell, Mon. Not. R. Astr. Soc., **182**, 443 (1978)
54. G.F. Krymsky, Dokl. Acad. Nauk. SSSR, **243**, 1306 (1977)
55. L. O'C. Drury, Space Sci. Rev., **36**, 57 (1983)
56. J.R. Jokipii, Ap. J., **313**, 842 (1987)
57. F.C. Jones & D.C. Ellison, Space Sci. Rev., **58**, 259 (1991)
58. P.O. Lagage & C.J. Cesarsky, A&A, **118**, 223 (1983)
59. E.G. Berezhko, Astropart. Phys., **5**, 367 (1996)
60. D.C. Ellison, M.G. Baring & F.C. Jones, Ap. J., **453**, 873 (1995)
61. H.J. Völk & P.L. Biermann, Ap. J., **333**, L65 (1988)
62. J.G. Kirk & P. Schneider, Ap. J., **315**, 425 (1987)
63. H.J. Völk, L.A. Zank and G.P. Zank, A&A, **198** 274 (1988)
64. E.G. Berezhko &H.J. Völk, Astropart. Phys., **7**, 183 (1997)
65. A.P. Szabo & R.J. Protheroe, Astropart. Phys., **2**, 375 (1994)
66. R.J. Protheroe & T. Stanev, Astropart. Phys., **10**, 185 (1999)
67. L.O'C. Drury et al., A&A, **347**, 370 (1999)
68. V.L. Ginzburg, Elementary Processes for Cosmic Ray Astrophysics, Gordon & Breach (New York) 1969
69. G.R. Blumenthal & R.J. Gould, Rev. Mod. Phys., **42**, 237 (1970)
70. V.S. Berezinsky et al., Astropart. Phys., **1**, 281 (1993)
71. D.L. Bertsch et al., Ap. J., **416**, 587 (1993)
72. P.P. Kronberg, Rep. Prog. Phys., **57**, 325 (1994)
73. R. Beck et al., Ann. Rev. Astron. Astrophys., **34**, 155 (1996)
74. R.J. Rand & A.G. Lyne, MNRAS, **268**, 497 (1994)
75. J.P. Vallee, Ap. J., **366**, 450 (1991)
76. T. Stanev, Ap. J., **479**, 290 (1997)

77. Y. Sofue & M. Fujimoto, Ap. J., **265**, 722 (1983)
78. R. Beck, Sp. Sci. Rev., **99**, 243 (2001)
79. J.L. Han et al., A&A, **228**, 98 (1997); J.L. Han, *astro-ph/0110319*
80. S. Chandrasekhar, Rev. Mod. Phys., **15**, 1 (1943)
81. A. Achterberg et al., *astro-ph/9907060*
82. M. Gupta & W.R. Webber, Ap. J., **340**, 1124 (1989)
83. M.M. Shapiro & R. Silberberg, Ann. Rev. Nucl. Sci., **20**. 323 (1970)
84. J.A. Simpson, Ann. Rev. Nucl. & Part. Sci., **33**, 323 (1983)
85. J.A. Simpson & M. Garcia-Munoz, Sp. Sci. Rev., **46**, 205 (1988)
86. A. Laurasia et al., Ap. J., **423**, 426 (1994)
87. J.J. Connel, Ap. J., **501**, L59 (1998)
88. J.A. Simpson & J.J. Connel, Ap. J., **479**, L85 (1998)
89. J.J. Connel, M.A. DuVernois & J.A. Simpson, Ap. J., **509**, L97 (1998)
90. W.R. Binns et al., Proc. 26th Int. Cosmic Ray Conf. (Salt Lake City), **3**, 21, (1999)
91. C.E. Fichtel et al., Ap. J., **198**, 163 (1975)
92. H. Mayer-Hasselwander et al., A&A, **105**, 164 (1980)
93. Hunter, S.D., et al., Ap. J., **481**, 205 (1997)
94. T.K. Gaisser, R.J. Protheroe & T. Stanev, Ap. J., **492**, 219 (1998)
95. R.J. Protheroe & T. Stanev, Proc. 25 Int. Cosmic Ray Conf. (Durban), **3**, 137 (1997)
96. T. Maeno et al., Astropart. Phys., **16**, 121 (2001)
97. J.M. Grunsfeld et al., Ap. J., **327**, L31 (1988)
98. N.L. Grigorov et al., Proc. 10th Int. Cosmic Ray Conf. (Calgary), **1**, 512 (1967)
99. T.H. Burnett et al., Phys. Rev. Lett., **27**, 1310 (1983)
100. A.V. Apanasenko et al., Astropart. Phys., **16**, 13 (2001)
101. E.N. Parker, Ap. J., **128**, 664 (1958)
102. The Swarthmore/Newark neutron monitor is constructed and operated by the Bartol Research Institute of the University of Delaware
103. E.S. Seo et al., Ap. J., 378 (1991) 763.
104. L.J. Gleeson & W.I. Axford, Ap. J., **154**, 1011 (1968)
105. L.A. Fisk, M.A. Forman & W.I. Axford, J. Geophys. Res., **78**, 995 (1973)
106. M. Garcia-Munoz et al., J. Geophys. Res., **91**, 2858 (1986)
107. C. Stoermer, Arch. Sci. Phys. Nat. Ser. 4, **32**, 117 (1911)
108. J. W. Bieber, P. Evenson and Z. Lin, Antarctic J., **27**, 318 (1992)
109. R.A. Langel, International Geomagnetic Reference Field, 1991 Revision, IAGA news, no. 38 (1991)
110. P. Lipari & T. Stanev, Proc. 24th Int. Cosmic Ray Conf. (Rome), **1** (1995)
111. J.P. Wefel, in *Cosmic Rays, Supernovae and the Interstellar Medium*, NATO ASI Series C, **337** (Dordrecht, Kluwer Academic Publishers) 1990
112. M. Casse, P. Goret & C.J. Cesarsky, Proc. 14th Int. Cosmic Ray Conf. (Munich), **2**, 646 (1975)]
113. M.M. Shapiro, Proc 20th Int. Cosmic Ray Conf. (Adelaide), **4**, 8 (1990)
114. J.P. Meyer, L.O'C. Drury & D.C. Ellison, Space Sci. Rev., **86**, 179 (1998)
115. R. Silberberg and C.H. Tsao, Ap. J., **352**, L49 (1990)
116. J.-P. Meyer, Ap. J. Suppl., **57**, 173 (1985)
117. P.L. Biermann, T.K. Gaisser & T. Stanev, Phys. Rev., D**51**, 3450 (1995)
118. J.J. Engelmann et al., A&A, **233**, 233 (1990)
119. D. Müller et al., Ap. J., **374**, 356 (1991)

120. K. Asakimori et al., Proc. 23rd Int. Cosmic Ray Conf. (Calgary), **2**, 25 (1993)
121. R. Silberberg et al., Ap. J., **363**, 265 (1990)
122. W.R. Webber, R.L. Golden & S.A. Stephens, Proc. 20th Int. Cosmic Ray Conf. (Moscow), **1**, 325 (1987)
123. R. Bellotti et al., Phys. Rev., **D60**:052002 (1999)
124. W. Menn et al., Ap. J., **533**, 281 (2000)
125. M. Boezio et al., Ap. J., **518**, 457 (1999)
126. J.Z. Wang et al., Ap. J., **564**, 244 (2002)
127. J. Alcaraz et al., Phys. Lett., B**490**, 27 (2000)
128. J. Alcaraz et al., Phys. Lett., B**494**, 193 (2000)
129. M.J. Ryan, J.F. Ormes & V.K. Balasubrahmanyan, Phys. Rev. Lett., **28** 985 (&E1497) (1972)
130. I.P. Ivanenko et al., Proc. 23rd Int. Cosmic Ray Conf. (Calgary), **2**, 25 (1993)
131. V.I. Zatsepin et al., Proc. 23rd Int. Cosmic Ray Conf. (Calgary), **2**, 13 (1993)
132. Y. Kawamura et al., Phys. Rev., D**40**, 729 (1989)
133. K. Asakimori et al., Ap. J., **502**, 278 (1998)
134. J. Buckley et al., Ap. J., **429**, 736 (1994)
135. M. Simon et al., Ap. J., **239**, 712 (1980)
136. M. Ichimura et al., Phys. Rev., D**48**, 1949 (1993)
137. J.A. Earl, Phys. Rev. Lett., **6**, 125 (1961)
138. R.J. Protheroe, Ap. J., **254**, 391 (1982)
139. I.V. Moskalenko & A.W. Strong, Ap. J., **496**, 694 (1998)
140. D. Müller & J. Tang, Ap. J., **312**, 183 (1987)
141. K.K. Tang, Ap. J., **278**, 881 (1984)
142. J. Nishimura, Adv. Sp. Res., **19**, 711 (1997)
143. M. Boezio et al., Ap. J., **532**, 635 (2000)
144. M.A. DuVernois et al., Ap. J., **559**, 296 (2001)
145. S. Torii et al., Ap. J., **559**, 973 (2001)
146. M. Boezio et al., Ap. J., **561**, 787 (2001)
147. J.W. Mitchel et al., Phys. Rev. Lett., **76**, 3057 (1996)
148. M. Hof et al., Ap. J., **467**, 33 (1996)
149. S. Orito et al., Phys. Rev. Lett., **87**, 1078 (2000)
150. J.W. Bieber et al., Phys. Rev. Lett., **83**, 674 (1999)
151. P. Lipari, Astropart. Phys., **1** 195 (1993)
152. K. Allkofer et al., Phys. Lett., **36**B, 425 (1971)
153. H. Jokish et al., Phys. Rev., D**19**, 1368 (1979)
154. V. Agrawal et al., Phys. Rev., D**53**, 1314 (1996)
155. M. Conversi, Phys. Rev., **79**, 749 (1950)
156. P. LeCoultre, Proc. 27th Int. Cosmic Ray Conf. (Hamburg), **3**, 974 (2001)
157. C.A. Ayre et al., J. Phys., G**1**, 584 (1975)
158. I.P. Ivanenko et al., Proc. 19th Int. Cosmic Ray Conf. (La Jolla), **8**, 210 (1985)
159. B.C. Rastin, J. Phys. G**10**, 1609 (1984)
160. M.P. DiPascale et al. J. Geophys. Res., **98**, 350 (1993)
161. M. Boezio et al., Phys. Rev., D**62**:032007 (2000)
162. M. Motoki et al., Astropart. Phys. **19**, 113 (2003)
163. R. Engel, T.K. Gaisser & T. Stanev, Proc. 27th Int. Cosmic Ray Conf. (Hamburg), **3**, 1029 (2001)
164. L.V. Volkova, G.T. Zatsepin & L.A. Kuzmichev, Yad. Phys., **29**, 1252 (1979)
165. R. Bellotti et al., Phys. Rev., D**53**, 35 (1996)

166. R. Engel et al., Proc. 27th Int. Cosmic Ray Conf. (Hamburg), **4**, 1381 (2001)
167. A.E. Brenner et al., Phys. Rev., D**26**, 1497 (1982)
168. M. Aguilar-Benitex et al., Z. Phys., C**50**, 405 (1991)
169. F. Reines et al., Phys. Rev. Lett., **15**, 429 (1965)
170. C.V. Achar et al., Phys. Lett., **18**, 196; **19**, 78 (1965)
171. M. Crouch, Proc. 20th Int. Cosmic Ray Conf. (Moscow), **6**, 165 (1987)
172. Yu.M. Andreev, V.I. Gurentzov & L.M. Kogai, Proc. 20th Int. Cosmic Ray Conf. (Moscow), **6**, 200 (1987)
173. M. Aglietta et al., Astropart. Phys., **3**, 311 (1995)
174. M. Ambrosio et al., Phys. Rev., D**52**, 3793 (1995)
175. Ch. Berger et al., Phys. Rev., D**40**, 2163 (1989)
176. P. Lipari & T. Stanev, Phys. Rev., D**44**, 3543 (1991)
177. W. Lohman, R. Kopp & R. Voss, CERN Yellow Report EP/85-03 (1985)
178. D.E. Groom, N.V. Mokhov & S.I. Striganov, At. Data Nucl. Data Tables, **78**, 183 (2001)
179. A.A. Petrukhin & V.V. Shestakov, Can. J. Phys., **46**, S377 (1968)
180. I.L. Rozental, Usp. Phys. Nauk, **94**, 91 (1968)
181. W.K. Sakumoto, Phys. Rev., D**45**, 3042 (1992)
182. R.P. Kokoulin & A.A, Petrukhin, Proc. 12th Int. Cosmic Ray Conference (Hobart), **6**, 2436 (1971)
183. L.B. Bezrukov & E.V. Bugaev, Yad. Phys., **33**, 1195 (1981) [Sov. J. Nucl. Phys., **33**, 635 (1981)]
184. S. Iyer Dutta et al., Phys. Rev., D**63**:094020 (2001)
185. T.K. Gaisser & T. Stanev, Phys. Rev., D**30**, 985 (1984)
186. T.K. Gaisser & T. Stanev, Phys. Rev., D**31**, 2770 (1985)
187. T.K. Gaisser, F. Halzen & T. Stanev, Phys. Reports, **258**, 174 (1995)
188. C.H. Lllewellyn Smith, Phys. Reports, **3**C, 261 (1972)
189. See the presentation of the DIS cross-sections in: R. Gandhi et al., Phys. Rev., D**58**: 093009 (1998)
190. P. Lipari, M. Lusignoli & F. Sartogo, Phys. Rev. Lett., **74**, 4384 (1995)
191. L.B. Okun, Leptons and Quarks (North Holland, Amsterdam) (1982)
192. M. Gluck, E. Reya & A. Vogt, Europ. Phys. J., C**67**, 433 (1995)
193. L.V. Volkova, Yad. Phys., **31**, 1510 (1980) [Sov. J. Nucl. Phys., **31**, 784 (1980)]
194. D.H. Perkins, Astropart. Phys., **2**, 249 (1994)
195. See Ref. [154] and T.K. Gaisser & T.Stanev, Proc. 24th Int. Cosmic Ray Conf. (Rome), **1**, 694 (1995)
196. M. Honda et al., Phys. Rev., D**54**, 4985 (1995)
197. G. Battistoni et al., Astropart. Phys., **12**, 315 (2000)
198. M. Honda et al., Phys. Rev., D**64**:053011 (2001)
199. P. Lipari, Astropart. Phys., **14**, 151 (2000)
200. K.S. Hirata et al., Phys. Lett., B**205**, 416 (1988); Phys. Lett., B**280**, 146 (1992)
201. D. Casper et al., Phys. Rev. Lett., **66**, 2561 (1991); R. Becker–Szendy et al., Phys. Rev., D**46**, 3720 (1992)
202. Y. Fukuda et al., Phys. Rev. Lett., **81**, 1562 (1998)
203. J. Kameda et al. (The Super-K collaboration), Proc. 27th Int. Cosmic Ray Conf. (Hamburg), **3**, 1057 (2001)
204. G.L. Fogli, E. Lisi & A. Marone, Phys. Rev., D**64**: 093005 (2001)
205. G.L. Fogli, E. Lisi & A. Marone, Phys. Rev., D**63**: 053008 (2001)

206. M. Goodman et al. (The Soudan 2 collaboration), Proc. 27th Int. Cosmic Ray Conf. (Hamburg), **3**, 1085 (2001)
207. M. Ambrosio et al., Phys. Lett., **B434**, 451 (1998)
208. T. Kajita & Y. Totsuka, Rev. Mod. Phys., **73**, 85 (2001)
209. B. Pontecorvo, Zh. Exp. Theor. Fiz., **53**, 1717 (1967) [Sov. Phys. JETP, **26**, 984 (1968)]
210. L. Wolfenstein, Phys. Rev., **D17**, 2369 (1978)
211. S.P. Mikheev & A.Yu. Smirnov, Yad. Phys., **42**, 1441 (1985) [Sov. J. Nucl. Phys., **42**, 913 (1985)]
212. S. Fukuda et al., Phys. Rev. Lett., **85**, 3999 (2000)
213. T. Montaruli et al. (The MACRO Collaboration), Proc. 27th Int. Cosmic Ray Conf. (Hamburg), **3**, 1069 (2001)
214. S. Fukuda et al., Phys. Rev. Lett., **86**, 5656 (2001)
215. Q.R. Ahmad et al., Phys. Rev. Lett., **89**:011301 (2002)
216. Q.R. Ahmad et al., Phys. Rev. Lett., **89**:011302 (2002)
217. See http://www.hep.anl.gov/ndk/longbnews for the latest news on long baseline neutrino oscillation detectors
218. S.H. Ahn et al., Phys. Lett., **B511**, 178 (2001)
219. P. Auger, Rev. Mod. Phys., **11**, 288 (1939)
220. W. Heitler, Quantum Theory of Radiation (Oxford University Press, 1944).
221. B. Rossi & K. Greisen, Revs. Mod. Phys., **13**, 240 (1941)
222. K. Greisen, Progr. Cosmic Ray Physics, **3**, 1 (1956)
223. J. Nishimura, Handbuch der Physik, **XLVI/2**, 1 (1967)
224. K. Kamata & J. Nishimura, Progr. Theor. Phys., Suppl. **6**, 93 (1958)
225. J.C. Butcher & H. Messel, Phys. Rev., **112**, 2096 (1958)
226. W.R. Nelson, H. Hirayama & D.W.O. Rogers, preprint SLAC-265 (UC-32), unpublished) 1985.
227. T. Stanev & H.P. Vankov, Comp. Phys. Comm., **4**, 47 (1972)
228. A.M. Hillas & J. Lapikens, Proc. 15th Int. Cosmic Ray Conf. (Plovdiv), **8**, 460 (1977)
229. J. Matthews, Proc. 27th Int. Cosmic Ray Conf. (Hamburg), **1**, 161 (2001)
230. J. Engel et al., Phys. Rev., **D46**,5013 (1992)
231. J. Alvarez-Muñoz et al., Phys. Rev., **D66**: 033011 (2002)
232. H.O. Klages et al., Nucl. Phys. (Proc. Suppl.), **52B**, 92 (1997)
233. P. Sokolsky, AIP Conf. Proc. **579**, 296 (2001)
234. J.W. Fouler et al., Astropart. Phys., **15**, 49 (2001)
235. K. Greisen, Ann. Revs. Nucl. Sci., **10**, 63 (1960)
236. A.M. Hillas et al., Proc 12th Int. Cosmic Ray Conf. (Hobart), **3**, 1001 (1971)
237. J.R. Patterson & A.M. Hillas, J. Phys. G: Nucl. Phys., **9**, 1433 (1983)
238. F. Arqueros et al., Astron. Astrophys., **359**. 682 (2000)
239. J.E. Dickinson et al., Proc. 26th Int. Cosmic Ray Conf. (Salt Lake City), **3**, 136 (1999)
240. D. Cline, F. Halzen & J. Luthe, Phys. Rev. Lett., **7**, 491 (1973)
241. T.K. Gaisser & F. Halzen, Phys. Rev. Lett., **54**, 1754 (1985)
242. G. Pancheri & Y. Srivastava, Phys. Lett., **159B**, 69 (1985)
243. L. Durand & H. Pi, Phys. Rev. Lett., **58**, 303 (1987)
244. T.K. Gaisser & T. Stanev, Phys. Lett., **219B**, 375 (1989)
245. A. Capella & A. Krzywicki, Phys. Rev., **D18**, 3357 (1978)
246. A. Capella & J. Tran Than Van, Z. Phys., **C10**, 249 (1981)

247. H. Bengtsson & T. Sjöstrand, Comp. Phys. Comm., **46**, 43 (1987)
248. A. Capella & J. Tran Than Van, Z. Phys., C**38**, 177 (1988)
249. R.S. Fletcher et al., Phys. Rev., D**50**, 5710 (1994)
250. C. Adlof et al., Nucl. Phys., B**497**, 3 (1997)
251. J. Breitweg et al., Phys. Lett., B**407**, 432 (1997)
252. L.V. Gribov et al., Phys. Rep., **100**, 1 (1983)
253. P.D.B. Collins, Introduction to Regge Theory and High Energy Physics, Cambridge University Press, 1977.
254. CORSICA: A Monte Carlo Code to Simulate Extensive Air Showers, Forschungszentrum Karlsruhe **FZKA 6019** (1998).
255. S.J. Sciutto, AIRES: A System for Air Shower Simulations, *astro-ph/9911331*
256. J. Ranft, Phys. Rev., D **21**, 64 (1995)
257. N.N. Kalmikov, S.S. Ostapchenko & A.I. Pavlov, Nucl. Phys. B (Proc. Suppl.), **52B**, 17 (1997).
258. R. Engel et al., Proc. 26th Int. Cosmic Ray Conf. (Salt Lake City), **1**, 415 (1999)
259. J.-N. Capdevielle, J. Phys. G, **20**, 637 (1994)
260. J. Knapp et al., Astropart. Phys., **19**, 77 (2003)
261. N.L. Grigorov et al., Yad. Phys. **11**, 1058 (1970) and Proc. 12th Int. Cosmic Ray Conf. (Hobart), **2**, 206 (1971)
262. T.V. Danilova et al., Proc. 15th Int. Cosmic Ray Conf. (Plovdiv), **8**, 129 (1977).
263. Yu.A. Fomin, Proc 22nd Int. Cosmic Ray Conf. (Dublin), **2**, 85 (1991)
264. M. Amenomori et al., Ap. J., **408** (1996)
265. M.A.K. Glasmacher et al., Astropart. Phys., **10**, 291 (1999)
266. M. Nagano et al., J. Phys., G**10**, 1295 (1984)
267. D.J. Bird et al., Ap. J., **441**, 144 (1994)
268. K.-H. Kampert et al., Invited, Rapporteur and Highlight papers of Int. Cosmic Ray Conf. 2001: 240 (Copernikus Gesellschaft), 2002.
269. B. Peters, Nuovo Cim. (Suppl.), **14**, 436 (1959)
270. B. Alessandro for the EASTOP Collaboration, Proc. 27th Int. Cosmic Ray Conf. (Hamburg), **1**, 124, (2001)
271. J.H. Weber et al., Proc 26th Int. Cosmic Ray Conf. (Salt Lake City), **1**, 347 (1999)
272. T. Abu-Zayyad et al., Ap. J., **557**, 686 (2001)
273. N.N. Kalmykov & G.B. Khristiansen, J. Phys. G, **21**, 1279 (1995)
274. J.W. Fowler et al., Astropart. Phys., **15**, 49 (2001)
275. S.P. Swordy & D.B. Kieda, Astropart. Phys., **13**, 137 (2000)
276. S.M. Paling et al., Proc. 25th Int. Cosmic Ray Conf. (Durban), **5**, 253 (1997)
277. J. Linsley, Phys. Rev. Lett., **10**, 146 (1963)
278. G. Cocconi, Nuovo Cim., **3** 1433 (1956)
279. A.A. Penzias & R. Wilson, Ap. J., **142**, 419 (1965)
280. G.T. Zatsepin & V.A. Kuzmin, JETP Lett. 4 78 (1966).
281. K. Greisen, Phys. Rev. Lett., **16**, 748 (1966)
282. J.C. Mather et al., Ap. J., **512**, 511 (1999)
283. A. Mücke et al., Comp. Phys. Comm., **124**, 290 (2000)
284. F.W. Stecker & M.H. Salamon, Ap. J., **512**, 521 (1999)
285. J.L. Puget, F.W. Stecker & J.H. Bredekamp, Ap. J., **205**, 638 (1976)
286. R.J. Protheroe & P.A. Johnson, Astropart. Phys., **4**, 253 (1996)

287. V.S. Berezinsky & S.I. Grigorieva, Astron. Astrophys., **199**,1 (1988)
288. M.J. Chodorowski, A.A. Zdziarski & M. Sikora, Ap. J., **400**, 181 (1992)
289. A. Achterberg et al., MNRAS, **328**, 393 (2001)
290. G. Bertone et al., Phys. Rev., D**66**:3003 (2002)
291. D.J. Bird et al., Ap. J., **441**, 144 (1995)
292. M. Nagano & A.A. Watson, Rev. Mod. Phys., **72**, 689 (2000)
293. M. Takeda et al., Phys. Rev. Lett., **81**, 1163 (1998); see *http://www-akeno.icrr.u-tokio.ac.jp* for update with events detected after this publication
294. T. Abu-Zayyad et al., *astro-ph/0208243*.
295. B.N. Afanasiev et al., in Proceedings of the Tokyo Workshop on Techniques for the Study of the Extremely High Energy Cosmic Rays, ed. M. Nagano (Institute for Cosmic Ray Research, University of Tokyo) p. 35
296. M.A. Lawrence, R.J.O. Reid & A.A. Watson, J. Phys. G, **17**, 733 (1991)
297. M. Ave et al., Astropart. Phys., **19**, 47 (20030 *astro-ph/0112253*
298. A.M. Hillas et al., Proc. 12th Int. Cosmic Ray Conf. (Hobart), **3**, 1001 (1971)
299. D.J. Bird et al., Phys. Rev. Lett., **71**, 3401 (1993)
300. *http://www-akeno.icrr.u-tokyo.ac.jp/AGASA/results.html*
301. M. Ave et al., Phys. Rev. Lett., **85**, 2244 (2000)
302. T.K. Gaisser et al, Phys. Rev., D**47**, 1919 (1993)
303. A.M. Hillas, Ann. Rev. Astron. Astrophys., **22**, 425 (1984)
304. C.A. Norman, D.B. Melrose & A. Achterberg, Ap. J., **454**, 60 (1995)
305. P. Kronberg, Rep. Prog. Phys., **21**, 325 (1994)
306. T.E. Clarke, P.P. Kronberg & H. Böringer, Ap. J., **547**, L111 (2001)
307. H. Kang, D.Ryu & T.W. Jones, Ap. J., **456**, 422 (1996)
308. J.P. Rachen & P.L. Biermann, Astron. Astrophys., **272**, 161 (1993)
309. F. Halzen & E. Zas, Ap. J., **488**, 669 (1997)
310. A.P. Szabo & R.J. Protheroe, Astropart. Phys., **2**, 375 (1994)
311. M. Milgrom & V. Usov, Ap. J., **449**, L37 (1995)
312. M. Vietri, Ap. J., **453**, 883 (1995)
313. E. Waxman, Ap. J., Phys. Rev. Lett., **386** (1995)
314. C. Cesarsky & V. Ptuskin, Proc. 23rd Int. Cosmic Ray Conf. (Calgary), **2**, 341 (1993)
315. P. Blasi, R.I. Epstein & A.V. Olinto, Ap. J., **533**, 123 (2000)
316. C.T. Hill, Nucl. Phys., B**224**, 469 (1983)
317. D.N.Schramm & C.T. Hill, Proc. 18th Int. Cosmic Ray Conf. (Bangalore), **2**, 393 (1983)
318. V.S. Berezinsky & M. Kachelriess, Phys. Rev., D**63**: 034007 (2001)
319. R.J. Protheroe & P.L. Biermann, Astropart. Phys., **6**, 45 (1996); err. ibid, **7**, 181 (1997)
320. P. Bhattacharjee & G. Sigl, Phys. Reports, **327**, 109 (2000)
321. C.T. Hill, D.N. Schramm & T.P. Walker, Phys. Rev., D**36**, 1007 (1987)
322. V.S. Berezinsky & A. Vilenkin, Phys. Rev. Lett., **79**, 5202 (1997)
323. M. Birkel & S. Sarkar, Astropart. Phys., **9**, 297 (1998)
324. T. Weiler, Astropart. Phys., **11**, 303 (1999)
325. D. Fargion, B. Mele & A. Salis, Ap. J., **517**, 517 (1999)
326. P. Blasi, S. Burles & A.V. Olinto, Ap. J., **512**, L79 (1999)
327. N. Hayashida et al., Astropart. Phys., **10**, 303 (1999)
328. G. de Vaucouleurs, Vistas. Astron., **2**, 1584 (1956)
329. T. Stanev et al., Phys. Rev. Lett., **75**, 3056 (1995)

330. M. Takeda et al., Ap. J., **522**, 225 (1999)
331. Y. Uchihori et al., Astropart. Phys., **13**, 151 (2000)
332. P.L. Biermann et al., Nucl. Phys. B (Proc. Suppl.), **87**, 417 (2000)
333. T. Stanev et al., Phys. Rev., D **62**:093005 (2000)
334. F. Halzen et al., Astropart. Phys., **3**, 151 (1995)
335. M. Ave et al., Phys. Rev., D **65**:063007 (2002)
336. K. Shinozaki et al., Ap. J., **571**, L117 (2002)
337. for information on the current status see *http://www.auger.org*
338. see *http://www-ta.icrr.u-tokyo.ac.jp*
339. see *http://www.euso-mission.org*
340. see *http://owl.gsfc.nasa.gov*
341. R. Ghandi et al., Phys. Rev., D **58**:093009 (1998)
342. M. Samorski & W. Stamm, Ap. J., **268**, 175 (1983)
343. W.T. Vestrand & D. Eichler, Ap. J., **261**, 251 (1982)
344. T.K. Gaisser & T. Stanev, Phys. Rev. Lett., **54**, 2265 (1985)
345. V.S. Berezinsky & O.F. Prilutski, Astron. Astrophys., **66**, 327 (1978)
346. L.O'C Drury, F. Aharonian & H.J. Völk, Astron. Astrophys., **285**, 995 (1994)
347. J.A. Esposito et al., Ap. J., **461** 520 (1996)
348. J.H. Buckley et al., Astron. Astrophys., **329**, 639 (1998).
349. M.G. Baring et al., Ap. J., **285**, 995 (1999)
350. P.L. Biermann, Sp. Sci. Rev., **74**, 385 (1995)
351. R. Mukherjee et al., Ap. J., **490**, 116 (1997)
352. M. Pohl, Invited, Rapporteur and Highlight Papers, 27th Int. Cosmic Ray Conf. (Hamburg), 2001; *astro-ph/0111552*
353. L. Marasci, G. Ghisellini, & A. Celotti, Ap. J., **397**, L5 (1992)
354. C.D. Dermer & R. Schlickeiser, Ap. J. Suppl., **90**, 945 (1994)
355. K. Mannheim & P.L. Biermann, Astron. Astrophys., **221** 211 (1989)
356. F.W. Stecker et al., Phys. Rev. Lett., **66**, 2697 (1991)
357. P. Meszaros & M.J. Rees, Ap. J., **405**, 278 (1993)
358. E. Waxman & J.N. Bahcall, Phys. Rev. Lett., **78**, 2292 (1997)
359. K. Mannheim, Astropart. Phys., **3**, 295 (1995)
360. E. Waxman & J.N. Bahcall, Phys. Rev., D**59**:023002 (1999)
361. K. Mannheim, R.J. Protheroe & J. Rachen, Phys. Rev., D**63:** 023003 (2001)
362. V.S. Berezinsky & G.T. Zatsepin, Phys. Lett., **28B**, 423 (1969)
363. F.W. Stecker, Astrophys. Sp. Sci., **20**, 47 (1973)
364. R. Engel, D. Seckel & T. Stanev, Phys. Rev., D**64**:093010 (2001)
365. S.L. Glashow, Phys. Rev., **118**, 316 (1960)
366. K.S. Capelle et al., Astropart. Phys., **8**, 321 (1998)
367. G.A. Askaryan, JETP, **14**, 441 (1962)
368. F. Halzen & D. Saltzberg, Phys. Rev. Lett., **81**, 4305 (1998)
369. J.G. Learned & S. Pakvasa, Astropart. Phys., **3**, 267 (1995)
370. *http://www.ifh.de/baikal.baikalhome.html*
371. *http://www.uoa.gr/nestor/intro.html*
372. *http://antares.in2p3.fr*
373. *http://nemoweb.lns.infn.it*
374. *http://amanda.physics.wisc.edu*
375. *http://icecube.wisc.edu*
376. *http://www.ps.uci.edu/%7Eanita/*
377. *http://kuhep4.phsx.ukans.edu/ iceman*

Index

Printing: Mercedes-Druck, Berlin
Binding: Stein + Lehmann, Berlin

RETURN TO: PHYSICS LIBRARY

351 LeConte Hall   510-642-3122

LOAN PERIOD 1	2	3
1-MONTH		
4	5	6

ALL BOOKS MAY BE RECALLED AFTER 7 DAYS.

Renewable by telephone.

DUE AS STAMPED BELOW.

This book will be held in PHYSICS LIBRARY until JUL 1 9 2004		
SEP 1 1 2004		
MAY 2 0 2005		
July 1 9. 005		
MAR 0 8 2006		
SEP 0 8 2009		

FORM NO. DD 22
500   4-03

UNIVERSITY OF CALIFORNIA, BERKELEY
Berkeley, California 94720–6000